Explorations
In Elementary
Mathematics

EXPLORATIONS

PRENTICE-HALL, INC., ENGLEWOOD CLIFFS, NEW JERSEY

Seaton E. Smith, Jr.

Faculty of Elementary Education
University of West Florida

IN ELEMENTARY

MATHEMATICS

2nd Edition

EXPLORATIONS IN ELEMENTARY MATHEMATICS, 2nd Ed.

Seaton E. Smith, Jr.

© 1971, 1966 by Prentice-Hall, Inc.,
Englewood Cliffs, New Jersey

Current printing (last digit): 10 9 8 7 6 5 4 3 2 1

13-296046-X

Library of Congress Catalog Card Number 78-138625

Printed in the United States of America

PRENTICE-HALL INTERNATIONAL, INC., *London*
PRENTICE-HALL OF AUSTRALIA, PTY. LTD., *Sydney*
PRENTICE-HALL OF CANADA, LTD., *Toronto*
PRENTICE-HALL OF INDIA PRIVATE LIMITED, *New Delhi*
PRENTICE-HALL OF JAPAN, INC., *Tokyo*

to Lucille, Jimmy, and Bobby

Foreword

The preparation and in-service training of elementary school teachers has been receiving increased attention in the last few years. There are so many new books to serve this purpose that a recent survey of texts for elementary school teachers listed forty-seven books. Among these forty-seven the first edition of *Explorations in Elementary Mathematics* was cited as one of the most elementary, and as suitable for self-study.

Seaton Smith developed his book from his experience as West Virginia State Supervisor of Mathematics working very closely with several thousand elementary school teachers. Since that time he has worked in pre-service and in-service training of elementary school teachers for three years at Ohio University and three years at the University of West Florida.

The first edition has been widely used and recognized as a book that would be understood by elementary school teachers and by prospective elementary school teachers with minimal mathematics background. The second edition is based upon six years of testing the first edition. In the second edition he has:

(1) updated the vocabulary,
(2) substantially increased the number of exercises and the types of exercises,
(3) included Chapter Tests for self-evaluation, and
(4) reorganized and expanded the material to provide separate chapters on number theory and the integers, two chapters on the rational numbers, and two chapters on informal geometry.

In addition practically all of the recently developed devices for introducing various topics have been included in the exposition. These will provide very useful background for teachers.

This book is designed primarily as an introductory course in modern mathematics for the in-service and pre-service training of elementary school teachers. It may also be studied by parents who are interested in gaining a knowledge of the kind of mathematics that their children are studying in the schools of today. The subject matter is also suitable for the general education of undergraduate college students with very limited secondary school training in mathematics.

Bruce E. Meserve

Preface

As we move into the decade of the seventies, the pre-service and in-service training of teachers in elementary school mathematics continues to be a major concern to many educators. We have seen the switch to "modern mathematics" take place in many school systems and often have been disappointed to find that the only real change was in the form of a new textbook adoption. We have seen many colleges and universities develop more rigorous mathematics courses for the prospective and in-service elementary school teacher to study. We have heard many complaints from students who do not have the mathematical background to comprehend mathematics presented in a rigorous manner. We have seen many cases of individuals who have successfully completed such rigorous mathematics courses only to question the relevance of the material to the teaching of elementary school mathematics.

The second edition of *Explorations in Elementary Mathematics* reflects the author's attempt to avoid some of the problems identified above. The material has been extensively field tested with thousands of elementary school teachers at both the in-service and pre-service level. Although mathematically precise language is used throughout the book, a major emphasis has been placed on the need to communicate with the reader who has a limited mathematical background. Many of the approaches used in the teaching of elementary school mathematics have been woven into the text in an attempt to have the student see the relevance of the material to teaching. The structure of the number systems is emphasized and many illustrations and exercises provide an opportunity for the reader to develop his understanding. The answers for the odd-numbered exercises are provided in the back of the book and the answers for the even-numbered exercises are available in a separate booklet.

The purpose of this book is to present the basic ideas of contemporary elementary school mathematics in such a manner that the reader will be able to understand these ideas and see their relevance to teaching.

The book is designed to serve as a text either in an in-service program or in a college course for prospective teachers or others interested in extending their knowledge of some basic mathematical concepts.

The author wishes to express his appreciation to all of the individuals who contributed toward the development of this book. A special note of gratitude is extended to Dr. Bruce E. Meserve of the University of Vermont for his valuable suggestions. Appreciation is also expressed to Dr. Patricia Spross of the U. S. Office of Education and to Dr. Carl A. Backman of the University of West Florida for their comments and suggestions. Special thanks are also due the author's wife, Lucille, for her dedicated efforts in typing the manuscript.

Seaton E. Smith, Jr.

Contents

chapter 5

Whole Numbers: Multiplication and Division   92

chapter 6

Elementary Number Theory   126

Explorations
In Elementary
Mathematics

chapter 1

Introduction to Mathematical Ideas

The use of the term "modern mathematics" in discussions of improvements in mathematics curricula seems to generate considerable misunderstanding. Although it is true that more mathematics has been created in the past half-century than in all previous times, the term "modern mathematics" more appropriately refers to an approach to the study of mathematics than to new mathematics as such. This approach emphasizes the importance of concepts, patterns, and mathematical structure, as well as the development of mathematical skills. The new applications and uses of today's mathematics demand that pupils learn the "why" as well as the "how" at each step in the learning process.

1-1 The Language of Sets

Many areas of mathematics that are called "modern" or "new" have come into prominence in the last decade. These terms are somewhat misleading because they imply that these areas of mathematics were developed or created in the last few years. Actually, the study of most of the new areas of mathematics began many years ago. For example, the theory of sets was introduced by George Cantor, a German mathematician, during the latter part of the nineteenth century. An Englishman by the name of George Boole was one of the first to find a use for the theory of sets in his algebra of sets, sometimes referred to as "Boolean Algebra."

There are two distinct aspects of sets. The first is *set theory*. This is beyond the level of consideration here. The second aspect is *set language*. Set language is being used extensively in programs in modern mathematics at all levels from kindergarten through graduate school. The notions of set and set language can be used to great advantage in clarifying and unifying many other mathematical concepts. An understanding of the concepts and language of sets is an important step in the orderly and meaningful development of mathematical ideas.

1-2 Concept of Sets

Mathematicians consider the term "set" to be undefined because it seems impossible to define the concept of a set in terms of simpler concepts. It is necessary in mathematics to begin with certain undefined terms, and the idea of a set is so intuitive and basic that it is a convenient starting place.

The idea of set is therefore described by synonyms rather than by a precise definition. The term **set** will be used to mean any well-defined collection, group, or class of objects or ideas. The phrase "well-defined" means that the set is described precisely enough for one to tell whether or not any given object belongs to it. We shall be primarily concerned with well-defined sets.

Some examples of sets are the letters in our alphabet, a baseball team, a set of dishes, a coin collection, and a group of children in a school play. In the case of a baseball team, for example, the set is well defined because the team is considered as a single entity and may be distinguished from any other team or group.

The objects contained in a set are called the **elements** of the set (they are also called **members** of the set). In a set of dishes, each dish is an element (member) of the set of dishes. In the case of a coin collection, each coin is a member of the set of coins.

There are several techniques for presenting sets. One technique is to show the members of the set in a loop or enclosure.

Another technique is to list the names of the members. The set of fruit consists of an apple, an orange, and a banana.

Sometimes members of a set are presented simply by spacing on a page.

            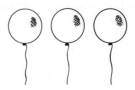

The children may be thought of as one set and the balloons as another set.

A capital letter may be used to name a given set, and the names of all the elements may be *represented* in some fashion within braces. Thus, $B = \{1, 2, 3\}$ means that the letter B is used to represent the same set that $\{1, 2, 3\}$ represents. It is read "B is the set whose elements are the numbers 1, 2, and 3." The statement $A = \{$Jones, Wilson$\}$ is read "A is the set whose elements are Jones and Wilson." Notice that commas are used to separate the elements of a set when they are listed between braces. The symbolism $1 \in B$ is used to mean "1 is a member of the set B" or, simply, "1 is in B."

It is also important to note that the order of listing the names of the members of a set is immaterial: the sets $\{1, 2, 3\}$, $\{2, 3, 1\}$, and $\{3, 1, 2\}$ are all the same. Thus, we can state that $\{1, 2, 3\} = \{2, 3, 1\} = \{3, 1, 2\}$. The "$=$" between two symbols for sets means that each symbol stands for the same set.

Exercises

1. Illustrate each of the four techniques for presenting sets.

2. Use braces to show each set.
   (a) The names of the days of the week.
   (b) The names of the months of the year that begin with the letter J.
   (c) The names of the four seasons in a year.
   (d) The first five letters in our alphabet.

3. Describe each set in words.
   (a)   (c) Set $G = \{1, 3, 5, 7, 9\}$

   (b)   (d) Set $H = \{$candy, cake$\}$

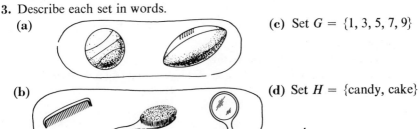

4. Indicate which of these sets are well defined and which are not. List the elements of each set if possible.
   (a) The set of months of the year that have 30 days.
   (b) The set of interesting books written this year.
   (c) The set of United States presidents from the State of Virginia.
   (d) The set of good basketball players in the State of Ohio.

5. Use braces and represent:
   (a) The set of whole numbers, ending in 0, between 20 and 90.
   (b) The set of the first two presidents of the United States.
   (c) The set of odd whole numbers greater than 10 and less than 18.
   (d) The set of whole numbers less than 100 that are divisible by 19.

6. Given set $K = \{1, 2, 3\}$; set $M = \{4, 6, 8, 10\}$; set $R = \{s, t\}$; and set $W = \{\triangle, \square, \otimes\}$. Copy and complete each statement.
   (a) $1 \in$ ____
   (b) $t \in$ ____
   (c) $3 \in$ ____
   (d) ____ $\in M$
   (e) ____ $\in K$
   (f) ____ $\in W$

1-3 Empty Set

The set that contains no members is another important concept. Consider the set whose elements are all the pink elephants pictured on this page. Clearly, there are no members in this set. The set that has no elements is called the **empty set,** also referred to as the **null set.** The empty set may be represented by the symbol $\{\ \}$. Another illustration of the empty set is the set of all United States presidents who were ten years of age when president. Frequently the symbol $\emptyset$ is used to represent the empty set. This cookie jar shown below is another familiar example of the empty set.

Exercises

1. Use braces to show
   (a) The set of all odd numbers that are even.
   (b) The set of names of the months of the year that begin with the letter B.

2. Which of these sets are empty sets?
   (a) The set of whole numbers greater than 7 and less than 9.
   (b) The set of whole numbers greater than 0 and less than 1.
   (c) The set of odd numbers divisible by an even number.
   (d) The set of even numbers divisible by an odd number.
   (e) The set of two-digit numerals ending in 3.
   (f) The set of squares of odd numbers that are even.

3. Give two examples of sets that have no elements.

4. Consider the set $A = \{1\}$. Is A the empty set? Explain.

5. Consider the set $B = \{0\}$. Is B the empty set? Explain.

6. Consider the set $C = \{\varnothing\}$. Is C the empty set? Explain.

1-4 One-to-One Correspondence

From earliest time man probably has been aware of simple numbers in counting. As man's possessions increased, it was necessary to devise a way to keep track of them. Among early methods of keeping count were scratching notches in sticks, tying knots in ropes, and putting pebbles in piles to represent the number of objects being counted. For instance, in the morning the shepherd would build a pile of stones by setting one stone for each sheep as it went out to pasture; at night, when the sheep came back, he would remove one stone from the pile for each sheep that returned. In this very simple way he knew whether or not the number of sheep that went out in the morning was the same as the number that came back at night. This pairing of two things, the sheep and the stone, is an example of one-to-one correspondence.

One-for-one matching of objects was a forerunner of counting. As soon as man began to match things in a definite order, he was counting. Then at the same time he had to invent words and symbols for his counting system in order to communicate his ideas.

The modern-day child, as he learns to count and communicate the concept of number, develops his mathematical insights in much the same way that his primitive ancestors did. The early number experiences of the child lead him through the ideas of "more than," "less than," and "the same amount as" until he learns that these ideas will not satisfy his needs because they are not precise enough. The child is then ready to deal with a further result of sets in one-to-one correspondence, the idea of number.

Two sets, C and D, are said to be in **one-to-one correspondence** when we have a matching, or pairing, of the elements of C with the elements of D, such that each element of C is matched to one and only one element of

D and each element of D is matched to one and only one element of C. For example, consider the sets

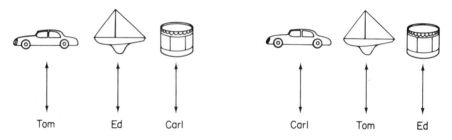

$$D = \left\{ \text{Tom, Ed, Carl} \right\}.$$

The following illustrations show that the sets C and D may be placed in one-to-one correspondence. Notice that this may be done in more than one way as shown in these two figures. Is there another way to place the two sets in one-to-one correspondence?

Another illustration of a one-to-one correspondence would be the set of names on a class roll and the set of students who are members of the class. Clearly, in a class there should be one name for each child and one child for each name.

It is important to note at this time that matchings of two sets are not always one-to-one correspondences. For example, consider the sets

$$E = \left\{ \text{Sylvia, Pat, Ingrid} \right\}.$$

These sets may be placed in a one-to-one correspondence, but they may also be matched in a manner that does not represent one-to-one correspondence. Observe in the first of the next two figures that Sylvia is matched with both pie and cake, while Pat is also matched with cake, and Ingrid is matched with ice cream. In the second illustration, pie is matched with both Sylvia and Pat, while cake is matched with Ingrid, and ice cream is not matched.

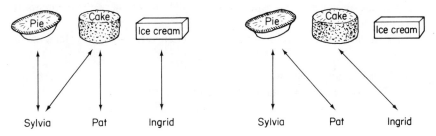

Another example of a correspondence that is not one-to-one is shown in the next figure. Since Bill is not matched with anything (or he could be matched with an element that has already been paired off), this is not a one-to-one correspondence.

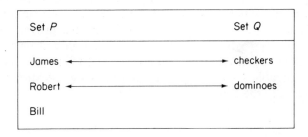

Exercises

1. In each part, can the sets be placed in one-to-one correspondence?
   (a) $\{h, a, r, d\}$ and $\{e, a, s, y\}$.
   (b) $\{@, \#, \$\}$ and $\{1, 2, 5, 9\}$.
   (c) {hammer, saw, rake, shovel} and {Dick, Bill, Al, Gordon}.
   (d) $\{3\}$ and $\{\varnothing\}$.

2. Indicate whether each pair of sets may be placed in one-to-one correspondence. If they do not match, state which contains the greater number of elements.
   (a) The fingers on your right hand and the fingers on your left hand.
   (b) The stripes in the American flag and the states in the United States of America.
   (c) The days in the month of July and the days in the month of August.
   (d) The set of even numbers and the set of odd numbers.

3. Consider the two sets $A = \{@, \#, \$, d\}$ and $B = \{*, \not{c}, \&, x, g\}$.

(a) If we match each element of A with one and only one element of B, how many elements of B have not been matched?

(b) Draw a diagram that represents one such matching of the elements of A and B.

(c) Can we make a one-to-one correspondence between the elements of A and B?

4. Make a drawing to show:

(a) Two sets placed in one-to-one correspondence.

(b) Two sets placed in a correspondence that is not one-to-one.

1-5 Counting

After a very long time of making one-to-one correspondences between sets of objects, man finally developed a more advanced kind of one-to-one correspondence by using symbols. Most of the ancient civilizations invented their own symbolic systems for counting, and after many centuries of development the number symbols and names that we use today were produced. Our number symbols 1, 2, 3, 4, . . . are of Hindu-Arabic origin. (*Note:* The three dots after "4" indicate that the numbers go on and on in the same manner.) A one-to-one correspondence between a set of stick men and the set of Hindu-Arabic symbols that are used to count the stick men might be illustrated as in the figure.

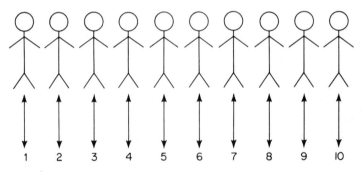

In counting, the number names beginning with 1 are matched in their order with the members of the set, and we know that the set contains as many elements as the last number named. The relationship between counting and one-to-one correspondence is very close, for essentially they are both matching processes. For example, to count the members of the set $\{\Box, \triangle, \Diamond, \bigcirc\}$ we would think about a pairing of the elements with numbers. The last number we use in this matching process is the one that tells us how many elements the set contains. Hence, we know that there are four elements in this set.

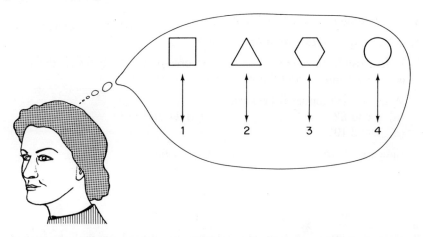

1-6 Counting Numbers (Natural Numbers)

Let us agree that the first counting number is 1. We then assume that each counting number is followed by another counting number that is 1 more than its predecessor. In this way, the set of **counting numbers** may be generated beginning with 1 and continually counting one more. Hence, the sequence of counting numbers continues without end, as illustrated by the three dots in the example

$$C = \{1, 2, 3, 4, \ldots\}.$$

The set C of counting numbers is also called the set N of **natural numbers**:

$$N = \{1, 2, 3, 4, \ldots\}.$$

Some textbooks refer to the numbers 0, 1, 2, 3, . . . as the natural numbers. In this text, the set of natural numbers will be denoted by the set N, and it will be understood to mean the set $N = \{1, 2, 3, \ldots\}.$

1-7 Whole Numbers

The set of **whole numbers** is the set of counting numbers and zero. We may represent the set of whole numbers as

$$W = \{0, 1, 2, 3, 4, \ldots\}.$$

We notice that the three dots are again used to mean that there are more numbers than we can name. In other words, 4 is the last number named in set W, but it is certainly not the last whole number.

Exercises

1. What is the smallest whole number?

2. What is the smallest counting number?

3. What is the largest whole number?

4. What is the largest counting number?

5. Is it possible to make a one-to-one correspondence between the set of counting numbers and the set of whole numbers?

6. What whole number is between
  (a) 6 and 8?   **(b)** 11 and 13?
  (c) 9 and 10?

7. Indicate whether each statement is true or false.
  (a) $0 \in W$ **(b)** $0 \in N$

  (c) $28 \in C$ **(d)** $\frac{1}{2} \in N$

  (e) $^{-}5 \in W$ **(f)** $.85 \in C$

8. If the elements of set D are in one-to-one correspondence with the elements of set E and if the elements of set E are in one-to-one correspondence with the elements of set R, what may we conclude about the elements of sets D and R?

1-8 Cardinal Numbers

Now let us use one-to-one correspondence to develop the meaning of number. The notions of set and of matching of sets are basic and must be a part of every child's experience.

Suppose we consider a set of two space capsules, a set of two books, a set of two airplanes, and a set of two apples. What is common to all of these sets? Obviously, these sets all have the same number of elements, namely two. We shall *imagine* that the same name tag or label is associated with each of the sets.

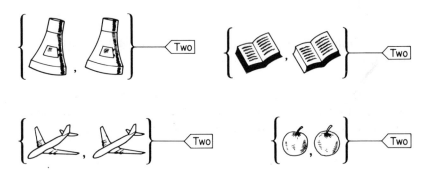

We shall call this mental label a **number** and shall attach this same number concept to any set that can be placed in one-to-one correspondence with any one of the given sets. The property common to all the given sets and to any other set that may be placed in one-to-one correspondence with the given set is "twoness." We naturally need a symbol to represent this property of twoness, and so we agree to use the symbol 2, which was passed down to us by our ancestors. It is important to realize, however, that our number system is man-made and the symbol 2 was invented to represent this number property of twoness, and that we could invent another symbol for the purpose of representing the number two if we so desired. In fact, the Romans attached the symbol II to sets having the twoness property.

The number of elements in a set is often referred to as the **cardinal number** of the set. Cardinal numbers are used to answer the question, "How many members are in the set?". In counting "one, two, three, four," the number names are matched in order with the elements of the set being counted, without regard for the arrangement of the elements, until each element has been placed in one-to-one correspondence with a counting number. For example, consider the set $A = \{\triangle, \hexagon, \bigcirc, \square\}$ and the set of counting numbers. Notice in each of the next two figures that the elements of set A may be arranged in more than one way for counting.

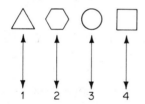   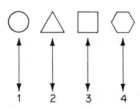

The number matched with the last element of the set is the cardinal number of the set. Hence, the cardinal number of set A is 4; that is, as shown in Section 1-5, there are 4 elements in the set. We write $n(A) = 4$, "the number of elements in set A is 4."

Exercises

1. Given set $B = \{a, b, c\}$; set $C = \{0, 2, 4, 6, 8, 10\}$; set $D = \{0, 5, 10, 15, \ldots, 35, 40\}$; and set $E = \{0, 6, 12, 18, \ldots, 72, 78\}$. Copy and complete.
   (a) $n(B) = $ ____
   (b) $n(C) = $ ____
   (c) $n(D) = $ ____
   (d) $n(E) = $ ____

2. Use braces to show an example for a set K with:
   (a) $n(K) = 0$
   (b) $n(K) = 2$
   (c) $n(K) = 5$
   (d) $n(K) = 18$

1-9 Ordinal Numbers

In counting to determine the position of an item in a group, or the order of the items, a number is used in an ordinal sense. Thus, a person may be seated in the fifth row of a theatre. Page numbers in a book are another illustration of numbers used in an ordinal sense. The common names for the ordinals are "first," "second," "third," "fourth," and so on.

In counting to determine the cardinal number of a set, we matched the counting numbers in order with the elements of the set being counted without regard for the arrangement of the elements in the set. If we impose an order on the elements of the set being counted, each element of the set has an **ordinal number.**

Ordinal numbers are used to answer such questions as, "Which element in the set is first?". We use numbers in an ordinal sense when the elements of a set have been arranged in order, and the numbers are used to identify in their order the elements of the set.

$$R = \left\{ \triangle, \diamondsuit, \bigcirc, \square \right\}.$$

$$S = \left\{ 1, 2, 3, 4 \right\}.$$

Consider this one-to-one correspondence between the elements of the sets R and S. When we say that R has 4 elements, the number 4 is used as a cardinal number. When we say that $\square$ is the fourth element in the set, the number 4 is used as an ordinal number.

Exercises

1. Indicate whether the number in each of these sentences is cardinal or ordinal.
   (a) The table has only four legs.
   (b) He is a member of Lodge 689.
   (c) Twenty-nine children are in Room 14.
   (d) "Name three words that are spelled with four letters."

2. What is the cardinal number of each of the following sets?
   (a) $\{@, \#\}$ (b) $\{x, y, z, g, h\}$
   (c) $\{ \ \}$ (d) $\{0\}$

3. Identify the number used in each of the following as either a cardinal or ordinal number.

(a) Charles IV   (b) 500-word essay
(c) 35-hour week   (d) First Prize

4. Show two sets:
   (a) That have the same cardinal number, three.
   (b) In which the cardinal number of one set is four greater than the cardinal number of the other.

1-10 Distinction between Number and Numeral

To a mathematician a **number** is an idea. It is an abstract concept that is used in thinking about the elements in a set. One cannot see, read, write, or draw a *number*.

Each of these sets contains different elements:

$$\{c, d, e\}, \quad \{@, \%, \phi\}, \quad \{\triangle, \square, \bigcirc\}, \quad \{\text{Bill, Tom, Harry}\}.$$

A common property of these sets is the property of threeness. This is the property that is meant when we speak of the number three. The property of number is abstract. All numbers are abstract ideas, and the symbols we write are the names for these abstract ideas. The symbols are called **numerals.**

It is important to understand that we *operate* with numbers but we *write* numerals for the purpose of communicating our results. The symbols 2, V, $4 - 3$, $1 \div 3$, and 2×3 are all numerals.

A NUMBER is an IDEA.

A NUMERAL is a SYMBOL.

Exercises

1. Which is larger, the number 6 or the number 7?

2. Which is larger, the numeral 3 or the numeral 4?

3. Which of the following numerals are names for the same number?
   (a) 2   (b) $11 - 8$
   (c) 2×0 (d) $7 - 5$
   (e) $1 + 3$ (f) $6 \div 3$

1-11 Finite and Infinite Sets

A set may contain any number of elements. In some cases, the number of members may be so large that it would be very tiresome to list them all; for example, the set of all natural numbers less than 500. It is therefore convenient to use three dots to indicate that some of the members are left out of a listing, as $\{1, 2, 3, \ldots, 498, 499\}$ for the set of natural numbers less than 500. The three dots denote that the natural numbers in sequential order from 3 to 498 are included in the set. This notation is obviously quite a time- and space-saver.

Sometimes there is no last number in a set and therefore no end to the list of elements of the set, as, for example, in the set of natural numbers and the set of whole numbers. We say that these sets have an infinite number of elements. They cannot be counted so that the counting comes to an end. Hence, we say that the set of natural numbers is an *infinite set*, and we represent it by $\{1, 2, 3, \ldots\}$. The three dots are used in this case to mean that the numbers continue in the same manner indefinitely; that is, we can always count one more. In general, any set that does not have a whole number as its cardinal number is an **infinite set.**

When a set is not infinite, we can count the elements in it (using whole numbers), and it is then said to be a **finite set.** Any finite set is either the empty set or has some cardinal number that we may call n for some natural number n. Some examples of finite sets are the set of children in a particular class, the set of players on a football team, the set of people living in West Virginia, and the set of all natural numbers less than 500.

Mathematics deals with both finite and infinite sets.

Exercises

In Exercises 1 through 10, indicate whether each set is finite or infinite.

1. The set of natural numbers.

2. The set of letters in our alphabet.

3. The set of people living in California.

4. The set of two-digit numerals ending in 9.

5. The set of odd numbers evenly divisible by 10.

6. The set of whole numbers.

7. The set of even whole numbers.

8. The set of counting numbers less than 999.

9. The set of players on a basketball team.

10. The set of grains of sand on Pensacola Beach.

11. Use braces to show each of these finite sets.
   (a) The set of natural numbers less than 3.
   (b) The set of whole numbers between 15 and 20.

(c) The set of even whole numbers less than 2.

(d) The set of multiples of 6 from 6 to 108 inclusive.

12. Use braces to show each of these infinite sets.

(a) The set of natural numbers greater than 10.

(b) The set of odd whole numbers.

(c) The set of multiples of 5.

(d) The set of natural numbers evenly divisible by 7.

1-12 Equal and Equivalent Sets

Two sets are said to be **equal** (they may also be referred to as **identical**) if every element of each set is an element of the other. Consider these sets:

$$A = \{\text{Elliott, Anne, James}\},$$

$$B = \{\text{Anne, Elliott, James}\}.$$

Then A is equal to B because "A" and "B" are different names for the same set. Recall that the order of the elements listed in a set is immaterial; thus, when the statement "$A = B$" is written, it means that there is one set that has been given two names. In reference to sets, the term "equal" is restricted to the meaning "same as," that is, "identical."

It is not always easy to determine the equality of sets. For example, consider the sets of numbers $D = \{2, 4, 6\}$ and $E = \{3 \times 2, 64 \div 32, 3 + 1\}$; it is possible to determine that $D = E$, but it may not be immediately obvious.

If two sets have the same cardinal number, their elements can be placed in one-to-one correspondence. Such sets are **equivalent sets.** Any two equivalent sets have the same number of elements, but the elements are not necessarily identical. For example, the sets $X = \{1, 2, 3\}$ and $Y = \{\triangle, @, \#\}$ are equivalent because each set contains the same number of elements, but, since the elements are not identical, set X is not equal to set Y.

It should be noted that equal sets are always equivalent sets but equivalent sets are not necessarily equal sets. Two commonly used symbols for denoting the equivalence of sets are $\leftrightarrow$ and $\sim$. In the previous example, the equivalence of sets X and Y may be denoted as $X \leftrightarrow Y$ and as $X \sim Y$ (read as "the set X is equivalent to the set Y"). It would be *incorrect* to say that $X = Y$.

An easy way to determine whether or not two sets are equivalent is to count the elements of each set. As we found in Section 1-8, the number of elements of a set T is the cardinal number of the set T and is written $n(T)$. The number of elements in the set of even natural numbers less than 10, $T = \{2, 4, 6, 8\}$, is 4, and we write $n(T) = 4$. The set M, where $M = \{\text{Mary,}$ Gene, Sandra, Allen$\}$, also has 4 members; thus, $(n)M = 4$. Since $n(T) = 4$ and $n(M) = 4$, the elements of the sets can be placed in one-to-one correspondence. Hence, the cardinal numbers of the sets are the same, and $T \leftrightarrow M$.

Exercises

1. Compare the sets of each pair and indicate whether they are equal sets, equivalent sets, or neither.
   (a) $\{d, e, f, g\}$ and $\{f, d, g, e\}$.
   (b) $\{2, 4, 6, 10\}$ and the set of even natural numbers less than 12.
   (c) $\{1, 3, 5, 7, \ldots\}$ and $\{2, 4, 6, 8, \ldots\}$.
   (d) $\{a, b, c, d, e\}$ and $\{c, e, a, f, b, d\}$.
   (e) $\{$Paul, Dave, George$\}$ and $\{$Pat, Lou, Mary$\}$.

2. What is the cardinal number of each set?
   (a) $A = \{\ \ \}$ (b) $B = \{*, @, \#\}$
   (c) $C = \{a, b, c, d, e, f\}$ (d) $D = \{0, 2, 3, 1, 7, 9, 6, 12\}$
   (e) $E = \{0\}$

3. Use braces to show a set that is equivalent to:
   (a) $\{r, s, t\}$ (b) $\{0, 3, 6, 9, \ldots, 24\}$
   (c) $\{1, 3, 5, 7, \ldots, 99\}$

4. Use braces to show a set that is equal to:
   (a) $\{n, c, t, m\}$ (b) $\{0, 4, 8, 12, \ldots, 36\}$
   (c) $\{0, 10, 20, 30, \ldots, 100\}$

5. The sets $R = \{p, a, b, g\}$, $S = \{b, p, g, a\}$, and $T = \{g, t, a, b\}$ are given.
   (a) Does $R = S$?
   (b) Does $R = T$?
   (c) Does $T = S$?
   (d) Are sets R, S, and T equivalent sets?

6. If $A = \{7, 9, 11\}$ and $B = \{7 + 4, 14 - 5, 21 \div 3\}$, does $A = B$?

7. List the elements of the set Y, when it is known that $X = Y$ and $X = \{$John, Ruth, Bob$\}$.

8. Consider the sets $A = \{2, 3, 4, 5, 6\}$, $B = \{2, 4, 6, 5, 3\}$, $C = \{7, 6, 5, 4, 3\}$, $D = \{5, 7, 4, 3, 6\}$, and $E = \{6, 4, 2, 5, 3\}$. Which of these sets are:
   (a) Equal to A?   (b) Equal to D?
   (c) Equal to E?   (d) Equivalent?

1-13 Relations of Equality and Inequality

In the study of mathematics we are continually involved with relations between numbers. A number may be **equal to, not equal to, less than,** or **greater than** a given number. Traditional elementary school mathematics courses have stressed equalities. The symbol $=$ is read "is equal to." We use this symbol between two numerals to indicate that both numerals represent the same number. For example,

$9 = 9$, nine is equal to nine;

$8 + 2 = 10$, eight plus two is equal to ten.

For convenience in representing a relation between two numbers that are not equal, the symbol $\neq$ is used, meaning "is not equal to." For example,

$6 \neq 7$, six is not equal to seven;

$8 + 4 \neq 5$, eight plus four is not equal to five.

The relation "is not equal to" often is not restrictive enough in comparing two numbers. Many times it is important to know which of the numbers is the larger. Then the symbol $>$, which is read "is greater than," may be used. For example,

$5 > 3$, five is greater than three;

$2 > 1$, two is greater than one.

It is just as important to be able to symbolize the relation "is less than." Symbolically, this relation is represented by $<$. For example,

$1 < 8$, one is less than eight;

$0 < 9$, zero is less than nine.

If a and b are whole numbers, $a > b$ means $b < a$ and $a = b + n$ for some natural number n. Consider the whole numbers 6 and 7. Since $7 = 6 + 1$,

$$7 > 6 \quad \text{and} \quad 6 < 7.$$

Observe that in the symbols $>$ and $<$ the pointed, or small, end is next to the numeral representing the smaller number, and the open, or large, end is next to the numeral representing the larger number. The symbols $\geq$ and $\leq$ are used to mean "is greater than or equal to" and "is less than or equal to," respectively.

Exercises

1. Use mathematical symbols to write each statement.
   (a) Seventeen is greater than eleven.
   (b) Twenty-five is less than fifty.
   (c) Nine is equal to six plus three.
   (d) Set G consists of the elements $\triangle$, $\square$, and $\diamondsuit$.
   (e) Two is not equal to three.
   (f) Eight is greater than or equal to seven.
   (g) Twelve is less than or equal to nine plus three.

2. Tell whether each statement is true or false.
   (a) $3 > 4$ (b) $8 > 7$
   (c) $9 + 4 \geq 13$ (d) $6 + 0 = 0$

    (e) $3 \times 6 > 21$ **(f)** $36 \div 3 \leq 9$
    (g) $7 + 9 < 12 + 5$ **(h)** $2 \times 6 \geq 7 + 5$
    (i) $6 \neq 2 + 4$

Copy and complete each statement. Write $<$ or $>$ in place of each blank.

3. If $6 < 10$, $6 + n$ ___ $10 + n$ for any whole number n.

4. If $294 < 492$, $294 + n$ ___ $492 + n$ for any whole number n.

5. If a, b, and n are whole numbers and $a < b$, $a + n$ ___ $b + n$.

6. If $10 > 6$, $10 \times n$ ___ $6 \times n$ for any natural number n.

7. If $41 > 40$, $41 \times n$ ___ $40 \times n$ for any natural number n.

8. If a, b, and n are natural numbers and $a > b$, $a \times n$ ___ $b \times n$.

9. If $5 < 6$ and $6 < 7$, 5 ___ 7.

10. If $28 < 29$ and $29 < 30$, 28 ___ 30.

11. If a, b, and c are whole numbers, where $a < b$ and $b < c$, a ___ c.

12. If $15 > 14$ and $19 > 18$, 15×19 ___ 14×18.

13. If $47 > 23$ and $52 > 47$, 47×52 ___ 23×47.

14. If a, b, c, and d are natural numbers, where $a > b$ and $c > d$, $a \times c$ ___ $b \times d$.

1-14 Number Sentences

When we discuss number relations and write, for example, that seven is less than eight, $7 < 8$, we are writing a **number sentence.** It is important to note that the statement represented by a number sentence may be either *true* or *false.* The statement "five plus four is equal to nine," $5 + 4 = 9$, is a true statement; the statement "five plus four is not equal to ten," $5 + 4 \neq 10$, is a true statement; the statement "three is greater than four," $3 > 4$, is a false statement.

    We may also have such number sentences as $3 + \square = 5$ and $7 + 4 = \triangle$. These are called **open number sentences,** and we cannot determine whether they express true or false statements until each of the symbols $\square$ and $\triangle$ is replaced by a symbol for a particular number. The symbol $\square$ or any other symbol used in place of a numeral is a **placeholder,** also called a **variable.**

    An open number sentence such as $4 + 3 = \square$ in terms of the symbol $=$ is usually called an **equation,** whereas sentences such as $4 + 3 \neq \triangle$, $4 + 3 > \bigcirc$, and $4 + 3 < \bigcirc$ in terms of the symbols $\neq$, $>$, and $<$ are called **inequalities.**

    Relations of equality and inequality are very important in mathematics;

they are being introduced in schools at the primary level and their use is being extended to all grade levels. The number line (Section 1-15) is an excellent device for illustrating these number relations.

Exercises

1. Use the correct symbol (= or ≠) to complete each number sentence.
    (a) 1 _____ 3   (b) 9 + 7 _____ 3 × 5
    (c) 5 + 3 _____ 4 + 3 + 1   (d) 4 + 6 _____ 36 ÷ 4
    (e) 21 − 8 _____ 7 × 2   (f) 29 × 7 _____ 34 × 6

2. Write <, >, or = in the blank so that each number sentence is a true statement.
    (a) 3 + 4 _____ 4 + 3   (b) 9 _____ 6
    (c) 2 + 3 _____ 7   (d) 2 × 3 _____ 3 × 2
    (e) 7 + 8 _____ 8 + 5   (f) 42 − 19 _____ 19 + 3

3. From the set of whole numbers select the number or numbers that will make each open number sentence a correct statement.
    (a) □ + 3 = 11   (b) △ + △ − 3 = 5
    (c) 3 + ○ < 4   (d) ◇ − 3 > 15
    (e) □ + △ = 10 and □ − △ = 2   (f) △ + 2 ≤ 13

4. Tell which of the following number sentences are open sentences. If possible, tell whether the number sentence is true or false.
    (a) 6 + 3 = 4 + 5   (b) 7 − 5 > 4
    (c) 2 + □ < 6   (d) 3 ≠ 2 + □
    (e) 3 × 4 > 3 + 8   (f) 2 + 3 + 4 < 15 − 2

1-15 Number Line

The **number line** is a widely used teaching device. It is a representation of numbers by points on a line, which is usually pictured horizontally and is thought of as extending without end both to the right and to the left. This endless extension of the line in both directions is sometimes indicated in drawings by arrowheads on the line.

$$\longleftarrow \rule{4cm}{0.4pt} \longrightarrow$$

    To represent (to *graph*) the whole numbers on the number line, an arbitrary point on the line is selected as the graph of the number zero; then a unit of length is chosen and marked off on the line, beginning at the graph of 0 and extending to the right. The right endpoint of this unit segment is the graph of 1.

We use the unit segment from the graph of 0 to the graph of 1 to mark off consecutive points to the right. In this manner, illustrated in the following figure, we locate evenly spaced points on the line as the graphs of the numbers 2, 3, 4, 5, 6, and so on.

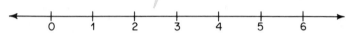

The number line is like a ruler, except that conceptually the number line extends without end. Hence, the number line is a straight line and includes points on it that can be placed in one-to-one correspondence with the set of whole numbers. Each number is the **coordinate** of the corresponding point on the number line; each of these points is the **graph** of the corresponding number. Thus, we say that the coordinate of the point F is the whole number 5 and that the graph of the whole number 5 is the point F.

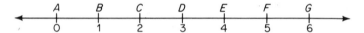

As previously mentioned, the number line may be used to illustrate number relations. For example, consider the numbers 3 and 5 graphed on the following number line:

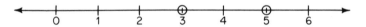

Here 5 is greater than 3, $5 > 3$; notice that the graph of 5 is to the right of the graph of 3 on this number line. 3 is less than 5, $3 < 5$; notice that the graph of 3 is to the left of the graph of 5 on the number line.

In our discussion of the number line we have assumed that the graph of 1 is located to the right of the graph of 0. From this assumption it follows in general that a number with a graph on the right of the graph of another number on the line is the greater number. A number with a graph on the left of the graph of another is the lesser number. For example,

> $9 > 6$, and the graph of 9 is on the right of the
> graph of 6 on the number line;

> $4 < 6$, and the graph of 4 is on the left of the
> graph of 6 on the number line.

The equality of two numbers may also be shown on the number line. Recall that equality of two numbers indicates that the numerals representing the numbers are actually names for the same number. Hence, two equal numbers are coordinates of the same point on the number line. An example, shown below, is the relation $\frac{6}{2} = 3$, in which $\frac{6}{2}$ and 3 are symbols for the same number.

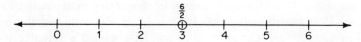

In the case of a relation between two numbers that are not equal, the number line may also be used. If two numbers are not coordinates of the same point on the number line, they are not equal. Notice that in the relation $4 \neq 3$ the graphs of 4 and 3 are not the same.

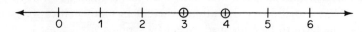

Exercises

1. What do we know about the graphs of two numbers on the number line
    (a) When one number is less than the other?
    (b) When one number is greater than the other?
    (c) When the two numbers are equal?

2. Use a number line to show:
    (a) $4 > 3$ (b) $2 < 5$
    (c) $2 \neq 3$ (d) $5 = \frac{10}{2}$

3. Refer to the following number line, in which letters are used to label points, and indicate which point is the graph of each of the numbers.
    (a) 3   (b) 0
    (c) 6   (d) 4
    (e) 2   (f) 1

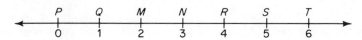

4. Refer to the number line given in Exercise 3 and give the coordinate of
    (a) Point *P*.
    (b) Point *Q*.
    (c) The point midway between *M* and *R*.
    (d) The point whose coordinate is twice the coordinate of point *N*.
    (e) The point whose coordinate is the sum of the coordinates of point *Q* and point *R*.

1-16 Precise Mathematical Language

One characteristic of mathematics is the precision of its language. Most of the modern programs in mathematics carefully develop correct usage of mathematical vocabulary. Mathematics is an exact science and requires an exact language. If students are trained in elementary mathematics to communicate ideas clearly and precisely, it is likely they will find the training useful in other areas. Certainly, they will have gained a valuable tool for seeing mathematical relationships in greater depth.

Every branch of learning has a vocabulary of its own, and mathematics is no exception. Definitions of terms in mathematics should always be kept in mind as definitions for this field only, regardless of dictionary definitions. For example, many dictionaries make little or no distinction between words such as *number* and *numeral*, and some even define them as synonyms. The word *set* has a dictionary definition, whereas in mathematics we generally consider the term to be undefined except by synonym.

The precise mathematical language that is used in most modern mathematics programs is well within the grasp of most elementary school children after they have had experience with the concepts involved. The use of mathematically correct language is important, and students should be encouraged to use it at the appropriate time.

1-17 Chapter Test

Indicate whether each statement is true or false. If false, tell why.

1. The number 4 is larger than the number 5.

2. The cardinal number of the set $\{0\}$ is zero.

3. If C is the set of counting numbers and N is the set of natural numbers, then $C = N$

4. Counting is finding "how many" by a matching process.

5. If $K = \{16\}$, then $1 \in K$.

6. Between any two whole numbers, there exists at least one whole number.

7. The number 3 is used as a cardinal number in this sentence: "She sits in seat number 3."

8. A number sentence is either true or false.

9. If $A = \{0, 2, 4, 6, \ldots, 24\}$, then A is an infinite set.

10. On the number line as we conventionally picture it, each point is the coordinate of a number.

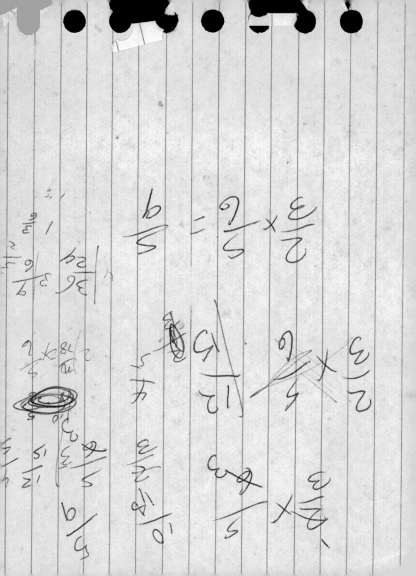

$$(X)(Y) = (Y)(X)$$

$(-)\times(-) = +$

$(-)\times(+) = -$ or $(+)\times(-)$ any time you multiply you can switch the signs.

$(+)\times(+) = +$

$(-)\div(-) = +$

$(-)\div(+) = -$

$(+)\div(+) = +$

$3 - (-4 \div 2 \times 3) - 4$

$-3 + 4$ $6 - 4$

$X - Y =$ The sign of the biggest

$X + Y =$ the sum or difference with the sign of $|x|$ or $|y|$

X or Y

Select the best possible answer.

11. Which of these statements apply to these two sets?

    I. They are equal.
   II. They are equivalent.
  III. They have the same cardinal number.
  (a) I only.
  (b) III only.
  (c) I and III only.
  (d) II and III only.

12. A number is:
  (a) An idea.
  (b) A numberal.
  (c) A symbol.
  (d) An operation.

13. The members of the set of whole numbers are:
  (a) 0, 1, 2, 3, 4, 5, 6, 7, 8, 9.
  (b) 0, 1, 2, 3, . . . , 99.
  (c) 1, 2, 3, 4, 5, ...
  (d) 0, 1, 2, 3, 4, 5, ...

14. Which statement best describes the number zero?
  (a) It is nothing.
  (b) It is a placeholder.
  (c) It is the cardinal number of the empty set.
  (d) It is the numeral found in the empty set.

chapter 2

Sets, Relations, and Operations

Set language is one of the important threads that unify many fundamental mathematical experiences. We have learned that a set is a well-defined collection of elements and that an empty set is a set containing no elements. The basic notions about sets are being used effectively to develop the understanding that supports the structure of mathematics. In this chapter we shall learn more about set language, relations among sets, and operations on sets.

2-1 Sets and Subsets

Often it is necessary to discuss sets that are part of a given set, that is, sets whose elements are also elements of the given set. Consider the following set of plates.

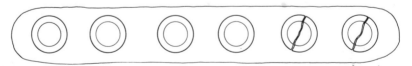

We can see that part of this set consists of broken plates and part of the set consists of plates that are not broken. We can say that the set of broken plates is a **subset** of the set of plates. We can also say that the set of plates that are not broken is a *subset* of the set of plates.

Now consider the set of students in a particular class. Assume that C represents the set containing all the members of the class and B represents the set consisting of all the boys in the class. Then each boy in the set B is a member of the set C and B is called a subset of C. In general, a set B is said to be a **subset** of a set C if each element of B is also an element of C. The symbolism $B \subseteq C$ is used to mean "B is a subset of C" or, as some people say, "B is included in C." If we let K represent the set of girls in the class who are ten feet tall, then K would be an empty set. However, the set G of girls in the class, regardless of height, is a subset of set C. Therefore, K is a subset of G and G is a subset of C. Thus $K \subseteq G$, $G \subseteq C$, and $K \subseteq C$. We will assume that the empty set is a subset of every set.

It should also be mentioned that a subset may contain all the elements of the set. For example, if you teach in a school for boys, and set T represents all the members of your class and set M represents all the boys in your class, then M is a subset of T. Since set M contains all the elements of set T, $M = T$. From this example, you may conclude that each set is a subset of itself.

It is important to recognize subsets of a given set that are different from the set itself. Thus, if set $L = \{a, b, c\}$, there are many subsets of L that are different from L. These subsets are

$$\{a\},\ \{b\},\ \{c\},\ \{a, b\},\ \{a, c\},\ \{b, c\},\ \text{and}\ \{\ \ \}.$$

Each subset of L that is different from L is called a **proper subset** of L. If $A = \{a, b\}$, we may write $A \subset L$, which is read "A is a proper subset of L." If $B = \{a, b, c\}$, we must use the symbolism $B \subseteq L$, which means "B is a subset of L." Note that the symbol used to represent a proper subset, $\subset$, is different from the symbol used to represent a subset, $\subseteq$. In the previous illustration there is at least one element of L that is not an element of A; therefore we use the symbolism $A \subset L$ to denote that A is a proper subset of L. Since the empty set is a subset of every set, there are seven proper subsets of L.

Perhaps the following problem will clarify some of the points previously discussed. Consider a child who goes to the toy box to select something to play with. Suppose there are only three toys available: a boat, a car, and a spin top.

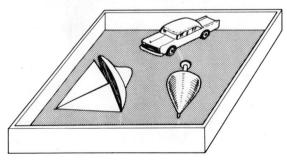

We will call the given set of choices C.

$$C = \left\{ \triangle , \; \text{🚗} , \; \triangledown \right\}$$

What possible selections can the child make from the toy box? The following choices come to mind:

$$P = \left\{ \triangle , \; \text{🚗} , \; \triangledown \right\}$$

$$Q = \left\{ \triangle , \; \text{🚗} \right\}$$

$$R = \left\{ \triangle , \; \triangledown \right\}$$

$$S = \left\{ \triangledown , \; \text{🚗} \right\}$$

$$T = \left\{ \triangle \right\}$$

$$V = \left\{ \text{🚗} \right\}$$

$$W = \left\{ \triangledown \right\}$$

$$X = \left\{ \quad \right\}.$$

Notice that each of these eight possible choices is a subset of the original set of choices:

$$C = \left\{ \triangle , \; \text{🚗} , \; \triangledown \right\}$$

Observe that the given set C is equal to P; $C = P$. Thus, we must write $P \subseteq C$ ("P is a subset of C"); since P is not a proper subset of C, $P \not\subset C$. Observe also that the other seven subsets are all proper subsets; $Q \subset C$, $R \subset C$, $S \subset C$, $T \subset C$, $V \subset C$, $W \subset C$, and $X \subset C$.

Exercises

1. Let $A = \{ \square, \bigcirc \}$; $B = \{ \bigcirc \}$; $C = \{ \quad \}$;

   and $D = \{ \square, \bigcirc, \star \}$.

   (a) Which of the sets are subsets of A?
   (b) Which of the sets are proper subsets of A?
   (c) Which of the sets are not subsets of A? Why?

2. Let $E = \{e, f, g, h\}$; $F = \{h, g, f, e\}$; $G = \{e, f, m\}$; $H = \{f, h, e\}$; and $J = \{g\}$. Tell whether each of these statements is true or false. If the statement is false, tell why.

   (a) $F = E$ (b) $F \subseteq E$
   (c) $F \subset E$ (d) $H \subset E$
   (e) $E \subset H$ (f) $G \subset E$
   (g) $G \subseteq E$ (h) $g \in E$
   (i) $J \in E$

3. Let $S = \{1, 2, 3\}$.
   (a) List all the possible subsets of S.
   (b) Which of the subsets listed in Exercise 3(a) is not a proper subset of S? Why?

4. Let $K = \{ \ \}$ and $N = \{0\}$. Tell whether each of these statements is true or false. If the statement is false, tell why.

   (a) $K = N$ (b) $0 \in N$
   (c) $0 \in K$ (d) $0 \subset K$
   (e) $\{ \ \} \subset K$ (f) $\{ \ \} \subseteq K$
   (g) $0 \subset N$ (h) $\{ \ \} \subset N$
   (i) $\{ \ \} \subseteq N$ (j) $\{0\} \subset N$
   (k) $\{0\} \subseteq N$ (l) $K \subset N$

5. Let $M = \{0, 5, 10, 15, 20, 25, 30, 35, 40, 45, 50\}$.
   (a) Find a subset of M which contains all the even numbers in M.
   (b) Find a subset of M which contains all the odd numbers in M.
   (c) List five proper subsets of M that have 4 members each.
   (d) List a subset of M that has 11 members. Is it a proper subset?

6. If set P is a proper subset of a finite set F, what can you conclude about the cardinal numbers of the sets P and F? Explain your answer.

7. Given a set with three elements, what is:
   (a) The greatest number of subsets that can be formed?
   (b) The greatest number of proper subsets that can be formed?

8. Complete this table.

Set	Subsets	Number of Elements in Original Set	Number of Subsets
{ }	{ }	0	1
{d}	{d}, { }	1	2
{d, e}	{d}, {e}, {d, e}, { }		
{d, e, f}			
{d, e, f, g}			

9. Use the information summarized in Exercise 8 to theorize a formula for the number S of subsets that can be formed from the original set consisting of x elements. Test your theory on a set consisting of five elements; on one consisting of six elements.

2-2 Universal Set

Whenever we discuss a set of elements, we have in mind a particular collection of objects or ideas that have some common characteristic. For example, we may be considering a set of jet airplanes or a set of pentagons. The common characteristic of the members of each of these sets is obvious: the airplanes are powered by jet engines and the pentagons are geometric figures with five sides. It should be noted that the elements of each of these sets can be thought of as elements of a larger set: the jet airplanes are elements also of the set of all airplanes and the pentagons are elements also of the set of all geometric figures.

In any situation the set from which all subsets under discussion are derived is called the **universal set,** also referred to as the **universe.** The universal set is denoted by the capital letter U. Thus, in a discussion about airplanes the universal set would be the set of all airplanes, and the set of jet airplanes would be a subset of it. You should note that the universal set is not the same for all situations.

Exercises

1. Describe a possible universal set for
   (a) The set of two-door automobiles.
   (b) The set of freshmen at George Wythe High School with B grades.

(c) The set of whole numbers between 16 and 23.

(d) The set of even numbers.

2. Given $U = \{1, 2, 3, 4\}$, indicate whether each statement is true or false. If the statement is false, tell why.

(a) $\{2, 3, 4\} \subseteq U$ (b) $\{2, 3, 4\} \subset U$

(c) $\{2, 3, 4, 1\} \subseteq U$ (d) $\{2, 3, 4, 1\} \subset U$

(e) $\{\ \ \} \subseteq U$ (f) $\{\ \ \} \subset U$

(g) $\{2, 3, 5\} \subseteq U$ (h) $\{2, 3, 5\} \subset U$

(i) $4 \in U$

3. Think of each of the following sets as a universal set, and select at least one proper subset of each.

(a) The set of all astronauts.

(b) The set of three-sided geometric figures.

(c) The set of natural numbers.

(d) The set of whole numbers less than 10.

2-3 Solution Sets

When we write an open number sentence such as $4 + \square < 10$, and then try to find a number or numbers that will make the sentence true, we are looking for the **solution set** with reference to some universal set. In fact, the solution set may vary, depending on the universal set.

In the example $4 + \square < 10$, the solution set is $S = \{0, 1, 2, 3, 4, 5\}$, if the universal set is the set of whole numbers. However, if the universal set is the set of counting numbers, then the solution set for $4 + \square < 10$ is $S = \{1, 2, 3, 4, 5\}$. Note that 0 is not a counting number.

Exercises

1. Let $U = \{0, 1, 2, 3, 4, \ldots\}$. Find the solution set for:

(a) $\triangle < 7$ (b) $\square \leq 7$

(c) $2 + \square > 8$ (d) $6 + \triangle = 15$

(e) $14 \geq 9 + \triangle$ (f) $12 + \triangle = 8$

2. Let $U = \{0, 2, 4, 6, 8, 10, \ldots\}$. Find the solution set for:

(a) $\square + 4 < 11$ (b) $\triangle + 8 \leq 8$

(c) $2 + \triangle \geq 6$ (d) $6 + \square < 5$

(e) $\square = 7 + 6$ (f) $15 > 1 + \triangle$

3. Repeat Exercise 2 for $U = \{2, 4, 6, 8, 10, \ldots\}$.

4. Repeat Exercise 2 for $U = \{0, 3, 6, 9, 12, \ldots\}$.

5. Find the solution set for $5 - \triangle \geq 0$ if the universal set is the set of:

(a) Counting numbers.

    (b) Whole numbers.

    (c) Even whole numbers.

    (d) Odd whole numbers.

6. Repeat Exercise 5 for $(2 \times \triangle) + 3 \leq 19$.

7. Repeat Exercise 5 for $\triangle + \square = 3$.

8. Repeat Exercise 5 for $\triangle + \square \leq 4$.

2-4 Union of Sets

In arithmetic, we are familiar with the processes of addition and multiplication. These are called binary operations, and for our present purposes a **binary operation** may be defined as one that assigns to *two* given elements a third element. The word binary implies two.

    Two of the operations on sets which we will discuss are union and intersection. Each of these operations is considered to be a binary operation because it is an operation on two sets which produces one new set. The new set may coincide with one of the given sets.

    Children intuitively discover what is meant by the union of sets when they combine the elements of two sets of things to form a single set; if the elements of $\left\{ \vcenter{} , \vcenter{} \right\}$ are joined with the element of $\left\{ \vcenter{} \right\}$, one obtains the new set $\left\{ \vcenter{} , \vcenter{} , \vcenter{} \right\}$. A teacher may think of the set of children in her class as the union of the set of girls in the class and the set of boys in the class. Consider two sets of boys:

$$R = \{\text{Russell, Dave, George}\},$$

$$G = \{\text{Gray, Bob, Paul, Bill}\}.$$

Suppose we wish to form a new set consisting of all the members of set R and all the members of set G. This new set is called the union of set R and set G and is written $R \cup G$. This is read "R union G." For this example, $R \cup G = \{\text{Russell, Dave, George, Gray, Bob, Paul, Bill}\}$.

    The **union** of any two sets S and T, written $S \cup T$, is the set consisting of all elements in either S or T or in both S and T. In other words, an element is a member of $S \cup T$ if the element is a member of at least one of the sets S and T. The following examples illustrate the meaning of the union of sets.

Example 1   If $S = \{2, 4, 6, 8\}$ and $T = \{10, 12\}$, then

$$S \cup T = \{2, 4, 6, 8, 10, 12\}.$$

$$n(S) = 4, \ n(T) = 2, \text{ and } n(S \cup T) = 6.$$

Example 2   If $D = \{$James, Sam$\}$ and $E = \{$Don, Joe, Sam$\}$, then

$$D \cup E = \{\text{James, Sam, Don, Joe}\}.$$

$$n(D) = 2, n(E) = 3, \text{ and } n(D \cup E) = 4.$$

Notice that in Example 2 we do not write the name "Sam" twice. As soon as the name is written once, we know that Sam is a member of the set $D \cup E$. Thus the set $\{$James, Sam, Don, Joe, Sam$\}$ is the same as the set $\{$James, Sam, Don, Joe$\}$, and it is unnecessary to write "Sam" twice. In Example 3, the 5 and 7 appear only once in $X \cup Y$:

Example 3   If $X = \{5, 6, 7, 8\}$ and $Y = \{1, 5, 7\}$, then

$$X \cup Y = \{1, 5, 6, 7, 8\}.$$

$$n(X) = 4, n(Y) = 3, \text{ and } n(X \cup Y) = 5.$$

Exercises

1. Given $G = \{2, 4, 6\}$ and $H = \{1, 3, 5\}$, find $G \cup H$ and $n(G \cup H)$.

2. Given $A = \{a, b, c\}$ and $B = \{b, c, d\}$, find $A \cup B$ and $n(A \cup B)$.

3. Given $R = \{\%, @, \#\}$ and $S = \{\text{¢}, \&, @\}$, find $R \cup S$ and $n(R \cup S)$.

4. Given $D = \{2, 3, 4\}$ and $D \cup E = D$, list each of the sets that can be used for E.

5. If $D \cup E$ is the same set as D, what could we conclude about sets D and E?

6. Given $T = \{w, x, y\}$, $K = \{l, m, n\}$, $\emptyset = \{\ \}$, and $U = \{a, b, c, d, \ldots, x, y, z\}$. Tell whether each statement is true or false. If the statement is false, tell why.

   (a) $T \cup T = U$
   (b) $T \cup T = T$
   (c) $T \cup U = U$
   (d) $T \cup U = T$
   (e) $T \cup K = K \cup T$
   (f) $\emptyset = T \cup \emptyset$
   (g) $T = \emptyset \cup T$
   (h) $T \cup \emptyset = \emptyset \cup T$
   (i) $\emptyset \cup U = \emptyset$
   (j) $\emptyset \cup U = U$
   (k) $T \cup U = \emptyset$
   (l) $\emptyset \cup \emptyset = \emptyset$

7. What observation can you make about a set in union with:

   (a) Itself?
   (b) The empty set?
   (c) The universal set?
   (d) Another set in either order?

8. What observations can you make about sets R and S if:

   (a) $n(R \cup S) = n(R) + n(S)$?
   (b) $n(R \cup S) = n(S)$?
   (c) $n(R \cup S) = n(R)$?

9. If the statement is possible, give an example; if the statement is impossible, tell why.

   (a) $n(X) = n(X \cup Y)$
   (b) $n(X) + n(Y) > n(X \cup Y)$
   (c) $n(X \cup Y) > n(X) + n(Y)$
   (d) $n(X \cup Y) < n(X)$

2-5 Intersection of Sets

It is frequently desirable to obtain the set of elements common to two given sets. Consider the sets D and E used in our discussion of the union of two sets:

$$D = \{\text{James, Sam, Charles}\},$$

$$E = \{\text{Don, Joe, Sam}\}.$$

Let us find the set of boys who belong to both D and E. Clearly, this set is {Sam}.

We have found the set of elements common to sets D and E. This set, $D \cap E$, is the intersection of D and E. The symbol used for intersection is $\cap$. For this example, $D \cap E$ is read "D intersection E," and $D \cap E =$ {Sam}.

The **intersection** of any two sets M and N, written $M \cap N$, is the set consisting of the elements that are in both M and N. The following examples illustrate the meaning of the intersection of sets.

Example 1 If $M = \{1, 3, 5\}$ and $N = \{0, 3, 4, 6\}$, then

$$M \cap N = \{3\}.$$

$$n(M) = 3, n(N) = 4, \text{ and } n(M \cap N) = 1.$$

Example 2 If $G = \{d, e, a, c, b\}$ and $H = \{l, m, a, b, n, d, h\}$,

$$G \cap H = \{a, b, d\}.$$

$$n(G) = 5, n(H) = 7, \text{ and } n(G \cap H) = 3.$$

Example 3 If $A = \{1, 3, 4, 7\}$ and $B = \{5, 6\}$, then

$$A \cap B = \{\ \}.$$

$$n(A) = 4, n(B) = 2, \text{ and } n(A \cap B) = 0.$$

Notice that in Example 3 the sets A and B have no elements in common, that is, their intersection is the empty set. This could also be represented by $A \cap B = \varnothing$, since $\varnothing = \{\ \}$.

Exercises

1. Given $A = \{a, b, c\}$ and $B = \{b, c, d\}$, find $A \cap B$ and $n(A \cap B)$.

2. Given $R = \{\%, @, \#\}$ and $S = \{\mathord{\text{¢}}, \&, @\}$, find $R \cap S$ and $n(R \cap S)$.

3. Given $G = \{2, 4, 6\}$ and $H = \{1, 3, 5\}$, find $G \cap H$ and $n(G \cap H)$.

4. Given $M = \{2, 3, 4\}$, $N \subseteq M$, and $M \cap N = M$, list each of the sets that can be used for N.

5. If $N \cup M = M$ and $M \cap N = M$, what could we conclude about sets M and N?

6. Given $D = \{a, b, c\}$ and $D \cap E = D \cup E$, list each of the sets that can be used for E.

7. If $D \cap E = D \cup E$, what do we know about sets D and E?

8. What is the intersection of the set of odd whole numbers less than 10 and the set of whole numbers from 1 to 10 inclusive?

9. Let $A = \{0, 1, 2, 3, \ldots\}$, $B = \{0, 2, 4, 6, \ldots\}$, $C = \{1, 3, 5, 7, \ldots\}$, and $D = \{5, 10, 15, 20, \ldots\}$; find:
   (a) $B \cup C$ (b) $B \cap C$
   (c) $B \cap D$ (d) $B \cup D$
   (e) $A \cap C$ (f) $A \cup C$
   (g) $A \cup D$ (h) $A \cap D$
   (i) $A \cup B$ (j) $A \cap B$
   (k) $C \cap D$ (l) $C \cup D$

10. Given $X = \{a, b, c\}$; $Y = \{d, e, f\}$; $\varnothing = \{\ \}$; and $U = \{a, b, c, d, e, \ldots, x, y, z\}$. Tell whether each statement is true or false. If the statement is false, tell why.
   (a) $X \cap X = U$ (b) $X \cap X = X$
   (c) $X \cap U = U$ (d) $X \cap U = X$
   (e) $X \cap Y = Y \cap X$ (f) $\varnothing = X \cap \varnothing$
   (g) $X = \varnothing \cap X$ (h) $\varnothing \cap U = U$
   (i) $\varnothing \cap U = \varnothing$ (j) $X \cap Y = X \cup Y$
   (k) $X \cup \varnothing = X \cap \varnothing$ (l) $Y \cup U = Y$

11. What observation can you make about a set in intersection with:
   (a) Itself?   (b) The empty set?
   (c) The universal set?   (d) Another set in either order?

12. What observation can you make about sets P and Q if:
   (a) $n(P \cap Q) = n(P)$?   (b) $n(P \cap Q) = n(Q)$?
   (c) $n(P \cap Q) = n(P \cup Q)$?   (d) $n(P \cap Q) = n(\varnothing)$?

13. If the statement is possible, give an example; if the statement is impossible, tell why.
   (a) $n(W) = n(W \cap K)$ (b) $n(W) + n(K) > n(W \cap K)$
   (c) $n(W) + n(K) < n(W \cap K)$ (d) $n(W \cap K) > n(W \cup K)$

2-6 Disjoint Sets

Any two sets with no elements in common are called **disjoint sets**. Thus, two sets are disjoint when their intersection is the empty set. Consider the sets $A = \{$Russell, Dave, George$\}$ and $B = \{$Gray, Bob, Paul, Bill$\}$. The intersection of sets A and B is the empty set; hence, we say that A and B are disjoint sets.

Example 1   If $S = \{2, 4, 6, 8\}$ and $T = \{10, 12\}$, then $S \cap T = \{\ \}$. The sets S and T are disjoint sets.

Example 2   If $X = \{5, 6, 7, 8\}$ and $Y = \{1, 5, 7, 9\}$, then $X \cap Y = \{5, 7\}$. The sets X and Y are *not* disjoint sets.

Exercises

1. $G = \{1, 3, 5, 7\}$ and $H = \{2, 4, 6, 3\}$; find $G \cap H$ and $n(G \cap H)$.

2. $X = \{@, \#, \%\}$ and $Y = \{*, \&, ¢\}$; find $X \cap Y$ and $n(X \cap Y)$.

3. $R = \{3, 6, 9, 12\}$ and $S = \{1, 5, 11, 12\}$; find $R \cap S$ and $n(R \cap S)$.

4. $A = \{7, 14, 21, 28, \ldots\}$ and $B = \{4, 8, 12, 16, \ldots\}$; find $A \cap B$ and $n(A \cap B)$.

5. Which pairs of sets in Exercises 1–4 are disjoint?

6. If $n(K \cap P) = 0$,
   (a) What do we know about the intersection of sets K and P?
   (b) Are sets K and P disjoint?

7. If W and Z are disjoint sets,
   (a) What do we know about the intersection of sets W and Z?
   (b) What does $n(W \cap Z)$ equal?

2-7 Complement of a Set

If A is a subset of a universal set U, the set of all elements of U that are not elements of A is called the **complement** of A with respect to U. We use the symbol A' (read "complement of A") to designate the complement of A. It is important to note that $A \subseteq U$ and $A' \subseteq U$.

If $U = \{a, e, i, o, u\}$ and $A = \{a\}$, then $A' = \{e, i, o, u\}$. In general, you must refer to a given set and a universal set in order to establish a complement of the given set.

If $U = \{1, 2, 3, \ldots, 10\}$ and $B = \{1, 9, 8, 3\}$, then $B' = \{2, 4, 5, 6, 7, 10\}$.

Exercises

1. Suppose that the universal set is the set of natural numbers less than 5 and that $A = \{2, 4\}$; find A'.

2. If the universal set is the set of all automobiles and C is the set of Chevrolets, describe C'.

3. Suppose $U = \{1, 2, 3, 4, 5, 6, 7, 8, 9\}$, $A = \{1, 2, 3\}$, $B = \{3, 5, 7, 9\}$, and $C = \{6, 7, 8\}$; find each of the following sets.

(a) $A \cap U$ (b) $A \cup U$
(c) B' (d) $B' \cap U$
(e) $C' \cup U$ (f) $C' \cap B'$
(g) $(A \cap C)'$ (h) $(B \cup C)'$

4. What observations can you make about a set in:
   (a) Union with its complement?
   (b) Intersection with its complement?

5. Let $U = \{1, 3, 5, 7, 9\}$; $M = \{1, 3, 5, 7\}$; and $N = \{5, 7, 9\}$. Find:
   (a) $(M \cup N)'$ (b) $M' \cup N'$
   (c) $(M \cap N)'$ (d) $M' \cap N'$

6. Use the information from Exercise 5 to tell whether each statement is true
   or false.
   (a) $(M \cup N)' = M' \cup N'$ (b) $(M \cup N)' = M' \cap N'$
   (c) $(M \cup N)' = (M \cap N)'$ (d) $(M \cap N)' = M' \cup N'$
   (e) $(M \cap N)' = M' \cap N'$

2-8 Cartesian Product of Sets

The Cartesian product of two sets S and T is quite different from the union
or intersection of the two sets. In the **Cartesian product** of sets S and T,
each element is an ordered pair (s, t) where $s \in S$ and $t \in T$. The Cartesian
product of sets S and T is denoted by $S \times T$ which is read "the Cartesian
product of S and T," or simply as "S cross T." If we let $S = \{a, b, c\}$ and
$T = \{g, h\}$, $S \times T$ is the set of all possible pairs (s, t) formed by selecting the
first element s from S and the second element t from T. Thus

$$S \times T = \{(a, g), (a, h), (b, g), (b, h), (c, g), (c, h)\},$$

and

$$T \times S = \{(g, a), (g, b), (g, c), (h, a), (h, b), (h, c)\}.$$

We should note that $n(S) = 3$, $n(T) = 2$, and $n(S \times T) = 6$. Does $n(S \times T) = n(T \times S)$?

Exercises

1. Given $A = \{1, 2\}$ and $B = \{e, f\}$, find:
   (a) $A \times B$ (b) $n(A \times B)$
   (c) $B \times A$ (d) $n(B \times A)$

2. In Exercise 1, does:
   (a) $A \times B = B \times A$?   (b) $n(A \times B) = n(B \times A)$?

3. Given $C = \{red\}$ and $D = \{white, blue\}$, find:
   (a) $C \times D$ (b) $n(C \times D)$
   (c) $D \times C$ (d) $n(D \times C)$

4. In Exercise 3, does:

    (a) $C \times D = D \times C$?   **(b)** $n(C \times D) = n(D \times C)$?

2-9 Venn Diagrams

Venn diagrams, named after the English mathematician John Venn, are frequently used to present a visual representation of relations and operations on sets. The points of a rectangular region (that is, the set of all points on and inside the rectangle) may be used to represent the elements of a universal set. In the next figure, the rectangular region represents the set U of all students attending Prince George High School.

A subset of the universal set U may be represented by a circular region within the rectangular region. For example, we may let B represent the set of boys attending Prince George High School. The set B is illustrated by the shaded region in the next figure. Note that the shaded region includes all points on and inside the circle.

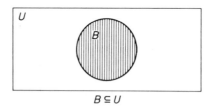

$B \subseteq U$

A proper subset of the set B may be represented by another circular region within the circular region B. For example, if F is the set of freshman boys attending Prince George High School, then the relationship may be represented as in the next figure.

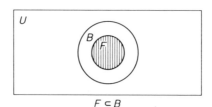

$F \subset B$

If B represents the set of boys attending Prince George High School, then B' denotes the set of girls attending Prince George High School. In the next figure we have represented B' by shading the points of U that are outside B. A dashed marking indicates the boundary of a shaded region when the boundary is not part of the region. Thus, the points on the circle are not included in the shaded region representing B'. Note that $B \cup B' = U$ and $B \cap B' = \varnothing$.

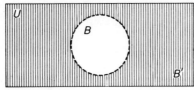

The union of sets D and E is shown in the next illustration. The elements of the set D are represented by the vertical shading; the elements of the set E, by the horizontal shading. The union of the two sets D and E is represented by the region that is shaded in either or both directions. Note that $D \cup E$ includes all the points of the set D as well as all those of the set E.

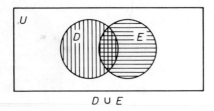

$D \cup E$

The intersection of sets D and E is pictured in the Venn diagram in the next picture. The elements of the set D are represented by vertical shading; the elements of the set E, by horizontal shading. The intersection of the two sets D and E is represented by the region that is shaded both vertically and horizontally. Note that $D \cap E$ includes only those points that are in both set D and set E.

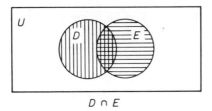

$D \cap E$

Venn diagrams may also be used to show relations and operations involving more than two sets. An illustration of $(J \cup K) \cap L$ is given in the following figure. The elements of $(J \cup K)$ are represented by horizontal shading and the elements of L are represented by vertical shading. Thus, $(J \cup K) \cap L$ is illustrated by the region that is shaded both vertically and horizontally.

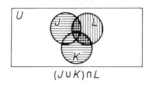

$$(J \cup K) \cap L$$

In the next figure, $K \cap L$ is represented by the region with vertical shading and J is represented by the region with horizontal shading. Thus, $J \cup (K \cap L)$ is illustrated by the region that is shaded in either or both directions.

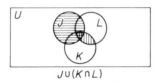

$$J \cup (K \cap L)$$

Does $(J \cup K) \cap L = J \cup (K \cap L)$? Study the previous two figures.

Exercises

1. Let $A = \{1, 3, 5, 7\}$ and $B = \{2, 4, 6, 8\}$.
   (a) Find $A \cup B$.   (b) Find $A \cap B$.

2. Copy the given figure and make Venn diagrams to show $A \cup B$ and $A \cap B$ for Exercise 1.

3. Let $M = \{\text{Louise, Ann}\}$ and $N = \{\text{Lucille, Louise}\}$.
   (a) Find $M \cup N$.   (b) Find $M \cap N$.

4. Copy the given figure and make Venn diagrams to show $M \cup N$ and $M \cap N$ for Exercise 3.

```
Ann              Lucille
        Louise
```

5. Draw Venn diagrams to represent two sets R and S such that $R \cap S \neq \varnothing$, and shade properly to show:
   (a) $R \cup S$ (b) $R \cap S$

6. Draw Venn diagrams to show $P \cup Q$ if:
   (a) $P \cap Q = \emptyset$ (b) $P \cap Q \neq \emptyset$
   (c) $P \subset Q$ (d) $Q \subset P$
   (e) $P = Q$ (f) $P \cup Q = Q$

7. Draw Venn diagrams to show $W \cap Y$ if:
   (a) $W \cap Y = \emptyset$ (b) $W \cap Y \neq \emptyset$
   (c) $W \subset Y$ (d) $Y \subset W$
   (e) $W = Y$ (f) $W \cup Y = Y$

8. If $G = H$, what relationship exists for $G \cup H$ and $G \cap H$?

9. Draw a Venn diagram to show $(A \cap B) \cup C$ where A, B, and C are subsets of U and the intersection of the three sets is not empty.

10. Draw Venn diagrams to show three sets:
    (a) A, B, and C such that $A \subset B$ and $B \subset C$.
    (b) X, Y, and Z such that all elements of X are elements of Y, all elements of Z are elements of Y, X and Z have no elements in common, and some elements of Y are not elements of either X or Z.

11. Draw Venn diagrams to represent two sets C and D such that $C \cap D \neq \emptyset$, and shade properly to illustrate:
    (a) C' (b) D'
    (c) $(C \cup D)'$ (d) $C' \cup D'$
    (e) $(C \cap D)'$ (f) $C' \cap D'$

12. In Exercise 11, does:
    (a) $(C \cup D)' = C' \cup D'$?   (b) $(C \cup D)' = C' \cap D'$?
    (c) $(C \cap D)' = C' \cap D'$?   (d) $(C \cap D)' = C' \cup D'$?
    (e) $(C \cup D)' = (C \cap D)'$?   (f) $C \cup D = C' \cap D'$?

13. Draw Venn diagrams to represent three sets R, S, and T such that $(R \cap S) \cap T \neq \emptyset$, and shade properly to illustrate:
    (a) $R \cup (S \cup T)$ (b) $(R \cup S) \cup T$
    (c) $R \cap (T \cup S)$ (d) $R \cap (T \cap S)$
    (e) $(R \cap T) \cup (R \cap S)$ (f) $(R \cup T) \cap (R \cup S)$

2-10 Chapter Test

Indicate whether each statement is true or false. If false, tell why.

1. If $A \subset U$, then $A \cap A' = \emptyset$.

2. Two sets H and W are disjoint if no member of H is a member of W.

3. G is a proper subset of R if every member of G is a member of R.

4. The empty set is a proper subset of every set.

5. If a set K has 3 elements, then there are $2 \times 3 = 6$ possible subsets of K.

6. If $U = \{0, 1, 2, 3, \ldots\}$, then the solution set for $\square \leq 5$ is $\{0, 1, 2, 3, 4\}$.

7. If $S = M$, then $S \cup M = S \cap M$.

8. If $n(D) = 4$ and $n(E) = 3$, then $n(D \times E) = 12$.

9. If $n(F) = 5$ and $n(Q) = 2$, then $n(F \cup Q) = 7$.

10. If $J = B$, then $J \subset B$.

Select the best possible answer.

11. How many proper subsets can be formed from a set with 4 elements?
   (a) 4^2 (b) 2^4
   (c) 15   (d) 14
   (e) 2^3

12. Which of the following represents the union of sets A and B?
   (a) $A = \{a, b\}$ (b) $a \in B$
   (c) $A \subset B$ (d) $A \cup B$
   (e) $A \cap B$

13. Given two sets R and S such that $R \cap S = \emptyset$. What does $R' \cap S$ equal?
   (a) $\emptyset$ (b) R
   (c) R' (d) S'
   (e) S

14. If $(J \cap K) \cap L \neq \emptyset$, then $J \cap (K \cup L) =$
   (a) $(J \cap K) \cup L$ (b) $(J \cup K) \cap (J \cup L)$
   (c) $(J \cup K) \cup (J \cap L)$ (d) $(J \cap K) \cup (J \cap L)$
   (e) $(J \cap K) \cup (K \cap L)$

15. If $A \cap B = \emptyset$, then $(A \cap B)' =$
   (a) U (b) A
   (c) B (d) $\emptyset$
   (e) $A' \cap B$

chapter 3

Systems of
Numeration

A **numeration system** is an organized procedure for arranging a set of symbols so that they effectively communicate the concept of number. Throughout the history of mathematics, many systems of numeration have been developed. Each system was organized on the basis of certain fundamental characteristics such as place value, base, additive, subtractive, multiplicative, repetitive, and so forth. In this chapter, we shall study the characteristics of a few numeration systems in the hope that we may develop a better understanding and appreciation of our decimal system.

3-1 Roman Numeration System

Most of us have had some acquaintance with the Roman system of numeration, because there are a few traces of it in our culture. For example, we often use Roman numerals to number the chapters in a book, to mark the hours on a clock, to designate dates on cornerstones, and so forth. Among the Roman numerals used were the following:

Roman Numeral	Our Numeral
I	1
V	5
X	10
L	50
C	100
D	500
M	1,000

The early Roman system was essentially an **additive system** of numeration. The number represented by a particular set of symbols was the sum of the numbers represented by each symbol in the set. Consider these examples:

Roman Numeral	Our Numeral
II	2
III	3
VII	7
XIII	13
XXXVIII	38
CCXXXVI	236

The Roman system also employed a **repetitive** characteristic, in that a particular symbol always represented the same value, regardless of position or frequency of use. In XXIII, the symbol X stands for 10 each time it is used, and the symbol I stands for 1 in each instance. Thus XXIII = 10 + 10 + 1 + 1 + 1 = 23.

Present-day use of the Roman numerals also involves a **subtractive** principle. This principle is used only for the numerals for four, nine, forty, ninety, four hundred, nine hundred, and so forth. For example, the numeral IV represents 5 − 1 or 4, the numeral IX represents 10 − 1 or 9, the numeral XL represents 50 − 10 or 40, and so forth.

We should note that the Roman system does not employ place value or a base. The position of a symbol may indicate whether to add or to subtract, but the position does not denote the value of the particular symbol. For example, in XC, the value of the X is 10, and the position of the X indicates that 10 is to be subtracted from 100 to give the value of XC. In CX, the value of the X is 10 and the position of the X indicates that 10 is to be added to 100 to give the value of CX.

It is also important to note that there was no symbol for zero in the Roman system.

The Roman system also used a **multiplicative** characteristic in order to represent large numbers. A bar was used over a symbol to indicate that the value of that symbol was to be multiplied by 1000. For example, C = 100, but $\overline{C} = 1000 \times 100 = 100,000$.

Adding and subtracting is relatively easy with Roman numerals. For example, the sum of 235 and 63 may be represented as follows:

$$\begin{array}{c} \text{CCXXXV} \\ \underline{\text{LXIII}} \\ \text{CCLXXXXVIII}; \end{array} \quad \text{this becomes CCXCVIII.}$$

We observe that LXXXX is symbolized by XC; hence the sum of 235 and 63 may be represented by CCXCVIII.

Multiplication with Roman numerals is much more cumbersome. For example, the product of 172 and 16 may be represented as follows:

```
                    CLXXII
                       XVI
          C   L   XX  II
      D  CC LLL     VV
  M  D   CC      XX
  MDDCCCCCLLLLXXXXVVII;    this becomes MMDCCLII.
```

We observe that the two V's may be symbolized by one X, the five X's by one L, the five L's by two C's and one L, the seven C's by one D and two C's, and the three D's by one M and one D. Hence the product of 172 and 16 is represented by MMDCCLII.

Exercises

1. Express each decimal numeral as a Roman numeral.
   (a) 13   (b) 52
   (c) 74   (d) 96
   (e) 149   (f) 387
   (g) 1971   (h) 100,498

2. Express each Roman numeral as a decimal numeral.
   (a) XXXI   (b) XXIX
   (c) X̄XIX   (d) CXCVI
   (e) CCXVI   (f) C̄CXVI
   (g) MMCD   (h) MMDC

3. Write the Roman numeral for the number that is 1 more than:
   (a) XV   (b) XVIII
   (c) CCXIX   (d) DCCXXXIII

4. Write the Roman numeral for the number that is 1 less than:
   (a) V   (b) X
   (c) C   (d) D

5. Use Roman numerals to find the sum of:
   (a) 12 and 53   (b) 74 and 95
   (c) 149 and 387   (d) 1,971 and 100,498

6. Use decimal numerals to find the sum of:
   (a) 12 and 53   (b) 74 and 95
   (c) 149 and 387   (d) 1,971 and 100,498

7. Compare and check the results obtained in:
   (a) Exercises 5(a) and 6(a).   (b) Exercises 5(b) and 6(b).
   (c) Exercises 5(c) and 6(c).   (d) Exercises 5(d) and 6(d).

8. Use Roman numerals to find the product of 12 and 53.

9. Use decimal numerals to find the product of 12 and 53.

10. Compare and check the results obtained in Exercises 8 and 9.

11. Express in Roman numerals.
   (a) 99   (b) 999
   (c) 88   (d) 888

3-2 Early Egyptian Numeration System

One of the earliest recorded systems of writing numerals is the Egyptian. Their hieroglyphic numerals have been traced back as far as 3300 B.C. The Egyptians developed a numeration system with which they could express numbers up to millions. Among the Egyptian numerals used were the following:

Egyptian Numeral	Name	Our Numeral
l	Stroke	1
∩	Arch or heel bone	10
9	Coiled rope or scroll	100
𝑖	Lotus flower	1,000
𝟋	Pointing finger	10,000
﹏	Burbot fish or tadpole	100,000
𝑥	Astonished man	1,000,000

The Egyptian numerals were carved on wood or stone. There was no symbol for zero, and the system was based on sets of ten with different symbols for ones, tens, hundreds, and so forth. These different symbols were needed because there was no use of place value. The Egyptian system was additive, since the number represented by a particular set of symbols was the sum of the numbers represented by each symbol in the set. The system was also repetitive since the same value was always assigned to a particular symbol. Consider these examples:

Egyptian Numeral	Our Numeral
IIII	4
∩∩∩IIIIIII	37
⁑99999∩∩∩∩III	2,453

Adding and subtracting is relatively easy with Egyptian numerals. For example, the sum of 2,453 and 85 may be represented as follows:

$$⁑99999∩∩∩∩III$$
$$∩∩∩∩∩∩∩IIIII$$
$$\overline{⁑99999∩∩∩∩∩∩∩∩∩∩∩IIIIIIII;}$$

this becomes   ⁑99999∩∩∩IIIIIIII.

We observe that the ten ∩ 's are symbolized by one 9 . Hence the sum of 2,453 and 85 is represented by   ⁑ 9 9 9 9 9 ∩ ∩ ∩ IIIIIIII.

Multiplication with Egyptian symbols is much more cumbersome. For example, the product of 12 and 365 may be represented as follows:

$$999∩∩∩∩∩IIIII$$
$$∩II$$
$$\overline{999∩∩∩∩∩IIIII}$$
$$999∩∩∩∩∩IIIII$$
$$⁑⁑⁑9999999∩∩∩∩∩$$
$$\overline{⁑⁑⁑99999999999999∩∩∩∩∩∩∩∩∩∩∩∩∩∩∩∩∩∩IIIIIIIIIIIIII;}$$

this becomes   ⁑⁑⁑⁑999∩∩∩∩∩∩∩.

We observe that the ten I 's are symbolized by one ∩ , then the eighteen ∩ 's are symbolized by one 9 and eight ∩ 's, and then the thirteen 9 's are symbolized by one ⁑ and three 9 's. Hence the product of 12 and 365 is represented by   ⁑⁑⁑⁑999∩∩∩∩∩∩∩.

Exercises

1. Express each decimal numeral as an Egyptian numeral.
   (a) 16   (b) 83
   (c) 201   (d) 879
   (e) 5,407   (f) 32,864
   (g) 253,478   (h) 1,040,104

2. Express each Egyptian numeral as a decimal numeral.
   (a) ∩∩I   (b) 99∩
   (c) 99∩∩II   (d) ∩⁑⁑⁑9IIII∩∩

(e) ꒒꒒꒒𓏲𓋹𓋹𓏴𓏴𓏴𓏴   **(f)** 𓂝𓏤𓆓𓂝𓂝𓆑‖‖‖‖‖

(g) 𓈖𓈖𓊖𓆓‖
𓈖𓈖𓊖𓆓‖

3. Write the Egyptian numeral for the number that is:
   (a) 1 more than ‖‖‖‖‖‖‖‖‖
   (b) 10 more than ∩∩∩∩∩∩∩∩∩
   (c) 100 more than 999999999
   (d) 1000 more than 𓆼𓆼𓆼𓆼𓆼𓆼𓆼𓆼𓆼

4. Write the Egyptian numeral for the number that is 1 less than:
   (a) ∩   **(b)** 9
   (c) 𓆼   **(d)** 𓂭

5. Write the Egyptian numeral for the number that is:
   (a) 10 less than 9   **(b)** 100 less than 𓆼
   (c) 1000 less than 𓂭   **(d)** 10,000 less than 𓂝

6. Use Egyptian numerals to find the sum of:
   (a) 16 and 83   **(b)** 201 and 879
   (c) 5,407 and 32,864   **(d)** 253,478 and 1,040,104

7. Use decimal numerals to find the sum of:
   (a) 16 and 83   **(b)** 201 and 879
   (c) 5,407 and 32,864   **(d)** 253,478 and 1,040,104

8. Compare and check the results obtained in:
   (a) Exercises 6(a) and 7(a).   **(b)** Exercises 6(b) and 7(b).
   (c) Exercises 6(c) and 7(c).   **(d)** Exercises 6(d) and 7(d).

9. Use Egyptian numerals to find the product of 16 and 83.

10. Use decimal numerals to find the product of 16 and 83.

11. Compare and check the results obtained in Exercises 9 and 10.

12. Express in Egyptian numerals.
    (a) 99   **(b)** 999
    (c) 9,999   **(d)** 99,999

3-3 Early Babylonian Numeration System

The Babylonians made their numerals by pressing a wedge-shaped instrument (stylus) into wet clay. One wedge shape, ▾, was used to represent 1. The symbol for ten, ⟨, was made in two steps by turning the stylus. Among the Babylonian numerals used were the following.

Babylonian Numeral	Our Numeral
▼	1
▼▼	2
▼▼▼	3
▼▼ ▼▼▼	5
▼▼▼ ▼▼▼	6
◀	10
◀▼▼	12
◀◀◀▼▼	34
◀◀ ▼▼▼ ◀◀◀▼▼▼	59

Notice that these numerals are based upon sets of ten but do not have place value for tens. Oddly enough, the Babylonians used a base of ten for naming only the numbers up to 60. For names of numbers 60 and greater they used the base 60. Actually their system contained a kind of place value, and the symbol ▼ represented 1, 60, 60 × 60, and so on, depending on its position. Since there was no symbol for zero, there was sometimes ambiguity in the value of the symbol as it was written. For example, 62 would be represented by the numeral▼ ▼ ▼, which could easily be confused with the numeral ▼ ▼ ▼ for 3. The Babylonian system also possessed an additive characteristic. The number 3,642 would be represented by the numeral ▼ ◀◀◀▼▼ . This numeral should be interpreted as $(1 \times 3,600) +$ $(0 \times 60) + (4 \times 10) + (2 \times 1)$.

Exercises

1. Express each decimal numeral as a Babylonian numeral.
   (a) 9   (b) 43
   (c) 59   (d) 61
   (e) 72   (f) 365
   (g) 3,624   (h) 4,107

2. Express each Babylonian numeral as a decimal numeral.
   (a) ◀▼▼▼   (b) ▼▼▼◀◀◀▼▼

   (c) ▼▼ ▼▼◀▼▼   (d) ▼ ▼

3. Write the Babylonian numeral for the number that is 1 more than:

(a) ▼   (b) ▼▼▼
                                                    ▼▼▼
                                                    ▼▼▼

(c) ◀◀ ▼▼▼   (d) ▼▼◀◀▼▼▼
    ◀◀ ▼▼▼   ▼▼◀◀▼▼▼
     ◀ ▼▼▼   ▼ ◀ ▼▼▼

4. Write the Babylonian numeral for the number that is 1 less than:

(a) ▼   (b) ◀

(c) ▼◀◀▼   (d) ▼▼◀◀
                                                         ◀◀

5. Use Babylonian numerals to find the sum of:
   (a) 9 and 43   (b) 59 and 61
   (c) 72 and 365   (d) 3,624 and 4,107

6. Use decimal numerals to find the sum of:
   (a) 9 and 43   (b) 59 and 61
   (c) 72 and 365   (d) 3,624 and 4,107

7. Compare and check the results obtained in:
   (a) Exercises 5(a) and 6(a).   (b) Exercises 5(b) and 6(b).
   (c) Exercises 5(c) and 6(c).   (d) Exercises 5(d) and 6(d).

3-4 Our Decimal System of Numeration

Like many other accomplishments of man, the development of our numeration system has gone through many stages to reach the present **decimal
system.** Many earlier civilizations contributed to the development of an
understandable, usable system. Characteristics such as the use of symbols,
place value, base, and zero should be considered when our system is compared with earlier systems. The decimal system of notation, which we sometimes take for granted, is one of the outstanding achievements of the human
mind.

The Hindu-Arabic system of numeration originated in ancient India and
was transmitted to Europe by traders and Arabic invaders. The present
forms of the symbols evolved for the most part in Europe, although the
historians have not agreed whether the Hindus or the Arabs contributed the
symbol for zero—which, it is generally agreed, was developed much later
than the other nine symbols. The following chart compares the tenth-century
form of the Hindu-Arabic numerals with the twentieth-century form.

Ancient Hindu–Arabic Numerals		I	?	?	?	Y	Ɫ	7	8	9
Modern Hindu–Arabic Numerals	0	1	2	3	4	5	6	7	8	9

Our decimal system of numeration, which is used in most of the world today, is based on sets of ten. Accordingly, the decimal system is often called the **base ten system.** Probably the base of ten was used because early man used his fingers for counting. When he had counted all ten fingers, he developed a procedure for starting the count over again, perhaps keeping some record of the number of tens.

In our system of counting, we start over when ten is reached and we use a **positional notation** to record the count. Thus, a count of ten is represented as 10, which means one set of ten and zero sets of one; a count of eleven is shown as 11, which means one set of ten and one set of one; and so on.

Ten Basic Number Symbols Used in the Decimal System

Set		*Numeral*	
{ }	0	is the numeral for a set containing zero elements.	
{/}	1	is the numeral for a set containing one element.	
{//}	2	is the numeral for a set containing two elements.	
{///}	3	is the numeral for a set containing three elements.	
{////}	4	is the numeral for a set containing four elements.	
{/////}	5	is the numeral for a set containing five elements.	
{//////}	6	is the numeral for a set containing six elements.	
{///////}	7	is the numeral for a set containing seven elements.	
{////////}	8	is the numeral for a set containing eight elements.	
{/////////}	9	is the numeral for a set containing nine elements.	

These ten numerals are called **decimal digits** (also referred to simply as **digits**).

The principle of **place value, also called positional notation,** is used in the decimal system. The value represented by each digit in a given numeral is determined by the position it occupies. For example, consider the numeral 35. The position of the 3 in this numeral indicates that it represents three sets of ten. In 53, on the other hand, the 3 represents three sets of one. The use of an additive characteristic is illustrated by the interpretation of 53 as 50 + 3.

The place values in the decimal system are based on *ten.* In the following place-value chart, beginning at the ones place (also called the "units" place) and reading to the left, we see that each place has a value ten times as large as the value of the place to its right.

Place – Value Chart For Whole Numbers

1 ten = 10 ones
1 hundred = 10 tens
1 thousand = 10 hundreds

Consider the numeral 2,345 in terms of the place value of each of its digits:

5 is in the ones place; 5 ones = 5 ones.

4 is in the tens place; 4 tens = 40 ones.

3 is in the hundreds place; 3 hundreds = 300 ones.

2 is in the thousands place; 2 thousands = 2,000 ones.

Thus the numeral 2,345 may be analyzed by using the principle of place value. It expresses the number two thousand three hundred forty-five. This may be represented in **expanded form** as

$$(2 \times 1{,}000) + (3 \times 100) + (4 \times 10) + (5 \times 1);$$

that is,

$$2{,}000 \quad + \quad 300 \quad + \quad 40 \quad + \quad 5.$$

If the digits of the numeral 2,345 were rearranged, the importance of the place-value principle would become very apparent. For example, consider 5,432:

$$(5 \times 1{,}000) + (4 \times 100) + (3 \times 10) + (2 \times 1);$$

that is,

$$5{,}000 \quad + \quad 400 \quad + \quad 30 \quad + \quad 2.$$

Although the same digits are used in the numerals 2,345 and 5,432, the two numerals do *not* name the same number.

Exercises

1. Some of the major characteristics of systems of numeration have been identified and discussed. Copy and complete this chart which summarizes

these characteristics for each system studied. Write yes or no in each box.

	Roman	Egyptian	Babylonian	Our Decimal
Place Value				
Base				
Additive				
Subtractive				
Multiplicative				
Repetitive				
Use of Zero				

2. In the decimal system, what value does the digit 3 represent in each of
   these numerals?
   (a) 612,734   (b) 3,281
   (c) 739,002   (d) 823

3. What is the smallest whole number you can represent by using each of the
   decimal digits 4, 5, and 1 exactly once? No other digits may be used.

4. What is the largest whole number you can represent by using each of the
   decimal digits 4, 5, and 1 exactly once? No other digits may be used.

5. What is the largest whole number represented by two decimal digits?

6. What is the smallest whole number represented by two decimal digits?

7. What is the smallest whole number represented by one decimal digit?

8. What is the largest whole number represented by one decimal digit?

9. Write an expanded form of:
   (a) 256   (b) 5,764
   (c) 97,438   (d) 385,061

3-5 Base of a System of Numeration

In general, the base of any system of numeration establishes the method of
grouping and the number of digits needed. In the decimal system, we collect
in sets of ten, and we can represent any number by using only the ten digits
0, 1, 2, 3, 4, 5, 6, 7, 8, and 9.

We may collect in sets other than ten. For example, the early Babylonians' system had a base of 60 and the Mayan Indians' system of numeration had a base of 20.

In the following illustrations a count of the elements in each produces the same number, but each illustration denotes a different way of grouping the elements. A subscript with a numeral denotes the base used to obtain that numeral, that is, the grouping. The general policy is to assume that a numeral is written in base 10 unless otherwise specified. However, in the first illustration the subscript "ten" is used for the decimal representation in order that it may easily be compared with the other two illustrations.

A decimal (base 10) grouping of the 16 elements in the figure indicates that the collection contains 1 set of ten elements and 6 more. This could be recorded as 16_{ten} and read as "one six, base ten," or "sixteen, base ten."

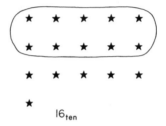

16_{ten}

If the same 16 elements were grouped in dozens (sets of twelve), there would be 1 set of twelve and 4 more. This grouping could be recorded as 14_{twelve} and read as "one four, base twelve."

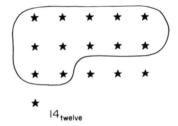

14_{twelve}

If the same 16 elements were grouped in sets of five, there would be 3 sets of five and 1 set of one. This could be recorded as 31_{five} and read as "three one, base five."

The three numerals 16_{ten}, 14_{twelve}, and 31_{five} are based upon different ways of collecting and representing the same number of elements; thus these numerals are names for the same number.

We have learned that the position a digit occupies in a numeral may determine the value of the digit. In the decimal system of numeration each place (position) is associated with a power of the base, ten. For example, the numeral 4,093 may be represented in expanded form as

$$(4 \times 1,000) + (0 \times 100) + (9 \times 10) + (3 \times 1).$$

In the base five system of numeration each place is associated with a power of the base, five. For example, the numeral 324_{five} may be represented in expanded form as

$$(3 \times 25) + (2 \times 5) + (4 \times 1).$$

Exercises

1. Represent each numeral in expanded form.
   - (a) 328
   - (b) 9,657
   - (c) 19,175
   - (d) 263,849

2. What is the value of the digit 7 in each numeral?
   - (a) 579
   - (b) 723
   - (c) 107
   - (d) 74,006

3. Copy the given set and group the elements in sets of:

   - (a) Ten
   - (b) Two
   - (c) Three
   - (d) Four
   - (e) Five

   31_{five}

4. Write an expanded form of the numeral for the grouping in each part of Exercise 3.

5. Copy the given set and group the elements in sets of:

   ★ ★ ★ ★ ★ ★   (a) Ten
                                (b) Two
   ★ ★ ★ ★ ★ ★   (c) Four
                                (d) Six
   ★ ★ ★ ★ ★ ★   (e) Eight

6. Write an expanded form of the numeral for the grouping in each part of Exercise 5.

7. Copy the given set and group the elements in sets of:

         (a) Ten
         (b) Three
         (c) Six
         (d) Nine
         (e) Twelve

8. Write an expanded form of the numeral for the grouping in each part of Exercise 7.

9. What is the base of the system of numeration when we group in sets of:
   (a) Ten?   (b) Twelve?
   (c) Nine?   (d) Five?
   (e) Two?   (f) Seven?
   (g) Three?   (h) Eight?
   (i) Four?

10. In general, how many digits are needed in order to be able to represent any number when we collect in sets of:
    (a) Ten?   (b) Twelve?
    (c) Nine?   (d) Five?
    (e) Two?   (f) Seven?
    (g) Three?   (h) Eight?
    (i) Four?

11. Study each of the following systems of numeration and tell which of these characteristics apply: additive, subtractive, multiplicative, repetitive, symbol for zero, place value, and base. If there is a base, tell what it is.

System	Counting numbers														
	1	2	3	4	5	6	7	8	9	10	11	12	13	14	15
I	a	aa	aaa	ab	b	ba	baa	baaa	abb	bb	bba	bbaa	bbaaa	abbb	bbb
II	▨	△	⊗▨	▨▨	△▨	⊗△	▨△	△△	⊗⊗▨	▨⊗▨	△⊗▨	⊗▨▨	▨▨▨	△▨▨	⊗△▨
III	A	B	C	BB	D	BC	E	BBB	CC	BD	F	BBC	G	BE	CD
IV	w	x	y	z	ww	wx	wy	wz	xw	xx	xy	xz	yw	yx	yy

12. For each system in Exercise 11, write the symbolism to represent the numbers 16 through 25.

3-6 Powers of Numbers

We have seen that in decimal notation each place has a value ten times as large as the value of the place on its right. For example,

$$1 \text{ ten} = 10 \text{ ones},$$
$$1 \text{ hundred} = 10 \text{ tens},$$
$$1 \text{ thousand} = 10 \text{ hundreds},$$
$$1 \text{ ten thousand} = 10 \text{ thousands},$$
$$1 \text{ hundred thousand} = 10 \text{ ten thousands}.$$

However, we are primarily interested in the value of each place in terms of ones. Then,

$$1 \text{ ten} = 10 \text{ ones},$$
$$1 \text{ hundred} = (10 \times 10) \text{ ones},$$
$$1 \text{ thousand} = (10 \times 10 \times 10) \text{ ones},$$
$$1 \text{ ten thousand} = (10 \times 10 \times 10 \times 10) \text{ ones},$$
$$1 \text{ hundred thousand} = (10 \times 10 \times 10 \times 10 \times 10) \text{ ones}.$$

Since the expressions of the form

$$10 \times 10 \times \ldots \times 10$$

become awkward when several 10's are multiplied together, we need a new symbol to indicate the number of 10's that are multiplied together.

Two or more numbers that are multiplied to form a **product** are called **factors** of the product. Thus, 4 and 6 are factors of the product 24, since $4 \times 6 = 24$; 3 and 8 are factors of 24, since $3 \times 8 = 24$; 2 and 12 are also factors of 24, since $2 \times 12 = 24$. Each of the numbers 1, 2, 3, 4, 6, 8, 12, and 24 is a factor of 24 since

$$24 = 1 \times 24 = 2 \times 12 = 3 \times 8 = 4 \times 6.$$

When the same factor is repeated, as in

$$10 \times 10, \quad 10 \times 10 \times 10, \quad \text{and} \quad 10 \times 10 \times 10 \times 10,$$

we use a small numeral above and to the right of the numeral representing the factor, to indicate the number of times the factor is repeated. For example,

$$10 \times 10 = 10^2,$$
$$10 \times 10 \times 10 = 10^3,$$
$$10 \times 10 \times 10 \times 10 = 10^4.$$

The repeated factor is the **base**, the number of times the base is used as a factor is the **exponent**, and the base with its exponent is sometimes called a **power.** In the case of 10^4 (read "ten to the fourth power"), 10 is the base, 4 is the exponent, and 10^4 is the power.

$$10^1 = (10) = 10,$$
$$10^2 = (10 \times 10) = 100,$$
$$10^3 = (10 \times 10 \times 10) = 1,000,$$
$$10^4 = (10 \times 10 \times 10 \times 10) = 10,000,$$
$$10^5 = (10 \times 10 \times 10 \times 10 \times 10) = 100,000.$$

Notice that the numeral for the exponent is usually omitted for the first power of ten; that is, ten is usually written 10 instead of 10^1. Exponents may be used with any number as a base. For example, $6 \times 6 \times 6$ is 6^3, where 6 is the base, 3 is the exponent, and 6^3 is the power.

Exercises

1. Represent each of the following as a power (that is, as a base with an exponent).
   (a) $8 \times 8 \times 8$ (b) $9 \times 9 \times 9 \times 9$
   (c) 3×3 (d) 5×25

2. Which numeral represents the larger number?
   (a) 5^6 or 6^5 (b) 2^7 or 7^2

3. What is the relation between the exponent of a power of ten and the number of zeros in the decimal numeral representing the number?

4. Represent the following as repeated factors.
   (a) 10^6 (b) 2^3
   (c) 4^5 (d) 5^4

5. In each of the following, identify the base, the exponent, and the power.
   (a) 3^6 (b) 7^2

3-7 Multiplication and Division of Powers

We have studied the exponent as a convention in notation that simplifies the representation of repeated factors. In general, if a and n are natural numbers, then a^n means $a \times a \times a \times \ldots \times a$, where there are n factors in the product. (*Note:* $a \times a$ may also be written $a \cdot a$.)

The use of exponents in the representation of repeated factors also simplifies the calculation of the product of two powers of the same base. For example,

$$4^2 \times 4^3 = (4 \times 4) \times (4 \times 4 \times 4) = (4 \times 4 \times 4 \times 4 \times 4) = 4^5,$$

$$6^3 \times 6^4 = (6 \times 6 \times 6) \times (6 \times 6 \times 6 \times 6)$$
$$= (6 \times 6 \times 6 \times 6 \times 6 \times 6 \times 6) = 6^7,$$
$$a^2 \cdot a^4 = (a \cdot a) \cdot (a \cdot a \cdot a \cdot a) = (a \cdot a \cdot a \cdot a \cdot a \cdot a) = a^6.$$

A careful observation of these three examples should lead us to the inference that

$$4^2 \times 4^3 = 4^{2+3} = 4^5,$$
$$6^3 \times 6^4 = 6^{3+4} = 6^7,$$
$$a^2 \times a^4 = a^{2+4} = a^6.$$

In general, any product of two powers of the same base is a power of that base, and the new exponent is the sum of the exponents of the factors; that is,

$$b^m \cdot b^n = b^{m+n}.$$

We may extend this idea to raising a power to a power. For example,

$$(4^2)^3 = (4 \times 4)^3 = (4 \times 4) \times (4 \times 4) \times (4 \times 4)$$
$$= (4 \times 4 \times 4 \times 4 \times 4 \times 4) = 4^6,$$

$$(6^3)^4 = (6 \times 6 \times 6)^4$$
$$= (6 \times 6 \times 6) \times (6 \times 6 \times 6) \times (6 \times 6 \times 6) \times (6 \times 6 \times 6)$$
$$= (6 \times 6 \times 6 \times 6 \times 6 \times 6 \times 6 \times 6 \times 6 \times 6 \times 6 \times 6) = 6^{12},$$

$$(a^2)^4 = (a \cdot a)^4 = (a \cdot a) \cdot (a \cdot a) \cdot (a \cdot a) \cdot (a \cdot a) = (a \cdot a \cdot a \cdot a \cdot a \cdot a \cdot a \cdot a) = a^8.$$

A careful observation of these three examples should lead us to the inference that

$$(4^2)^3 = 4^{2 \times 3} = 4^6,$$
$$(6^3)^4 = 6^{3 \times 4} = 6^{12},$$
$$(a^2)^4 = a^{2 \times 4} = a^8.$$

In general, any power of a power of a base is a power of that base, and the new exponent is the product of the exponent of the given power and the exponent that indicated the power to which it was raised; that is,

$$(b^m)^n = b^{m \cdot n}.$$

The use of exponents in the representation of repeated factors also simplifies the calculation of the quotient of two powers of the same base. For example,

$$\frac{4^5}{4^3} = \frac{4 \times 4 \times 4 \times 4 \times 4}{4 \times 4 \times 4} = \frac{4}{4} \times \frac{4}{4} \times \frac{4}{4} \times 4 \times 4$$
$$= 1 \times 1 \times 1 \times 4 \times 4 = 4 \times 4 = 4^2,$$

$$\frac{6^7}{6^4} = \frac{6 \times 6 \times 6 \times 6 \times 6 \times 6 \times 6}{6 \times 6 \times 6 \times 6} = \frac{6}{6} \times \frac{6}{6} \times \frac{6}{6} \times \frac{6}{6} \times 6 \times 6 \times 6$$

$$= 1 \times 1 \times 1 \times 1 \times 6 \times 6 \times 6 = 6 \times 6 \times 6 = 6^3,$$

$$\frac{a^6}{a^2} = \frac{a \cdot a \cdot a \cdot a \cdot a \cdot a}{a \cdot a} = \frac{a}{a} \cdot \frac{a}{a} \cdot a \cdot a \cdot a \cdot a = 1 \cdot 1 \cdot a \cdot a \cdot a \cdot a = a \cdot a \cdot a \cdot a = a^4.$$

A careful observation of these three examples should lead us to the inference that

$$\frac{4^5}{4^3} = 4^{5-3} = 4^2,$$

$$\frac{6^7}{6^4} = 6^{7-4} = 6^3,$$

$$\frac{a^6}{a^2} = a^{6-2} = a^4.$$

In general, any quotient of two powers of the same base is a power of that base, and the new exponent is the difference of the exponents of the given powers; that is,

$$b^m \div b^n = b^{m-n}.$$

Perhaps special attention should be given to the next example, in which a represents any natural number:

$$\frac{a^4}{a^4} = \frac{a \cdot a \cdot a \cdot a}{a \cdot a \cdot a \cdot a} = \frac{a}{a} \cdot \frac{a}{a} \cdot \frac{a}{a} \cdot \frac{a}{a} = 1 \cdot 1 \cdot 1 \cdot 1 = 1.$$

However, using the method of subtracting exponents previously discussed, we have

$$\frac{a^4}{a^4} = a^{4-4} = a^0.$$

Therefore, to insure consistency in the results, we must agree that

$$a^0 = 1,$$

where a represents any natural number.

Another special case of division involving exponents must be considered. If a represents any natural number, then

$$\frac{a^2}{a^4} = \frac{a \cdot a}{a \cdot a \cdot a \cdot a} = \frac{a}{a} \cdot \frac{a}{a} \cdot \frac{1}{a} \cdot \frac{1}{a} = 1 \cdot 1 \cdot \frac{1}{a} \cdot \frac{1}{a} = \frac{1}{a^2}.$$

However, using the method of subtracting exponents, we have

$$\frac{a^2}{a^4} = a^{2-4} = a^{-2},$$

where the -2 is read "negative two" and is called a negative exponent. Therefore, to insure consistency in the results, we must agree that

$$a^{-2} = \frac{1}{a^2} \quad \text{and, in general,} \quad a^{-n} = \frac{1}{a^n}$$

where a and n represent natural numbers.

Observe the pattern of the exponents in the following illustration of powers of ten:

$$10^3 = (10 \times 10 \times 10) = 1,000,$$

$$10^2 = (10 \times 10) = 100,$$

$$10^1 = (10) = 10,$$

$$10^0 = (1) = 1,$$

$$10^{-1} = \left(\frac{1}{10}\right) = \frac{1}{10},$$

$$10^{-2} = \left(\frac{1}{10} \times \frac{1}{10}\right) = \frac{1}{100},$$

$$10^{-3} = \left(\frac{1}{10} \times \frac{1}{10} \times \frac{1}{10}\right) = \frac{1}{1,000},$$

$$10^{-4} = \left(\frac{1}{10} \times \frac{1}{10} \times \frac{1}{10} \times \frac{1}{10}\right) = \frac{1}{10,000},$$

$$10^{-5} = \left(\frac{1}{10} \times \frac{1}{10} \times \frac{1}{10} \times \frac{1}{10} \times \frac{1}{10}\right) = \frac{1}{100,000}.$$

Exercises

1. Use exponents and represent each expression as a single power.
   (a) $2^4 \times 2^3$ (b) $3^{-4} \times 3^7$
   (c) $6^7 \div 6^2$ (d) $4^2 \times 4^7 \div 4$
   (e) $2^0 \times 3^0$ (f) $(5^2)^6$

2. Use exponents and represent each expression as a single power where b represents a natural number.
   (a) $b^3 \cdot b^5$ (b) $b^9 \div b^4$
   (c) $b^2 \cdot b$ (d) $b^7 \div b^7$
   (e) $b^2 \div b^5$ (f) $(b^4)^3$

3. Use exponents and represent each expression as a single power.
   (a) $(3^7)^2$ (b) $3^7 \times 3^2$
   (c) $5^0 \times 25^2$ (d) $(4^3)^{-2}$
   (e) $8^2 \div 2$ (f) $21^{13} \div 21^{19}$

4. In the decimal system, the place value of the tens place is how many times as large as that of the:
   (a) Ones place?   (b) Tenths place?
   (c) Hundredths place?   (d) Ten-thousandths place?

5. In the decimal system, the place value of the hundredths place is what fraction of that of the:

(a) Tenths place?   (b) Ones place?
(c) Tens place?   (d) Hundreds place?

3-8 Chapter Test

Indicate whether each statement is true or false. If false, tell why.

1. The set of digits used to write numerals in our decimal system is an infinite set.

2. A numeration system is said to be repetitive if the positioning of the symbol changes the value of the individual symbol.

3. In our decimal system, the hundreds place has a value that is 1,000 times the value of the tenths place.

4. The notation a^0 is another name for 1 where a represents any whole number.

5. $27^2 = 3^6$.

6. An expanded form of the numeral 2,056 is $(2 \times 10^4) + (0 \times 10^3) + (5 \times 10^2) + (6 \times 10^1)$.

7. In a base three system of numeration, we generally need only 3 different digits to write numerals.

8. In 42 the digit 4 has a value which is twenty times the value of the digit 2.

9. An exponent is a numeral.

10. If a numeration system has a base, it must employ the idea of place value.

Select the best possible answer.

11. In 2,756, the digit 7 represents:

(a) 7×10^4 (b) 7×10^3
(c) 7×10^2 (d) 7×10^1
(e) 7×10^0

12. The expression 5^3 means:

(a) $3 \times 3 \times 3 \times 3 \times 3$ (b) $5 \times 5 \times 5$
(c) 5×3 (d) $5 \div 3$
(e) 5 multiplied by itself three times.

13. Which of these groupings of elements would be indicated by the numeral 23_{five}?

(a)

(b)

(c)

(d)

(e)

14. The meaning of 10^{-3} is:

   (a) $^-1,000$ (b) $\dfrac{1}{100}$

   (c) $(^-10) \times (^-10) \times (^-10)$ (d) $\dfrac{1}{10} \times 3$

   (e) $\dfrac{1}{1,000}$

15. The Egyptian system of numeration differed from our decimal system in which of the following ways?
   (a) It had a base of five.
   (b) Zero was used to denote quantity but not as a place holder.
   (c) Position of symbols denoted additive or subtractive characteristics.
   (d) Groupings were made as successive powers of 2.
   (e) The number represented did not depend on the relative position of the symbols.

16. In our decimal system, the method of grouping is in sets of:
   (a) Twos   (b) Fives
   (c) Tens   (d) Hundreds
   (e) Thousands

17. Why do we teach expanded form of numerals?
   (a) To give a better understanding of exponents.
   (b) To give a better understanding of different bases.
   (c) To give a better understanding of place value in computation.
   (d) To give a better understanding of ancient systems of numeration.
   (e) None of these.

18. A system of numeration is:
   (a) A procedure used in joining sets.

(b) An innovation of modern arithmetic programs.
(c) An unchangeable law of mathematics.
(d) A method of adding whole numbers.
(e) A systematic method of naming numbers.

19. $(27^2)^{14} =$
   (a) 27^{16} (b) 27^7
   (c) 27^{-12} (d) 27^{28}
   (e) 27^{12}

20. How would you represent 17 in this system of numeration?

$$1 \quad 2 \quad 3 \quad 4 \quad 5 \quad 6 \quad \ldots \quad 10$$

$$a \quad b \quad c \quad ak \quad k \quad ka \quad \ldots \quad kk$$

   (a) *kkbcb*   (b) *akakk*
   (c) *kkkb*   (d) *kkkaa*
   (e) None of these

chapter 4

Whole Numbers: Addition and Subtraction

Increased emphasis is being placed at the elementary school level on some of the basic patterns and properties of mathematics. Three major purposes in this emphasis are to give the child an understanding of the structure of mathematics, to give him tools for building new ideas and concepts from what he already knows (that is, for making generalizations), and to develop in him an ability to apply generalizations to specific cases. There are many opportunities for the use and application of the basic properties of mathematics from the very beginning of systematic instruction in Grade One. Too often we fail to direct the student's attention to the way these properties operate and to what they permit him to do.

In this chapter some of the properties of the operations of addition and subtraction of whole numbers are identified. Modern courses in elementary school mathematics emphasize these basic properties as unifying concepts that enable us to explain logically why the familiar rules and procedures for adding and subtracting numbers really do work. The greater insight and understanding that the student will gain is also most helpful in providing a firm foundation for work encountered at the secondary school level.

In this chapter the operation of addition on the set of whole numbers is defined, and the basic properties are presented as assumptions. These assumptions are referred to by different writers as postulates, axioms, principles, laws, and properties.

4-1 Addition of Whole Numbers

Addition of two whole numbers may be understood clearly in terms of the union of two sets. Children intuitively discover the meaning of the union of sets when they put the blocks of two sets together to form a single set. For example, the blocks of

$$\left\{ \boxed{A}, \boxed{B} \right\} \text{ joined with those of } \left\{ \boxed{C}, \boxed{D}, \boxed{E} \right\}$$

$$\text{produce a new set } \left\{ \boxed{A}, \boxed{B}, \boxed{C}, \boxed{D}, \boxed{E} \right\}$$

The cardinal number (in this case 5) of the union of the two sets is the sum of the cardinal numbers (2 and 3) of the two given sets. This illustrates the statement

$$2 + 3 = 5.$$

Addition of two whole numbers may be defined as finding the cardinal number of a set formed by the union of two disjoint sets. Recall (Section 2-6) that disjoint sets are those with no elements in common.

The importance of the word *disjoint* in the definition of addition is shown in the next example. Let

$$A = \{\text{Tom, Beth, Sam}\},$$
$$B = \{\text{Carl, Tom, Gene, George}\};$$

then

$$A \cup B = \{\text{Tom, Beth, Sam, Carl, Gene, George}\}.$$

The cardinal number of set A is 3, and the cardinal number of set B is 4. However, the cardinal number of $A \cup B$ is 6 rather than $3 + 4$. In order that the cardinal number of $A \cup B$ may be the sum of the cardinal numbers of sets A and B, disjoint sets must be selected. Therefore, *only the union of disjoint sets is used to describe addition of two whole numbers.*

Since set union is an operation on *two* sets, it is called a binary operation. Similarly, the addition of two whole numbers may be described as a binary operation on the cardinal numbers of two disjoint sets. Addition is called a binary operation because it is fundamentally an operation performed on two numbers at a time. In the language of mathematics, $5 + 3$ indicates a binary operation to be performed in a *definite order* on two whole numbers, 5 and 3; that is, $5 + 3$ indicates that the operation is to be completed by adding 3 to 5, which results in the sum 8. The same definite order is indicated in the example below.

$$
\begin{array}{r}
5 \\
+3 \\
\hline
8
\end{array}
$$

In general, it may be stated that addition is an operation that combines a first number and a second number to produce a unique third number. The first and second numbers are called **addends,** and the result of combining the addends is called the **sum.**

A number line frequently is used as an aid in visualizing the operation of addition. Remember (Section 1-15) how a number line was constructed. A line was drawn and an arbitrary point on the line was selected to represent the graph of 0. Then a unit of length was chosen and marked off on the line, beginning at the graph of 0 and extending to the right. By marking off this unit of length again and again and labeling the endpoints of the segments 1, 2, 3, 4, and so on, the whole-number line may be pictured as follows.

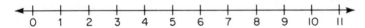

The whole number at each of the indicated points represents the number of segments of unit length that are measured off from the graph of 0 to that point. We must realize also that a given whole number may be represented on the number line by any line segment that has the given whole number of unit lengths as the distance between its endpoints. For example, consider the different illustrations of the number 2 represented by a line segment of two units' length on this number line.

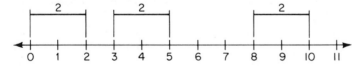

The addition of whole numbers may be pictured as an operation moving to the right on the number line. To add 2 and 4 on the number line, start with the graph of 0 and mark off to the right a segment with two units of length; then start at that point and mark off to the right a segment with four units of length. These two line segments have determined a new segment whose left end is at the graph of 0 and whose right end is at the graph of 6; the length of the new line segment is the sum of the lengths of the other two segments. Thus, the picture on the following number line represents the sum of 2 and 4.

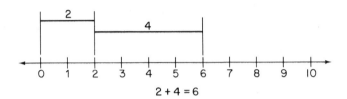

$$2 + 4 = 6$$

Exercises

1. Find the cardinal number for each set.
   (a) $A = \{\triangle, \square, \diamondsuit, \triangledown\}$, $n(A) =$ _____
   (b) $B = \{\star, \oslash\}$, $n(B) =$ _____
   (c) $C = \{x, y, z\}$, $n(C) =$ _____
   (d) $D = \{\ \}$, $n(D) =$ _____
   (e) $E = \{\diamondsuit, \star, y, \bigcirc, \#\}$, $n(E) =$ _____

2. Use the sets in Exercise 1 and find:
   (a) $A \cup B$ (b) $B \cup C$
   (c) $A \cup E$ (d) $C \cup D$
   (e) $C \cup E$ (f) $D \cup E$
   (g) $B \cup D$ (h) $B \cup E$

3. Use the sets in Exercise 2 and find:
   (a) $n(A \cup B)$ (b) $n(B \cup C)$
   (c) $n(A \cup E)$ (d) $n(C \cup D)$
   (e) $n(C \cup E)$ (f) $n(D \cup E)$
   (g) $n(B \cup D)$ (h) $n(B \cup E)$

4. Use the sets in Exercise 1 and tell whether each statement is true or false. If false, tell why.
   (a) $n(A) + n(B) = n(A \cup B)$ (b) $n(B) + n(C) = n(B \cup C)$
   (c) $n(A) + n(E) = n(A \cup E)$ (d) $n(C) + n(D) = n(C \cup D)$
   (e) $n(C) + n(E) = n(C \cup E)$ (f) $n(D) + n(E) = n(D \cup E)$
   (g) $n(B) + n(D) = n(B \cup D)$ (h) $n(B) + n(E) = n(B \cup E)$

5. Use sets to illustrate each addition equation.
   (a) $2 + 5 = 7$ (b) $3 + 6 = 9$
   (c) $4 + 3 = 7$ (d) $6 + 0 = 6$
   (e) $4 + 4 = 8$ (f) $0 + 0 = 0$

6. Write the equation represented on each number line.

(a)

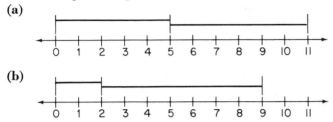

(b)

7. Represent each equation on a number line.

(a) $3 + 5 = 8$ (b) $4 + 7 = 11$

(c) $5 + 5 = 10$

4-2 Closure and Uniqueness Properties

Consider each of these sums of whole numbers: $1 + 3$, $4 + 7$, $9 + 13$, $18 + 25$, and $60 + 245$. Would you expect each of these sums to be a whole number? The answer to this question is dependent upon a very important mathematical property called **closure**. The closure property applied to addition of whole numbers means that the sum of two numbers in the set of whole numbers is a number that is also an element of the set of whole numbers. In other words, when one whole number is added to another whole number, the sum is also a whole number. Thus, the set of whole numbers is **closed** with respect to the operation of addition.

In the example $5 + 9 = 14$, note that the addends and the sum are all elements of the set of whole numbers. Is 14 the only possible sum when 5 and 9 are added? The answer to this question is dependent upon an important mathematical property called **uniqueness**. The uniqueness property applied to addition of whole numbers means that there is only one possible sum when two whole numbers are added. Thus, 14 is the *unique* sum of 5 and 9.

A generalized statement of these two properties is:

If a and b are whole numbers, then there is one and only one whole number that is called the sum of a and b. This sum is written a + b.

Not all sets are closed *under* addition. It is important to understand that a set is not closed under addition if there exist two elements of the set whose sum is not an element of the set. To illustrate this, one may consider the finite set $S = \{1, 2, 3, 4, 5, 6\}$. This set is not closed under addition because two numbers can be found in the set such that their sum is not in the set; for example, $5 \in S$, $6 \in S$, $5 + 6 = 11$, and 11 is not in set S. In other words, if there is in S at least one pair of elements whose sum is not in S, then S is not closed under the operation of addition.

Exercises

1. State whether or not each set is closed under the operation of addition.
   (a) The set of natural numbers greater than 100.
   (b) The set of counting numbers between 565 and 1,000.
   (c) The set of odd whole numbers less than 100.
   (d) The set of even natural numbers.
   (e) The set of whole numbers whose one's digit is 0.
   (f) The set of counting numbers.
   (g) The set of whole numbers.

2. State whether or not each sum is unique.
   (a) $3 + 2$ (b) $10 + 6$
   (c) $19 + 13$ (d) $57 + 84$

4-3 The Commutative Property of Addition

In the union of two sets, the order in which the elements and sets are joined does not affect the union of the two sets. Consider the disjoint sets $A = \{a, b, c\}$ and $D = \{d, e, f, g\}$:

$$A \cup D = \{a, b, c, d, e, f, g\},$$
$$D \cup A = \{d, e, f, g, a, b, c\}.$$

It may be observed in this illustration that $A \cup D$ is the same set as $D \cup A$, since they both contain identical elements; that is, $A \cup D = D \cup A$.

Since the addition of whole numbers may be defined in terms of the union of two disjoint sets, a change in the order of the addends will not affect the sum. Observe that in the above example the cardinal number of set A is 3 and the cardinal number of set D is 4. Note also that 7 is the cardinal number of both $A \cup D$ and $D \cup A$. In other words, $3 + 4 = 7$ and $4 + 3 = 7$. We conclude that $3 + 4 = 4 + 3$.

This example illustrates the **commutative property of addition** (sometimes called the order property). The commutative property of the addition of whole numbers may be described by saying that the sum of two whole numbers does not depend on the order of the addition. For each pair of whole numbers a and b,

$$a + b = b + a.$$

The commutative property of addition should be clear when one thinks in terms of combining sets. However, if addition is developed as counting, it is not quite so obvious that addition has the commutative property. For example, if we ask a child to add 8 and 5, he may, if he has not memorized the sum, begin with one of the numbers and count forward the other number. He may think, "I'll begin with 8 and count forward 5." This is illustrated on the following number line.

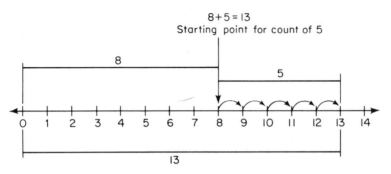

It is indeed important to understand that with this approach to addition one would get the same result if he started with 5 and counted forward 8. This is illustrated on the next number line.

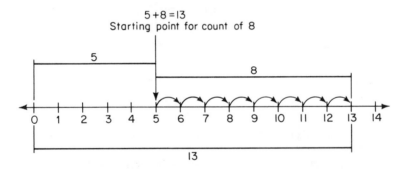

Thus, we conclude that $8 + 5 = 5 + 8 = 13$.

Elementary school children frequently use the commutative property of addition when they reverse the order of the addends to check their work.

<table>
<tr><td></td><td>Check</td></tr>
<tr><td>36</td><td>25</td></tr>
<tr><td>+25</td><td>+36</td></tr>
<tr><td>61</td><td>61</td></tr>
</table>

We do the same when we add a long column of figures down and check by adding up.

Exercises

1. State whether or not each activity is commutative.
 (a) To give a test, then grade the papers.
 (b) To salt, then pepper, your eggs.
 (c) To pour black paint into white paint.
 (d) To strike a match, then light the fire.
 (e) To put on socks, then put on shoes.

2. Given $A = \{a, b, c\}$ and $B = \{d, e, f, g\}$. Use these sets to illustrate:
 (a) $n(A) + n(B) = n(A \cup B)$ (b) $n(B) + n(A) = n(B \cup A)$
 (c) $n(A \cup B) = n(B \cup A)$ (d) $n(A) + n(B) = n(B) + n(A)$

3. Use sets to illustrate:
 (a) $2 + 4 = 4 + 2$ (b) $7 + 8 = 8 + 7$

4. Write the equation for this illustration of the commutative property.

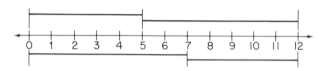

5. Use a number line to represent each of the following:
 (a) $4 + 5 = 5 + 4$ (b) $6 + 7 = 7 + 6$

6. Complete each statement to obtain an example of the commutative property of addition.
 (a) $1 + 2 = \underline{\quad} + \underline{\quad}$ (b) $19 + 6 = \underline{\quad} + \underline{\quad}$
 (c) $27 + 63 = \underline{\quad} + \underline{\quad}$ (d) $108 + 325 = \underline{\quad} + \underline{\quad}$

4-4 The Associative Property of Addition

Recall that addition was defined as a *binary* operation; that is, an operation upon *two* numbers. What is meant by $2 + 3 + 4$? It is necessary to establish a procedure for adding more than two numbers. We agree that when more than two numbers are to be added, we must first find the sum of two numbers, then find the sum of this sum and the third number, and then repeat the operation two numbers at a time. In the example $2 + 3 + 4$, there are only two ways (excluding use of the commutative property) to associate (group) the numbers for addition. Either the 2 and 3 are added first and then the 4 is added to their sum,

$$(2 + 3) + 4 = 5 + 4 = 9,$$

or the 3 and 4 are added first and then their sum is added to 2,

$$2 + (3 + 4) = 2 + 7 = 9.$$

Notice that in this example the way the three numbers are associated (grouped) for addition does not affect the sum.

The two methods of associating (grouping) the three numbers 2, 3, and 4 for addition may be illustrated on the number line. The sum $(2 + 3) + 4$ is represented on the first number line. Notice that we determine $2 + 3$ first and then add 4 to this sum.

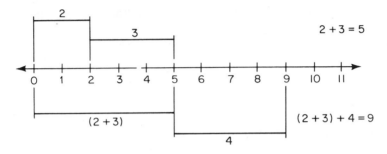

The sum $2 + (3 + 4)$ is represented on the next number line. Notice that we determine $3 + 4$ first and then add this sum to 2.

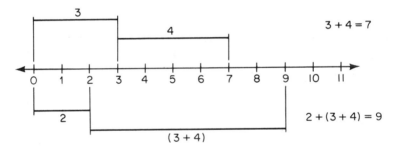

Thus we see that $(2 + 3) + 4 = 2 + (3 + 4)$.

The idea that grouping the three numbers differently, without changing their order, does not change their sum is called the **associative property of addition** for whole numbers. This property may be used to make addition easier. For example, in adding 12 and 6 it may be easier to consider $12 + 6 = (10 + 2) + 6 = 10 + (2 + 6) = 10 + 8 = 18$, and in the example $33 + 74 + 26$ it may be easiest to group $33 + (74 + 26) = 33 + 100 = 133$.

The associative property of addition for whole numbers a, b, and c may be stated as

$$(a + b) + c = a + (b + c).$$

Exercises

1. Given $C = \{g, h, j\}$, $D = \{k, l, m, n\}$, and $E = \{e, f, a, b, w\}$. Use these sets to illustrate:

   (a) $n(C \cup D) + n(E) = n((C \cup D) \cup E)$
   (b) $n(C) + n(D \cup E) = n(C \cup (D \cup E))$
   (c) $n((C \cup D) \cup E) = n(C \cup (D \cup E))$
   (d) $n(C \cup D) + n(E) = n(C) + n(D \cup E)$

2. Use sets to illustrate:
   (a) $(1 + 2) + 3 = 1 + (2 + 3)$ (b) $(2 + 8) + 5 = 2 + (8 + 5)$

3. Write the equation for this illustration of the associative property.
   (a)

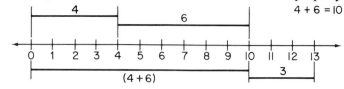

   (b)

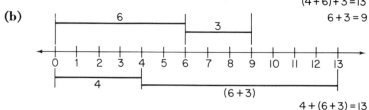

4. Use number lines as in Exercise 3 to represent each statement.
   (a) $(4 + 7) + 3 = 4 + (7 + 3)$ (b) $(9 + 1) + 5 = 9 + (1 + 5)$

5. Show how the associative property of addition can be used to find each sum in two different ways.
   (a) $15 + 3 + 42 \cdot$ (b) $22 + 7 + 18$
   (c) $109 + 23 + 11$ (d) $237 + 165 + 35$

6. Tell which property of addition, commutative or associative, is illustrated by each equation.
   (a) $15 + 6 = 6 + 15$ (b) $(4 + 5) + 6 = 4 + (5 + 6)$
   (c) $2 + (3 + 4) = (3 + 4) + 2$ (d) $d + (e + f) = (d + e) + f$

7. Use the commutative and associative properties of addition to find each sum.
   (a) $27 + 36 + 13 + 64$ (b) $135 + 81 + 15 + 9$
   (c) $58 + 24 + 32 + 16$ (d) $19 + 17 + 11 + 83$

8. Add without the use of paper and pencil. Did you mentally apply the associative property of addition?
   (a) $1 + (999 + 27)$ (b) $60 + (40 + 39)$
   (c) $(783 + 89) + 11$ (d) $(176 + 80) + 20$

9. Indicate whether each statement is true or false.
   (a) $18 + 25 > 25 + 18$ (b) $(4 + 8) + 7 = (4 + 7) + 8$
   (c) $505 + 98 = 98 + 505$ (d) $(34 + 9) + 10 < 10 + (9 + 34)$

10. Group the numbers 5, 6, 7, and 8 for addition in at least four different ways. Find the sum for each of the groupings. Are all the sums equal?

11. Use the commutative and associative properties of addition to justify or explain each of the following. Copy each step and give the reason when needed.

(a) Show that $(a + c) + b = a + (b + c)$

$(a + c) + b = a + (c + b)$ associative property of addition

$a + (c + b) = a + (b + c)$ _____

(b) Show that $(c + b) + a = a + (b + c)$

$(c + b) + a = a + (c + b)$ commutative property of addition

$a + (c + b) = a + (b + c)$ _____

(c) Show that $(c + a) + b = a + (b + c)$

$(c + a) + b = (a + c) + b$ _____

$(a + c) + b = a + (c + b)$ _____

$a + (c + b) = a + (b + c)$ _____

4-5 Identity Element for Addition

The number zero is the number of elements in the empty set. It is most important that zero be recognized as a specific number that is just as useful as any of the counting numbers 1, 2, 3, ... The number zero plays a very special role in the addition of whole numbers. The sum of any whole number and the number zero is always the original whole number. For example,

$$0 + 0 = 0,$$
$$1 + 0 = 1 \quad \text{and} \quad 0 + 1 = 1,$$
$$2 + 0 = 2 \quad \text{and} \quad 0 + 2 = 2,$$
$$3 + 0 = 3 \quad \text{and} \quad 0 + 3 = 3,$$

and a general rule is that for each whole number a,

$$a + 0 = a \quad \text{and} \quad 0 + a = a.$$

This property is indicated by calling zero the **identity element for addition** of whole numbers, that is, the **additive identity**.

Exercises

1. Given $R = \{r, s, t\}$ and $S = \{\ \}$. Use these sets to illustrate:

(a) $n(R) + n(S) = n(R \cup S)$

(b) $n(R \cup S) = n(R)$

(c) $n(R) + n(S) = n(R)$

(d) $n(R) + n(S) = n(S) + n(R) = n(R)$

2. Use sets to illustrate:

    (a) $4 + 0 = 0 + 4 = 4$ (b) $5 + 0 = 0 + 5 = 5$

3. Complete each statement to make a true number sentence.

    (a) $0 + 8 = $ ____

    (b) ____ $+ 0 = 27$

    (c) $0 + $ ____ $= 0$

    (d) $x + $ ____ $= $ ____ $+ x = x, x \in W$

4-6 The Table of Basic Addition Facts

Addition facts for decimal digits are sometimes summarized in a table like the following one.

+	0	1	2	3	4	5	6	7	8	9
0	0	1	2	3	4	5	6	7	8	9
1	1	2	3	4	5	6	7	8	9	10
2	2	3	4	5	6	7	8	9	10	11
3	3	4	5	6	7	8	9	10	11	12
4	4	5	6	7	8	9	10	11	12	13
5	5	6	7	8	9	10	11	12	13	14
6	6	7	8	9	10	11	12	13	14	15
7	7	8	9	10	11	12	13	14	15	16
8	8	9	10	11	12	13	14	15	16	17
9	9	10	11	12	13	14	15	16	17	18

    Many of the important properties of addition may be observed from the table. A line has been drawn diagonally through the table from the upper left corner to the lower right. The elements drawn through are called *diagonal elements* and form the *main diagonal* of the table. Notice that, except for the diagonal elements, this line divides the table into two equivalent parts. When the table is folded over its main diagonal, the blocks containing 1 fall upon each other, as do the blocks containing 2, and so on. This is due to the commutative property of addition, which justifies conclusions such as $4 + 3 = 3 + 4$. An understanding of the commutative property of addition for whole numbers reduces the number of basic facts that must be learned by nearly one-half.

A table of all addition facts would be endless. However, it is interesting to realize that a knowledge of decimal notation makes it sufficient that one learn the facts for the ten decimal digits, a knowledge of the commutative property of addition makes it sufficient that one learn the facts on and below the main diagonal of the table, and a knowledge of the property of adding zero provides the basis for the facts in the first column of the table. Accordingly, it is enough to know the sums shown in the following table.

+	0	1	2	3	4	5	6	7	8	9
0										
1		2								
2		3	4							
3		4	5	6						
4		5	6	7	8					
5		6	7	8	9	10				
6		7	8	9	10	11	12			
7		8	9	10	11	12	13	14		
8		9	10	11	12	13	14	15	16	
9		10	11	12	13	14	15	16	17	18

Thus it is possible for students to learn all of the 100 basic addition facts by learning only 45 facts and two properties, the commutative and the identity. The associative property may be used to explain many of the facts. For example,

$$9 + 9 = 9 + (1 + 8) = (9 + 1) + 8 = 10 + 8 = 18,$$
$$9 + 8 = 9 + (1 + 7) = (9 + 1) + 7 = 10 + 7 = 17,$$
$$9 + 7 = 9 + (1 + 6) = (9 + 1) + 6 = 10 + 6 = 16,$$
$$8 + 8 = 8 + (2 + 6) = (8 + 2) + 6 = 10 + 6 = 16,$$
$$8 + 7 = 8 + (2 + 5) = (8 + 2) + 5 = 10 + 5 = 15,$$
$$8 + 6 = 8 + (2 + 4) = (8 + 2) + 4 = 10 + 4 = 14.$$

The doubles of numbers ($2 + 2 = 4, 3 + 3 = 6, 4 + 4 = 8$, and so on) seem to be easier for most children to master than some of the other facts. Other sums may be related to the doubles. For example,

$3 + 4$ is one less than $4 + 4$,

$3 + 4$ is one more than $3 + 3$,

$2 + 3$ is one more than $2 + 2$,

$2 + 3$ is one less than $3 + 3$.

This procedure calls attention to a very basic approach to the learning of mathematics at all levels. One does not need to memorize a great many unrelated details. Instead, one should learn a few important basic ideas, and then learn how to use these basic ideas to reason out new facts for himself. All the ideas of mathematics can be derived from a few fundamental ideas and the use of logical reasoning. These fundamental ideas and the ability to reason are the tools that every student should develop.

Here is another example of the power of basic ideas in determining the sum of any pair of numbers. Consider $539 + 427$. From the concept of place value (expanded notation),

$$539 = 500 + 30 + 9$$
$$+427 = 400 + 20 + 7.$$

Now, using the previously summarized addition facts along with the concept of place value and the commutative and associative properties, we have

$$500 + 30 + 9$$
$$+400 + 20 + 7$$
$$\overline{900 + 50 + 16}$$

9 ones $+$ 7 ones $=$ 16 ones

3 tens $+$ 2 tens $=$ 5 tens

5 hundreds $+$ 4 hundreds $=$ 9 hundreds.

From the concept of place value, $16 = 10 + 6$, and

$$900 + 50 + 16 = 900 + 50 + (10 + 6).$$

Using the associative property we have

$$900 + 50 + (10 + 6) = 900 + (50 + 10) + 6.$$

Then, since $50 + 10 = 60$,

$$900 + (50 + 10) + 6 = 900 + 60 + 6.$$

This expanded form may be written in place-value notation as

$$900 + 60 + 6 = 966.$$

Therefore,

$$539 + 427 = 966.$$

By applying the commutative property of addition, it may also be concluded that

$$427 + 539 = 966.$$

The regrouping for this example is illustrated as follows.

$$539 = 5 \text{ hundreds, 3 tens, and } 9 \text{ ones}$$
$$\underline{+427} = 4 \text{ hundreds, 2 tens, and } 7 \text{ ones}$$
$$966 = 9 \text{ hundreds, 5 tens, and } 16 \text{ ones}$$

$$(5 \times 100) + (3 \times 10) + (9 \times 1)$$
$$(4 \times 100) + (2 \times 10) + (7 \times 1)$$
$$= (9 \times 100) + (5 \times 10) + (16 \times 1)$$

The 16 ones are grouped into one set of 1 ten and one set of 6 ones, and then the 1 ten is grouped with the 5 tens. This produces a sum of 9 hundreds, 6 tens, and 6 ones.

Exercises

1. In the addition table for decimal digits, describe briefly the pattern found as one proceeds from element to element in
   (a) The horizontal rows from left to right.
   (b) The vertical columns from top to bottom.
   (c) The diagonal blocks from upper left to lower right.
   (d) The diagonal blocks from lower left to upper right.
   (e) The horizontal rows from right to left.
   (f) The vertical columns from bottom to top.

2. Tell why each pattern noted in Exercise 1 occurs in the addition table.

3. From the addition table, what observation can you make about the sum of:
   (a) Two odd numbers?
   (b) Two even numbers?
   (c) An odd number and an even number?
   (d) An even number and an odd number?

4. What other patterns do you observe in the addition table?

5. Show how the associative property may be used to find each sum by renaming one addend to make a ten.
   (a) $9 + 6$ (b) $8 + 5$
   (c) $7 + 6$

6. Explain how each sum could be found by relating to a "doubles fact."
   (a) $5 + 6$ (b) $9 + 8$
   (c) $8 + 6$

7. Use the concept of place value and the associative property to find the sum of 157 and 365.

$$157 = 100 + 50 + 7$$
$$+365 = 300 + 60 + 5$$
$$\overline{? = \ ? + \ ? + \ ?}$$

(a) At each step, illustrate the indicated concept by filling in the expanded notation.

Place value	$400 + 110 + (? + 2)$
Associative property	$400 + (110 + ?) + \ ?$
Addition	$400 + \ ? + \ ?$
Place value	$400 + (100 + ?) + \ ?$
Associative property	$(400 + ?) + \ ? + \ ?$
Addition	$? + \ ? + \ ?$
Place-value notation	$?$

(b) Use a similar procedure to explain the following.

$$249$$
$$+657$$

8. How does the addition table for whole numbers illustrate the:
    (a) Commutative property?   (b) Identity property?

9. Can the associative property of addition be observed directly from the addition table? Explain.

4-7 Understanding the Addition Algorithm

Man has always looked for ways and methods of making his work easier. The procedures for adding numbers are the result of our efforts to simplify the operation of addition. Generally, as one becomes more and more familiar with addition facts, he finds it convenient to arrange the addends in column form and to use mentally the properties of addition of whole numbers and the concept of place value to determine the sum. A method of arranging numerals so as to reduce the number of steps necessary to determine the correct result is called an **algorithm, or algorism.** It is most important to realize that the short-cut method, the algorithm, produces the correct answer through numerical manipulations based on the reasoned use of number properties and concepts. Consider these examples.

Example 1	$\overset{1\,1}{239}$		**Example 2**	$\overset{1\,2}{328}$
	$+462$			227
	$\overline{701}$			$+789$
				$\overline{1{,}344}$

The examples illustrate the familiar addition algorithm. The use of this technique is highly recommended; the student *must*, however, understand why the process works before he concentrates on developing speed and accuracy in manipulative skills. Even though the student's calculations may become purely automatic, he should be capable of explaining what he is really doing. In Example 1 the calculator is not actually "carrying a one." He understands that the sum of 9 and 2 is 11, which is $10 + 1$. Then the 10 is added to 30, and the sum 40 is added to 60, and so forth. Thus, it should be clear that all "carrying" is based on an application of place-value notation, commutativity, and associativity.

The next three examples illustrate the "partial sums" method of addition. This method was popular in the sixteenth century and it illustrates the properties involved whenever addition is performed. Notice that the renaming occurs during addition.

Example 3	14	**Example 4**	47	**Example 5**	47
	$+\ 7$		$+21$		26
	11		8		$+59$
	10		60		22
	21		68		110
					132

Exercises

1. Find each sum. Use the familiar addition algorithm when applicable.

   (a)   2   (b)   20
        $+3$ $+30$

   (c)   24   (d)   27
       $+35$ $+38$

2. The examples in Exercise 1 are arranged in a possible sequence of difficulty for teaching. Analyze each example and explain what skills a student must have developed to be able to work each example successfully.

3. Find each sum. Use the familiar addition algorithm when applicable.

   (a)   300   (b)   320
      $+400$ $+460$

   (c)   390   (d)   397
      $+450$ $+456$

4. The examples in Exercise 3 are arranged in a possible sequence of difficulty for teaching. Analyze each example and explain what skills a student must have developed to be able to work each example successfully.

5. Find each sum, demonstrating the renaming that occurs when you use the partial-sums method of addition.

(a)   27
    +34

(b)   278
    +165

(c)   2,597
    +4,036

(d)   396
    222
    +137

4-8 Subtraction of Whole Numbers

It is imperative that children discover and understand the relationship between subtraction and addition. This relationship may be studied in the following statements.

If $2 + 7 = 9$,   then $9 - 7 = 2$.

If $7 + 2 = 9$,   then $9 - 2 = 7$.

If $3 + 5 = 8$,   then $8 - 5 = 3$.

If $5 + 3 = 8$,   then $8 - 3 = 5$.

If $\square + \triangle = \bigcirc$,   then $\bigcirc - \triangle = \square$.

If $\triangle + \square = \bigcirc$,   then $\bigcirc - \square = \triangle$.

The commutative property for the addition of whole numbers allows us to state that $7 + 2 = 2 + 7$; therefore, if we have $7 + 2 = 9$, we may change it to $2 + 7 = 9$ and get both $9 - 7 = 2$ and $9 - 2 = 7$ from the original equation, $7 + 2 = 9$. Thus, if $7 + 2 = 9$, then $9 - 2 = 7$ and $9 - 7 = 2$. This may also be expressed with a, b, and c representing whole numbers.

If $a + b = c$,   then $c - b = a$ and $c - a = b$.

The whole number n for the missing addend that makes $n + 3 = 8$ a true statement is called the **difference,** $8 - 3$ (read "eight minus three"), and the operation of finding n is called **subtraction.** The difference $8 - 3$ is another name for the number 5; hence, $8 - 3$ may be replaced by 5:

If $8 - 3 = n$,   then $n + 3 = 8$ and $n = 5$.

Subtraction is a binary operation, since it combines a first number and a second number to produce a unique third number. Frequently, we say that the first number is the **sum,** the second number is an **addend,** and the number to be determined is the **missing addend.** In the example $8 - 3 = n$, 8 is the sum, 3 is an addend, and the variable n stands for the missing addend that will produce 8 when 3 is added to it. If x and y are whole numbers and $y \geq x$, then the operation $y - x$ of subtracting x from y is finding some whole number d such that $d + x = y$, and the operation $y - d$ of subtracting d from y is finding some number x such that $x + d = y$.

The number line may be used to help visualize the operation of subtraction of whole numbers. Consider the following number line for $6 - 4 = (?)$.

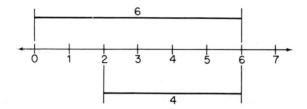

Notice that we start at the graph of 0 and find a point 6 units to the right of the graph of 0. This point has a coordinate of 6. Then we start at the graph of 6 and find a point 4 units to its left. This point has a coordinate of 2. Hence, we have a visualization of the statement $6 - 4 = (2)$.

Another way of using the number line to visualize the problem $6 - 4 = (?)$ is to think of the relation between addition and subtraction. Then the difference $6 - 4$ is the whole number n that 4 must be added to in order to get the sum 6.

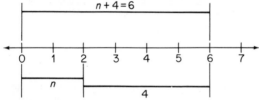

From the illustration on the number line it should be clear that n is 2. Since $6 - 4 = n$ means $n + 4 = 6$, then $n = 2$.

Subtraction of whole numbers may also be illustrated by *removing a subset from a given set*. Observe in this example that 2 blocks are being removed from the set of 5 blocks. The remaining set contains 3 blocks.

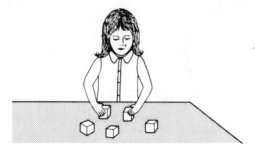

The cardinal number of the remaining set (in this case 3) is the result of subtracting the cardinal number of the subset being removed (2) from the cardinal number of the original set (5). This illustrates the statement:

$$5 - 2 = 3.$$

Subtraction of whole numbers may also be represented by *comparing two sets*. Consider the two sets of toy cars pictured.

Tom's cars   Mike's cars

A comparison or matching of the elements reveals an answer to the question "How many more toy cars does Mike have than Tom?". Since all of Tom's cars are matched and 2 cars are left in Mike's set that are not matched, we conclude that Mike has 2 more cars than Tom. This is a possible illustration of

$$6 - 4 = 2.$$

Exercises

1. Write each number sentence as an equation involving subtraction.
   (a) $7 + 6 = 13$ (b) $21 + 38 = 59$
   (c) $4 + 1 = 5$ (d) $k + h = m$

2. Write each number sentence as an equation involving addition.
   (a) $27 - 13 = 14$ (b) $16 - 7 = 9$
   (c) $145 - 65 = 80$ (d) $r - s = t$

3. Write two equations, one involving addition and one involving subtraction, that are represented on each number line.
   (a)

   (b)

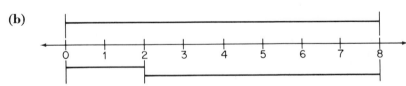

4. Use a number line to illustrate each equation.
   (a) $5 - 3 = 2$ (b) $2 + 3 = 5$
   (c) $7 - 4 = 3$ (d) $3 + 4 = 7$
   (e) $9 + 2 = 11$ (f) $11 - 2 = 9$

5. Write the subtraction equation represented by each of the following.
   (a)

   ★   ★   ★   ★ (★ ★ ★)

   (b) ★ ★ ★ ★ ★ ★   (★ ★ ★ ★)

6. Illustrate each equation using set removal.
   (a) $7 - 1 = 6$ (b) $4 - 4 = 0$
   (c) $9 - 6 = 3$

7. Write the subtraction equation represented by each of the following.
   (a)   (b)

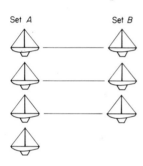

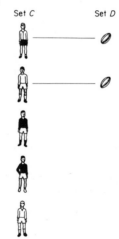

8. Illustrate each equation using set comparison.

   (a) $4 - 1 = 3$ (b) $6 - 5 = 1$
   (c) $5 - 5 = 0$

4-9 Inverse Operations: Addition and Subtraction

Consider the statements $8 + 3 = 11$ and $11 - 3 = 8$, that is, $(8 + 3) - 3 = 8$. Notice that adding 3 and subtracting 3 "undo" each other. That is, we start with 8, add 3 to obtain 11, then subtract 3 and get 8. The net effect of the two operations is the same as the operation of adding zero. Therefore, we state that adding a given number and then subtracting that same number are **inverse operations**. Likewise, subtracting a given number and then adding the same number back again are inverse operations.

Observe that $12 - 4 = 8$ and $8 + 4 = 12$, that is, $12 - 4 + 4 = 12$. For many years we have checked subtraction problems by addition, that is, by an application of this property of inverse operations, as in the following example.

$$\begin{array}{cc} & \text{Check} \\ 9 & 6 \\ \underline{-3} & \underline{+3} \\ 6 & 9 \end{array}$$

In general, we say that addition is the inverse operation of subtraction and subtraction is the inverse operation of addition.

Understanding that addition and subtraction are inverse operations allows us to determine unknown subtraction facts from our knowledge of addition facts, by finding missing addends. In other words, a subtraction problem may always be thought of as an addition problem with the sum and one addend known:

$$\begin{array}{lll} 10 - 2 = 8 & \text{because} & 8 + 2 = 10, \\ 13 - 7 = 6 & \text{because} & 6 + 7 = 13, \\ 35 - 19 = 16 & \text{because} & 16 + 19 = 35, \\ 41 - 23 = 18 & \text{because} & 18 + 23 = 41. \end{array}$$

Exercises

1. What is the inverse of each of the following?
   (a) Closing a door.   (b) Depositing $6.00 in the bank.
   (c) Putting on your shoes.   (d) Placing a pie in the oven.

2. Indicate whether each statement is true or false.
   (a) If $7 + 3 = 10$, then $10 - 3 = 7$ and $(7 + 3) - 3 = 7$.
   (b) If $6 + 8 = 14$, then $14 - 6 = 8$ and $(6 + 8) - 6 = 6$.
   (c) If $12 - 5 = 7$, then $7 + 5 = 12$ and $(12 - 5) + 5 = 12$.
   (d) If $26 + 9 = 35$, then $35 - 9 = 26$ and $(26 + 9) - 9 = 35$.

3. If you add 6 to a whole number n, what must you do to the sum to obtain n?

4. If you subtract 7 from a whole number n, what must you do to the difference to obtain n?

4-10 Properties of Subtraction of Whole Numbers

Although addition and subtraction are closely related operations, a distinction between them becomes evident when subtraction is considered with respect to some of the basic properties that hold for addition.

Commutative property   Since $9 - 4 \neq 4 - 9$, subtraction of whole numbers is not commutative.

In general, if a basic property of mathematics is applied to a given operation on the elements of a set of numbers and it leads us to a false conclusion in at least one case, then the property does not hold for the given operation on the elements of the specified set of numbers. The case in which the property does not hold is often called a *counterexample*. Note that $9 - 4 \neq 4 - 9$ is a counterexample that may be used to prove that subtraction of whole numbers is not commutative. Other counterexamples such as $4 - 0 \neq 0 - 4$ and $3 - 10 \neq 10 - 3$ may be used for the same purpose.

Associative property   We may observe that in the following example the associative property does not hold for subtraction of whole numbers:

$$7 - (6 - 5) \neq (7 - 6) - 5, \text{ because } 7 - 1 \neq 1 - 5.$$

Identity element   One property that does hold in a limited way for the subtraction of whole numbers is the special property of operating with zero. We have already agreed, for any whole number a, that $a + 0 = 0 + a = a$. It is true that $a - 0 = a$, although in general $0 - a \neq a$. Thus we may think of 0 as a *right identity* that is not a left identity for subtraction of whole numbers.

Closure and uniqueness properties   Recall that one whole number added to another whole number always produces a whole number as the sum. This property does not hold for subtraction. Certainly $3 - 5$ is not a whole number. The difference $3 - 5$, if it exists, must be a number that satisfies the equation $\square + 5 = 3$. There is no such number $\square$ in the set of whole numbers, since 5 added to any whole number produces a whole number greater than or equal to 5, and 3 is less than 5. Therefore, we conclude that differences of whole numbers do not always exist in the set of whole numbers. This is illustrated in the following diagram. The set of whole numbers is represented as being inside a circle, and a set of other numbers, represented by differences of whole numbers, as being outside the circle.

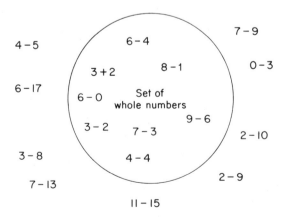

When any two whole numbers are subtracted, is there only one possible difference? We have seen that in some cases, no differences between whole numbers exist in the set of whole numbers. We say that the difference between two whole numbers, if it exists, is *unique*.

Exercises

1. Give a counterexample to show that each of the following properties does not hold for subtraction of whole numbers.
   (a) Commutative   (b) Associative
   (c) Closure

2. Find each answer in the set of whole numbers. Indicate those examples for which no answer exists.
   (a) $9 - 2$ (b) $3 - 4$
   (c) $5 - (4 + 2)$ (d) $9 + 5 - 8$
   (e) $7 - (9 - 6)$ (f) $8 - (10 - 2)$

3. Indicate whether each statement is true or false.
   (a) $11 - (6 - 3) = (11 - 6) - 3$ (b) $12 - 9 \neq 9 - 12$
   (c) $3 - (2 - 1) > (3 - 2) - 1$ (d) $8 - 7 = 7 - 8$
   (e) $5 + 0 = 5 - 0$ (f) $(11 - 0) - 3 = 11 - (0 - 3)$

4-11 Understanding the Subtraction Algorithm

A thorough understanding of the addition algorithm, the properties of addition, and the definition of subtraction makes clear the meaning of the subtraction algorithm. For example, to find the difference $43 - 17$, we first use the place-value concept and write

$$43 = 40 + 3$$
$$-17 = -(10 + 7).$$

Since 7 is greater than 3, $40 + 3$ must be renamed. Using addition, 40 can be renamed $30 + 10$, and so $40 + 3 = (30 + 10) + 3$. By the associative property of addition, $(30 + 10) + 3 = 30 + (10 + 3)$; then, by place value, $30 + (10 + 3) = 30 + 13$. Hence,

$$43 = 40 + 3 = 30 + 13$$
$$-17 = -(10 + 7) = -(10 + 7).$$

Now we must determine what number when added to 7 produces the sum 13, and likewise what number added to 10 produces a sum of 30; $\square + 7 = 13$ and $\triangle + 10 = 30$. Since $6 + 7 = 13$ and $20 + 10 = 30$,

$$43 = 30 + 13$$
$$-17 = -(10 + 7)$$
$$\overline{20 + 6 = 26.}$$

Most people do not think through all of the steps listed in the previous example. The familiar subtraction algorithm (short-cut method) is used, and the problem appears as follows.

$$\begin{array}{r} \overset{3\;1}{\cancel{4}3} \\ -17 \\ \hline 26 \end{array}$$

In using the algorithm in this example one must realize that he does not "borrow" a 1; he really "borrows" a 10. This means that he renames 43 as $30 + 13$ and then subtracts 7 from 13 and 10 from 30.

Actually, the word "borrow" is not precise, because it implies paying back. It is suggested that the process be taught as "renaming" or "regrouping" instead of "borrowing."

Another example of renaming may improve the student's understanding of the concepts that are actually applied in subtraction. From the concept of place value we have

$$\begin{array}{rl} 534 = & 500 + 30 + 4 \\ -245 = & -(200 + 40 + 5). \end{array}$$

Since 5 is greater than 4, it is necessary to rename $500 + 30 + 4$. With our knowledge of addition, 30 can be renamed $20 + 10$ and $500 + 30 + 4$ can be renamed $500 + (20 + 10) + 4$. By the associative property of addition, $500 + (20 + 10) + 4 = 500 + 20 + (10 + 4)$. Since $10 + 4 = 14$, $500 + 20 + (10 + 4) = 500 + 20 + 14$. We now have the problem

$$\begin{array}{r} 500 + 20 + 14 \\ -(200 + 40 + 5) \end{array}$$

and must consider the question of subtracting 40 from 20.

Since 40 is larger than 20, we must rename $500 + 20$. With the use of addition, 500 is renamed $400 + 100$ and $500 + 20$ is renamed $(400 + 100) + 20$. By the associative property of addition, $(400 + 100) + 20 = 400 + (100 + 20)$ and then, by place value, $400 + (100 + 20) = 400 + 120$. Hence,

$$\begin{array}{rlllll} 534 = & 500 + 30 + 4 = & 500 + 20 + 14 = & 400 + 120 + 14 \\ -245 = & -(200 + 40 + 5) = & -(200 + 40 + 5) = & -(200 + 40 + 5) \\ & & & \overline{200 + 80 + 9} \\ & & & = 289. \end{array}$$

An interesting method of subtraction taught in many countries of the world is known as the "Austrian," or "additive," method. Since to subtract 245 from 534 is to determine the missing addend n such that

$$245 + n = 534,$$

students are taught to find the addend by adding to 245 rather than by subtracting from 534. The following illustration shows how this is done.

	245	245
First, we determine what must be added to 245 to produce a 4 in the ones column.	$+\ \ ?$ $\overline{?4}$	$+\ \ 9$ $\overline{254}$

	254	254
Having added 9, we now have 254 and are working to obtain a 3 in the tens column.	$+\ \ ?$ $\overline{?34}$	$+\ 80$ $\overline{334}$

	334	334
We now have 334 and are looking for an addend to produce a 5 in the hundreds column.	$+\ \ ?$ $\overline{534}$	$+200$ $\overline{534}$

By adding 9, 80, and 200 to 245, we obtain 534.

$$\begin{array}{r} 200 \\ 80 \\ +\ \ 9 \\ \hline 289 \end{array}$$

Thus, we have

$$534 - 245 = 289.$$

Exercises

1. Find each difference. Use the familiar subtraction algorithm when applicable.

   (a) $\begin{array}{r} 7 \\ -4 \\ \hline \end{array}$

   (b) $\begin{array}{r} 70 \\ -40 \\ \hline \end{array}$

   (c) $\begin{array}{r} 76 \\ -42 \\ \hline \end{array}$

   (d) $\begin{array}{r} 75 \\ -48 \\ \hline \end{array}$

2. The examples in Exercise 1 are arranged in a possible sequence of difficulty for teaching. Analyze each example and explain what skills a student must have developed to be able to work each example successfully.

3. Find each difference. Use the familiar subtraction algorithm when applicable.

   (a) $\begin{array}{r} 500 \\ -200 \\ \hline \end{array}$

   (b) $\begin{array}{r} 560 \\ -210 \\ \hline \end{array}$

   (c) $\begin{array}{r} 520 \\ -230 \\ \hline \end{array}$

   (d) $\begin{array}{r} 527 \\ -239 \\ \hline \end{array}$

4. The examples in Exercise 3 are arranged in a possible sequence of difficulty for teaching. Analyze each example and explain what skills a student must have developed to be able to work each example successfully.

5. Work each example and show enough steps to illustrate the process used to calculate your answer.

   (a)   67   **(b)**   50
           − 19   − 13

   (c)   236   **(d)**   444
          − 158   − 355

6. Use the "Austrian" method of subtraction to find each missing addend.

   (a)   368   **(b)**   286
       + ?   + ?
        495   421

   (c)   172   **(d)**   537
       + ?   + ?
        301   729

7. Is it possible to use an addition table summarizing addition facts for whole numbers to find the solution to a subtraction problem involving whole numbers? If so, are there limitations? Explain.

4-12 Summary of Properties for Addition and Subtraction

The following table presents a summary of general properties for addition and subtraction of whole numbers. The letters a, b, and c represent arbitrary whole numbers where $a \neq b$, and $b \neq c$.

Property	*Addition*	*Subtraction*
Closure	$a + b$ always produces a whole number for the sum	$a - b$ does not always produce a whole number for the difference
Uniqueness	$a + b$ is a unique whole number	$a - b$, if it exists, is a unique whole number
Commutative	$a + b = b + a$	$a - b \neq b - a$
Associative	$a + (b + c) = (a + b) + c$	$a - (b - c) \neq (a - b) - c$ $(c \neq 0)$
Identity element	$a + 0 = a$ $0 + a = a$	$a - 0 = a$ $0 - a \neq a$ $(a \neq 0)$ (right identity only)
Inverse Operation	$a + b = c$ if and only if $c - b = a$	$a - b = c$ if and only if $c + b = a$

4-13 Chapter Test

Indicate whether each statement is true or false. If false, tell why.

1. The associative property of addition permits the grouping of three numbers in any order for addition.

2. Addition is a binary operation which is to be performed in a definite order.

3. The statement $2 + (3 + 4) = (3 + 4) + 2$ illustrates an application of the commutative property of addition.

4. The subtraction of whole numbers is associative since $(5 - 4) - 0 = 5 - (4 - 0)$.

5. A good way to reason an answer for $7 + 6$ is to think first that $7 + 7 = 14$, then subtract 1.

6. When two whole numbers are added, the sum is larger than either addend.

7. If $n(R) + n(T) = n(S)$ for sets R, T, and S, then $n(S) - n(R) = n(T)$.

8. Zero is the inverse element for addition of whole numbers.

9. The set of numbers $\{0, 4, 8, 12, 16, \ldots\}$ is closed under the operation of subtraction.

10. People learn, retain, and apply mathematics better if the material is learned by rote.

Select the best possible answer.

11. Let x, y, and z be whole numbers. $x - y = z$. If the same whole number is added to x and y, what can we say about z?
    (a) It remains unchanged.   (b) It increases.
    (c) It decreases.   (d) It is never a whole number.
    (e) It cannot be determined.

12. Which of the following charts could be used to illustrate an interpretation of subtraction?

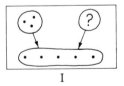   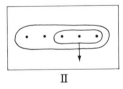   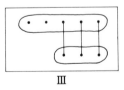

I   II   III

(a) I only   (b) II only
(c) III only   (d) I, II, and III
(e) The correct answer is not listed

13. Which of the following generalizations could *not* be developed directly
    from patterns appearing in a table of basic addition facts?
    (a) Identity property   (b) Closure property
    (c) Commutative property   (d) Associative property
    (e) None of the answers a–d is correct

14. How many basic subtraction facts are there?
    (a) Fifty   (b) Fifty-five
    (c) One hundred   (d) One hundred eighty
    (e) An infinite number

15. Which of the following illustrates an inverse relationship?
    (a) If $17 - 3 = 14$, then $17 - 14 = 3$
    (b) If $17 - 3 = 14$, then $14 + 3 = 17$
    (c) If $17 - 3 = 14$, then $14 + 17 = 3$
    (d) If $17 - 3 = 14$, then $14 - 3 = 11$
    (e) None of these

16. In teaching that $(30 + 2) + (40 + 3) = (30 + 40) + (2 + 3)$, one has
    made use of which of the following?

                    I. A commutative property.

                    II. An associative property.

                    III. A distributive property.

    (a) I only   (b) II only
    (c) I and III only   (d) I and II only
    (e) I, II, and III

17. Addition is to sum as subtraction is to:
    (a) Minuend   (b) Subtrahend
    (c) Addend   (d) Difference
    (e) Sum

chapter 5

Whole Numbers: Multiplication and Division

In this chapter the operation of multiplication of whole numbers is defined, and some of the basic properties of multiplication are presented as assumptions. The operation of division of whole numbers is presented as the inverse of multiplication, and some of the basic properties of division are studied.

5-1 Multiplication of Whole Numbers

Multiplication is a *binary* operation since it combines two numbers to produce a unique third number. Recall (Section 3-6) that the two numbers that are multiplied are called *factors* and the resulting number is called the *product* of the factors. In the example $3 \times 5 = 15$ the numbers 3 and 5 are factors of 15, and 15 is the product of 3 and 5. Multiplication within the set of whole numbers may be related to repeated addition. In other words, 3×5 may be considered as $5 + 5 + 5$, and 6×2 is equal to

$$2 + 2 + 2 + 2 + 2 + 2.$$

These examples may be illustrated by joining equivalent sets as shown in the next picture.

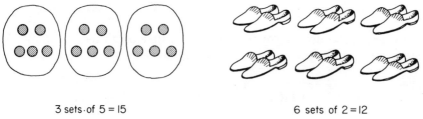

3 sets of 5 = 15   6 sets of 2 = 12
5 + 5 + 5 = 15   2 + 2 + 2 + 2 + 2 + 2 = 12
3 x 5 = 15   6 x 2 = 12

The repeated-addition interpretation is a good intuitive way to introduce multiplication; however, there are several limitations to this approach. Consider the following:

(1) When one of the factors is 0, it is meaningless to think of 0×4 as "add four, zero times" since addition is a binary operation.

(2) When one of the factors is 1, it is also meaningless to think of 1×4 as "add four, one time" since the binary operation of addition suggests two addends.

(3) When the factors are very large numbers (for example, 125×50), the use of the addition process to obtain the product is very time-consuming.

Therefore, it is desirable to deal with multiplication as a number operation in its own right, as soon as we become familiar with it as an abstract process.

It should be noted at this point that there are many different notations for multiplication. For example, each of the following denotes the product "3 times 2":

(a) 3×2 (b) 3 groups of 2   (c) $\begin{array}{r} 2 \\ \times 3 \end{array}$ (d) $3 \cdot 2$

(e) 3 twos   (f) 2 multiplied by 3   (g) $(3)(2)$ (h) Multiply 2 by 3

Multiplication of whole numbers may also be represented on a number line. For example, $3 \times 2 = 6$ may be represented as $2 + 2 + 2 = 6$.

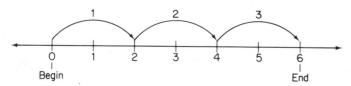

Notice that the representation of multiplication on the number line involves jumps of equal size. Begin at the graph of 0 and count off three jumps of 2 (that is, 3 twos) to determine that the coordinate of the terminal point is 6; $3 \times 2 = 6$.

A rectangular dot array may be used to relate multiplication to addition. Study this picture of a 3×6 array on a pegboard.

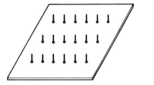

3 rows of 6 = 18

3 sets of 6 = 18

3 x 6 = 18

Dot arrays may also be shown on graph paper, or simply by arranging objects in rows and columns.

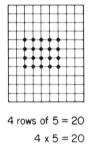

4 rows of 5 = 20

4 x 5 = 20

2 rows of 9 = 18

2 x 9 = 18

In order that we may be consistent, we shall agree that a 4×5 array has four (horizontal) rows and five (vertical) columns, and that a 2×9 array has two rows and nine columns. In other words, the first number indicates the number of rows, and the second number gives the number of columns.

The multiplication of whole numbers may also be described in terms of the Cartesian product of sets. Consider a set of three coats and a set of four hats. No two coats or two hats are of the same color. How many different combinations of coats and hats can be made? It may be observed in the following diagram that each coat can be paired with each of the four hats and that a total of twelve different color combinations can be formed.

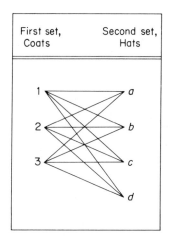

If we denote the set of coats by {1, 2, 3} and the set of hats by {a, b, c, d}, a convenient notation for the set of all possible coat–hat combinations is illustrated by listing the set of ordered pairs: {(1, a), (1, b), (1, c), (1, d), (2, a), (2, b), (2, c), (2, d), (3, a), (3, b), (3, c), (3, d)}. Note that a set of twelve combinations (pairs of elements) has been constructed from a set of three objects and a set of four objects. Each of these pairs is called an **ordered pair**, because we have assigned a significance to the order. In the ordered pair (1, a), coat 1 is the **first element**, and hat a is the **second element**.

Exercises

1. Write an addition and a multiplication equation for each of the following.
   (a)

   (b)

2. Make a set picture to illustrate each multiplication equation, if possible.
   (a) 2 × 7 = 14   (b) 6 × 1 = 6
   (c) 3 × 0 = 0   (d) 8 × 3 = 24
   (e) 0 × 4 = 0   (f) 1 × 9 = 9

3. For each part of Exercise 2, write an addition equation that is equivalent to the given multiplication equation, if possible.

4. Write the multiplication equation represented on each number line.

(a)

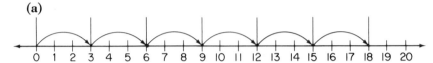

(b)

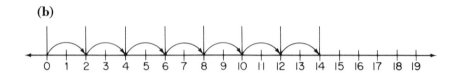

5. Draw a number line to represent each equation, if possible.
   (a) $4 \times 2 = 8$ (b) $5 \times 1 = 5$
   (c) $0 \times 3 = 0$ (d) $7 \times 0 = 0$

6. Write a multiplication equation for each of the following arrays.

(a)
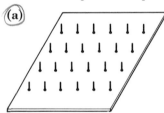

(b)

(c)
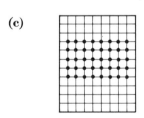

(d)

7. Draw an array to represent each equation, if possible.
   (a) $6 \times 4 = 24$ (b) $8 \times 1 = 8$
   (c) $3 \times 0 = 0$ (d) $2 \times 10 = 20$

8. Write a multiplication equation for each of the following.

(a)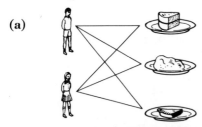

(b) $\{(a, 1), (a, 2), (a, 3), (a, 4), (a, 5)\}$

9. Jack has five different-appearing sport shirts and two different-appearing pairs of pants. Assume that each shirt may be worn with each pair of pants. How many different outfits (combinations) can Jack make?

5-2 Closure and Uniqueness Properties

The property of **closure** applied to the multiplication of whole numbers means that the product of any two whole numbers is also a whole number. The property of **uniqueness** indicates that there is only one possible product when two given whole numbers are multiplied. In the example $7 \times 3 = 21$ both the factors and the product are elements of the set of whole numbers. It should also be noted that there is one and only one possible product when 3 is multiplied by 7. Hence, 21 is the *unique* product 7×3.

The properties of closure and uniqueness for the multiplication of whole numbers may be stated as

$$a \times b = c$$

where a and b may be any whole numbers, and the whole number c is the unique product of a and b.

The property of closure under multiplication does not hold for all sets. In the case of a finite set $X = \{1, 3, 5, 7, 9\}$ a product of two elements of the set may not be a member of the set. For example, $3 \in X$, $9 \in X$, and $3 \times 9 = 27$, but 27 is not an element of the set X. Hence, set X is not closed with respect to the operation of multiplication.

Exercises

1. State whether or not each set is closed under the operation of multiplication.
   (a) The set of even whole numbers.
   (b) The set of odd whole numbers.
   (c) The set of natural numbers greater than 100.
   (d) The set of counting numbers between 25 and 385.
   (e) The set of whole numbers whose one's digits are 0.

(**f**) The set of odd natural numbers less than 888.

(**g**) The set of counting numbers whose one's digits are 5.

2. State whether or not each product is unique.

(**a**) 9×8 (**b**) 3×6

(**c**) 25×36 (**d**) 234×197

5-3 The Commutative Property of Multiplication

The order of multiplication of two numbers (factors) in the set of whole numbers does not affect the product. For example, $3 \times 4 = 4 \times 3$, since $3 \times 4 = 12$ and $4 \times 3 = 12$. This example illustrates the **commutative property of multiplication** for whole numbers. If a and b are any two whole numbers, then

$$a \times b = b \times a.$$

Arrays may be used in leading children to the discovery of the commutative property of multiplication. A 3×4 array has the same number of elements as a 4×3 array, but the arrays look different as shown in these pictures.

3 x 4

4 x 3

The number line also may be used to illustrate the commutative property of multiplication of whole numbers.

The next picture shows that the effect of three jumps of 4 units' length is the same as four jumps of 3 units' length.

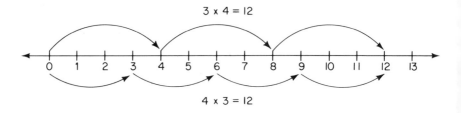
3 x 4 = 12

4 x 3 = 12

Exercises

1. Make set pictures to show that $2 \times 6 = 6 \times 2$.

2. Given $A = \{\triangle, \square, \otimes\}$ and $B = \{p, q\}$, find:

(a) $A \times B$ (b) $n(A \times B)$
(c) $B \times A$ (d) $n(B \times A)$

3. In Exercise 2, does:
(a) $A \times B = B \times A$?   (b) $n(A \times B) = n(B \times A)$?

4. Which statement in Exercise 3 illustrates the commutative property of multiplication for whole numbers?

5. Draw array pictures to show that $8 \times 4 = 4 \times 8$.

6. Draw a number line to show that $4 \times 5 = 5 \times 4$.

7. Copy and complete each statement to obtain an example of the commutative property of multiplication.
(a) $5 \times 3 = \underline{\quad} \times \underline{\quad}$ (b) $7 \times 9 = \underline{\quad} \times \underline{\quad}$
(c) $143 \times \underline{\quad} = \underline{\quad} \times 143$ (d) $\underline{\quad} \times 4 = 4 \times \underline{\quad}$

5-4 The Associative Property of Multiplication

Multiplication, like addition, is a binary operation. When more than two factors are to be multiplied, the factors are grouped for binary operations. For example, $3 \times 4 \times 5$ may be grouped for multiplication as $(3 \times 4) \times 5$ or as $3 \times (4 \times 5)$. The expression $(3 \times 4) \times 5$ indicates 12×5, and the expression $3 \times (4 \times 5)$ indicates 3×20. Both 12×5 and 3×20 equal 60. Notice that the associative property does *not* provide for a change in the order of the factors. It provides only for a change in the grouping.

Consider this example illustrating the idea of the associative property of multiplication. Given 4 sets of 3 buttons with 2 holes in each button, how many buttonholes are there in all?

Interpretation A:

> 4 sets of 3 buttons each = 12 buttons
>
> 12 buttons with 2 holes each = 24 buttonholes
>
> $(4 \times 3) \times 2 = 12 \times 2$
> $= 24$

Interpretation B:

> 3 buttons with 2 holes each = 6 buttonholes
>
> 4 cards with 6 buttonholes each = 24 buttonholes

$$4 \times (3 \times 2) = 4 \times 6$$
$$= 24$$

We can see clearly from this example that $(4 \times 3) \times 2 = 4 \times (3 \times 2) = 24$.
The **associative property of multiplication** may be stated as

$$(a \times b) \times c = a \times (b \times c),$$

where a, b, and c represent any whole numbers.

The associative and commutative properties of multiplication provide us with a flexibility in rearranging and grouping the factors in a problem involving only multiplication.

Exercises

1. Show how the associative property of multiplication can be used to find each product in two different ways.
   (a) $2 \times 5 \times 46$ (b) $3 \times 6 \times 9$
   (c) $50 \times 20 \times 7$ (d) $4 \times 8 \times 25$

2. Tell which property, commutative or associative, is illustrated by each equation.
   (a) $(7 \times 8) \times (9 \times 6) = (9 \times 6) \times (7 \times 8)$
   (b) $2 \times (3 \times 4) = (2 \times 3) \times 4$
   (c) $(e \times f) \times g = e \times (f \times g)$
   (d) $(5 \times 8) \times 10 = 10 \times (5 \times 8)$

3. Multiply without the use of paper and pencil. Did you mentally apply the associative property?
   (a) $2 \times (5 \times 13)$ (b) $5 \times (20 \times 17)$
   (c) $4 \times (25 \times 16)$ (d) $8 \times (125 \times 7)$

4. Indicate whether each statement is true or false.
   (a) $(26 \times 8) \times 3 < (8 \times 26) \times 3$
   (b) $101 \times 14 \neq 14 \times 101$
   (c) $(17 \times 20) \times 13 = (17 \times 13) \times 20$
   (d) $7 \times (8 \times 9) = (8 \times 9) \times 7$

5. Group the numbers 3, 4, 5, and 6 for multiplication in at least four different ways. Find the product for each of the groupings. Are all the products equal?

6. Use the commutative and associative properties of multiplication to justify or explain each of the following. Copy each step and give the reason when needed.
   (a) Show that $(a \times b) \times c = a \times (c \times b)$.
   $(a \times b) \times c = a \times (b \times c)$ <u>associative property of multiplication</u>
   $a \times (b \times c) = a \times (c \times b)$ <u>   </u>

(b) Show that $a \times (b \times c) = c \times (b \times a)$.

$a \times (b \times c) = (b \times c) \times a$ commutative property of multiplication

$(b \times c) \times a = (c \times b) \times a$ _____

$(c \times b) \times a = c \times (b \times a)$ _____

(c) Show that $a \times (b \times c) = b \times (a \times c)$.

$a \times (b \times c) = (a \times b) \times c$ _____

$(a \times b) \times c = (b \times a) \times c$ _____

$(b \times a) \times c = b \times (a \times c)$ _____

5-5 Identity Element for Multiplication

The product of any whole number and the number 1 is the original whole number. For example, $2 \times 1 = 1 \times 2 = 2$, $3 \times 1 = 1 \times 3 = 3$, $4 \times 1 = 1 \times 4 = 4$, $5 \times 1 = 1 \times 5 = 5$, and so on. Each of the statements may be illustrated by arrays as shown in the next figure. A 2×1 array consists of two rows of one member each, a 1×2 array consists of one row of two members, and consequently each array contains only two members. A similar situation exists for 3×1 and 1×3 arrays, 4×1 and 1×4 arrays, 5×1 and 1×5 arrays, and so on.

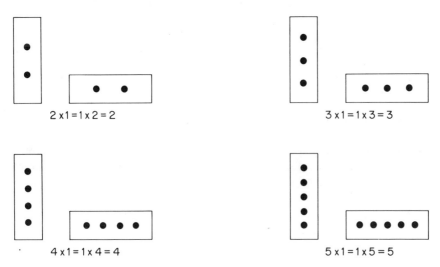

$2 \times 1 = 1 \times 2 = 2$ $3 \times 1 = 1 \times 3 = 3$

$4 \times 1 = 1 \times 4 = 4$ $5 \times 1 = 1 \times 5 = 5$

For any whole number a,

$$a \times 1 = 1 \times a = a.$$

Thus the number 1 is called the **identity element for multiplication,** that is, the **multiplicative identity.** Note that the role of 1 in multiplication is similar to that of 0 in addition.

Exercises

1. Given $A = \{a\}$ and $B = \{\triangle, \otimes, \square\}$. Use these sets to illustrate:
   (a) $n(A) \times n(B) = n(A \times B)$
   (b) $n(A \times B) = n(B)$
   (c) $n(A) \times n(B) = n(B)$
   (d) $n(A) \times n(B) = n(B) \times n(A) = n(B)$

2. Use sets to illustrate:
   (a) $5 \times 1 = 1 \times 5 = 5$ (b) $6 \times 1 = 1 \times 6 = 6$

3. Write the multiplication equation shown by each pair of arrays.
   (a)   (b)

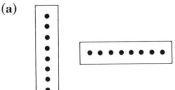

4. Draw arrays to illustrate:
   (a) $7 \times 1 = 1 \times 7 = 7$ (b) $1 \times 9 = 9 \times 1 = 9$

5. Complete each statement to make a true equation.
   (a) $1 \times 16 = $ _____   (b) _____ $\times 1 = 23$
   (c) $1 \times $ _____ $= 1$ (d) _____ $\times 1 = 0$
   (e) $d \times $ _____ $= $ _____ $\times d = d, d \in W$

5-6 Property of Multiplication by Zero

The number zero has a special characteristic when used as a factor. Consider 5×0 as the number of ordered pairs that can be formed by selecting a first object from a set of five things and a second object from a set of zero things. We can then see that $5 \times 0 = 0$. If we recall that a coat–hat combination requires one element from each set, there are exactly 0 coat–hat combinations possible from a set of 5 coats and a set of 0 hats.

The product of any whole number and zero is always zero. For example,

$$0 \times 0 = 0,$$
$$1 \times 0 = 0 \text{ and } 0 \times 1 = 0,$$
$$2 \times 0 = 0 \text{ and } 0 \times 2 = 0,$$
$$3 \times 0 = 0 \text{ and } 0 \times 3 = 0,$$
$$4 \times 0 = 0 \text{ and } 0 \times 4 = 0;$$

for each whole number a,

$$a \times 0 = 0 \times a = 0.$$

Exercises

1. Given $D = \{d, e, f, g\}$ and $H = \{\ \}$. Use these sets to illustrate:
   (a) $n(D) \times n(H) = n(D \times H)$
   (b) $n(D \times H) = n(H)$
   (c) $n(D) \times n(H) = n(H)$
   (d) $n(D) \times n(H) = n(H) \times n(D) = n(H)$

2. Use sets to illustrate:
   (a) $2 \times 0 = 0 \times 2 = 0$ (b) $4 \times 0 = 0 \times 4 = 0$

3. Illustrate 0×7 using:
   (a) Pictures of sets.   (b) A number line.
   (c) An array.   (d) Cartesian product of sets.

4. Complete each statement to make each equation true.
   (a) $8 \times \underline{\quad} = 0$ (b) $\underline{\quad} \times 0 = 0$
   (c) $0 \times \underline{\quad} = 1$ (d) $7 \times 8 \times 0 = \underline{\quad}$
   (e) $(2 \times 3) \times \underline{\quad} = (7 \times 0) \times 9$ (f) $r \times \underline{\quad} = \underline{\quad} \times r = 0, r \in W$

5-7 Distributive Property of Multiplication over Addition

The operations of addition and multiplication have been discussed, and
some of the properties of these operations have been considered. The distributive property, however, may involve both addition and multiplication.

Consider the example 3×7 and recall from previous discussions that
there are many numerals that represent 7. In other words, 3×7 may also
be represented as $3 \times (5 + 2)$. The question that is immediately posed is,
"Does $3 \times (5 + 2)$ equal 3×7 or $(3 \times 5) + (3 \times 2)$?" Since 3×7 equals
21 and $(3 \times 5) + (3 \times 2)$ equals $15 + 6 = 21$, we have $3 \times (5 + 2) =
3 \times 7 = (3 \times 5) + (3 \times 2)$. This may be illustrated by arrays as in the
next figure.

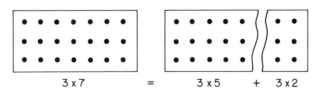

$$3 \times 7 \qquad = \qquad 3 \times 5 \quad + \quad 3 \times 2$$

The **distributive property of multiplication over addition** is a very
basic and important one. It may be stated in general terms as

$$a \times (b + c) = (a \times b) + (a \times c),$$

where a, b, and c are any whole numbers.

It is important to note that since multiplication is commutative, we may distribute multiplication over addition from both the left and the right. In other words, we have a *left distributive property*,

$$a \times (b + c) = (a \times b) + (a \times c),$$

and also a *right distributive property*,

$$(b + c) \times a = (b \times a) + (c \times a).$$

We speak of "the distributive property of multiplication over addition" in referring to whichever one of these is applicable.

This distributive property is the basis of many of our short-cut methods of computation. When we work the following multiplication problem we apply a distributive property. In the problem

$$\begin{array}{r} 31 \\ \times\ 2 \\ \hline \end{array}$$

we multiply 1 by 2 and multiply 30 by 2. We then add these products and write the answer 62. Actually $31 = 30 + 1$, and we multiply this sum by 2. Thus we have

$$2 \times (30 + 1) = (2 \times 30) + (2 \times 1) = 60 + 2 = 62.$$

Application of the distributive property of multiplication over addition is further illustrated in these examples:

$$(8 \times 27) + (8 \times 73) = 8 \times (27 + 73) = 8 \times 100 = 800,$$
$$(16 \times 12) + (34 \times 12) = (16 + 34) \times 12 = 50 \times 12 = 600,$$
$$30 \times 17 = 30 \times (10 + 7) = (30 \times 10) + (30 \times 7) = 300 + 210 = 510.$$

There is also a distributive property of multiplication over subtraction (see Exercise 7).

Exercises

1. Use a distributive property and insert numerals to make each number sentence true.
   (a) $7 \times (8 + 29) = (7 \times \underline{\quad}) + (\underline{\quad} \times 29)$
   (b) $(5 \times 3) + (5 \times \underline{\quad}) = \underline{\quad} \times (\underline{\quad} + 26)$
   (c) $128 \times (64 + 216) = (\underline{\quad} \times \underline{\quad}) + (\underline{\quad} \times \underline{\quad})$
   (d) $(785 \times 38) + (785 \times 59) = \underline{\quad} \times (\underline{\quad} + \underline{\quad})$

2. Indicate whether each statement is true or false.
   (a) $2 \times (3 + 4) = (2 \times 3) + (3 \times 4)$
   (b) $(7 + 8) \times 5 = (7 \times 5) + (8 \times 5)$
   (c) $(26 \times 8) + (17 \times 8) = (26 \times 17) + 8$
   (d) $(5 \times 11) + (16 \times 5) = (16 + 11) \times 5$

3. Apply a distributive property to compute each of the following.
   (a) $(78 \times 43) + (78 \times 57)$ (b) $(39 \times 5) + (39 \times 5)$

4. We have illustrated that multiplication is distributive over addition. Is addition distributive over multiplication? That is, if a, b, and c represent any three whole numbers, is it always true that $a + (b \times c) = (a + b) \times (a + c)$?

5. State the property that justifies each step illustrated in the following example.

$(30 \times 13) + (37 \times 30) = (30 \times 13) + (30 \times 37)$ ⎯⎯⎯⎯⎯⎯
$(30 \times 13) + (30 \times 37) = 30 \times (13 + 37)$ ⎯⎯⎯⎯⎯⎯
$30 \times (13 + 37) = 30 \times 50$ ⎯⎯⎯⎯⎯⎯
$30 \times 50 = 1{,}500$ ⎯⎯⎯⎯⎯⎯

6. Given $A = \{a, b\}$, $B = \{g, h, j, k\}$, and $C = \{\triangle, \square, \otimes\}$. Use these sets to illustrate the following.
   (a) $A \times (B \cup C) = (A \times B) \cup (A \times C)$
   (b) $n[A \times (B \cup C)] = n[(A \times B) \cup (A \times C)]$
   (c) $n(A) \times n(B \cup C) = n(A \times B) + n(A \times C)$
   (d) $n(A) \times [n(B) + n(C)] = n(A) \times n(B) + n(A) \times n(C)$

7. Is multiplication distributive over subtraction? That is, if a, b, and c represent any three whole numbers, is it always true that $a \times (b - c) = (a \times b) - (a \times c)$? Is it always true that $(b - c) \times a = (b \times a) - (c \times a)$?

8. Apply a distributive property for multiplication over subtraction to compute each of the following.
   (a) $(49 \times 17) - (9 \times 17)$ (b) $(28 \times 15) - (28 \times 5)$

9. State the property that justifies each step illustrated in the following examples.
   (a) $(3 + 5) \times (4 + 2) = 3 \times (4 + 2) + 5 \times (4 + 2)$ ⎯⎯⎯⎯⎯⎯
   $3 \times (4 + 2) + 5 \times (4 + 2) = (3 \times 4) + (3 \times 2) + (5 \times 4)$
   $+ (5 \times 2)$ ⎯⎯⎯⎯⎯⎯
   (b) $(d + e) \times (g + h) = d \times (g + h) + e \times (g + h)$ ⎯⎯⎯⎯⎯⎯
   $d \times (g + h) + e \times (g + h) = (d \times g) + (d \times h)$
   $+ (e \times g) + (e \times h)$ ⎯⎯⎯⎯⎯⎯

10. Each example involves the application of one of the properties of multiplication or addition. Name the property illustrated by each example.
   (a) $3 + 4 = 4 + 3$ (b) $2 \times (5 + 6) = (5 + 6) \times 2$
   (c) $2 \times 3 \times 4 \times 0 \times 6 = 0$ (d) $(7 \times 8) \times 9 = 7 \times (8 \times 9)$
   (e) $(7 + 5) \times (0 + 6) = (7 + 5) \times 6$
   (f) $(9 \times 12) \times 1 = (9 \times 12)$
   (g) $(2 + 8) \times (7 + 4) = (2 + 8) \times 7 + (2 + 8) \times 4$
   (h) $9 \times (2 + 3 + 4 + 5) = (9 \times 2) + (9 \times 3) + (9 \times 4) + (9 \times 5)$

5-8 The Table of Basic Multiplication Facts

As in addition, a table may be used to summarize multiplication facts. The following table is a summary of the products of the decimal digits.

×	0	1	2	3	4	5	6	7	8	9	
0	0	0	0	0	0	0	0	0	0	0	
1	0	1	2	3	4	5	6	7	8	9	
2	0	2	4	6	8	10	12	14	16	18	⟵ (2 × 9)
3	0	3	6	9	12	15	18	21	24	27	
4	0	4	8	12	16	20	24	28	32	36	
5	0	5	10	15	20	25	30	35	40	45	
6	0	6	12	18	24	30	36	42	48	54	⟵ (6 × 7)
7	0	7	14	21	28	35	42	49	56	63	
8	0	8	16	24	32	40	48	56	64	72	
9	0	9	18	27	36	45	54	63	72	81	

This table may be used to emphasize many of the important properties that apply to the multiplication of whole numbers. The special characteristics of multiplying by 0 and by 1 are shown in the first two rows and the first two columns. A line has been drawn diagonally through the table from the upper left-hand corner to the lower right. The elements drawn through are called *diagonal elements* and form the *main diagonal* of the table. Notice that except for the diagonal elements this line divides the table into two equivalent parts. When the table is folded over its main diagonal, the blocks containing 0 fall upon each other, as do the blocks containing 2, and so on. This is due to the commutative property of multiplication, which justifies such conclusions as $2 \times 3 = 3 \times 2$.

Further examination of the table will reveal other patterns that are most interesting. For example, study the diagonal elements from upper left to lower right, $\{1, 4, 9, 16, \ldots, 81\}$. These diagonal elements are consecutive squares; that is, each diagonal element is obtained by squaring a number (using the number as a factor two times). What are the differences between the consecutive squares? Look for other patterns and try to explain why they develop.

Special attention should be given to the fact that an understanding of the commutative property and the special characteristics of 0 and 1 in multiplication make it unnecessary to learn the 100 basic facts in the multiplication table. It is sufficient to learn the 36 facts shown in the next table.

×	0	1	2	3	4	5	6	7	8	9
0										
1										
2			4							
3			6	9						
4			8	12	16					
5			10	15	20	25				
6			12	18	24	30	36			
7			14	21	28	35	42	49		
8			16	24	32	40	48	56	64	
9			18	27	36	45	54	63	72	81

Actually, some of the simple products in this table could be considered as repeated addition, and this would reduce the number of needed multiplication facts even further. An understanding of the distributive property and the basic addition facts could also reduce the number of needed multiplication facts. For example, if we wish to multiply 6 by 9, we could think in terms of 6 as $3 + 3$ and then use the distributive property to determine

$$9 \times 6 = 9 \times (3 + 3) = (9 \times 3) + (9 \times 3) = 27 + 27 = 54.$$

Another example may improve the understanding of concepts that are actually applied in the multiplication process. For example, consider 6×18. From the concept of place value, $18 = 10 + 8$. Hence,

$$6 \times 18 = 6 \times (10 + 8).$$

Applying the distributive property, we have

$$6 \times (10 + 8) = (6 \times 10) + (6 \times 8).$$

Using multiplication, we have

$$(6 \times 10) + (6 \times 8) = 60 + 48,$$

and, from the concept of place value,

$$60 + 48 = 60 + (40 + 8).$$

Then, using the associative property of addition, we have

$$60 + (40 + 8) = (60 + 40) + 8,$$

and, since $60 + 40 = 100$,

$$(60 + 40) + 8 = 100 + 8 = 108.$$

Therefore,

$$6 \times 18 = 108.$$

If we use the expanded notation in vertical form, we obtain

$$
\begin{array}{r}
18 = 10 + 8 \\
\times\ 6 = \underline{\times\ 6} \\
60 + 48 = 60 + (40 + 8) = (60 + 40) + 8 = 108.
\end{array}
$$

Exercises

For Exercises 1 through 8, consider a multiplication table for the whole numbers from 0 to 1,000.

1. What property of multiplication is illustrated by the top row of facts and the left-hand column of facts?

2. What property of multiplication is illustrated by the facts in the second row and those in the second column from the left?

3. It was pointed out that in the multiplication table for the decimal digits the diagonal elements from upper left to lower right form the set of consecutive squares,

$$\{1, 4, 9, 16, 25, 36, 49, 64, 81\}.$$

   What pattern seems to be developing in this set of numbers? Predict the next five numbers that would be represented on the main diagonal of the multiplication table for the whole numbers from 0 to 25.

4. How does the multiplication table for whole numbers illustrate the commutative property?

5. From the multiplication table, what observation can you make about the product of:
   (a) Two odd numbers?
   (b) Two even numbers?

    (c) An odd number and an even number?

    (d) An even number and an odd number?

6. What other patterns do you observe in the multiplication table?

7. What are the first four entries in the row named 700? In the column named 700?

8. Is each number represented in the row named 700 matched by an equal number represented in the column named 700?

9. Can the associative property of multiplication be observed directly from the multiplication table? Explain.

10. Show how the distributive property of multiplication over addition may be used to find each product by renaming one factor as two addends.

    (a) 5×9 (b) 7×8 (c) 8×12

5-9 Understanding the Multiplication Algorithm

There are several conventional algorithms that aid in solving multiplication problems. Each of these algorithms can be explained in terms of the properties of multiplication and addition. It is extremely important that children understand *why* these algorithms give the correct result when used properly.

    The following examples demonstrate a familiar multiplication algorithm.

$$
\textbf{Example 1} \quad
\begin{array}{r}
{\scriptstyle 2} \\
14 \\
\times\ 7 \\
\hline
98
\end{array}
\qquad\qquad
\textbf{Example 2} \quad
\begin{array}{r}
{\scriptstyle 1} \\
24 \\
\times 13 \\
\hline
72 \\
24 \\
\hline
312
\end{array}
$$

    Using the algorithm in Example 1, we think "7 times 4 is 28"; put an 8 in the ones column and a 2 above the 1 in 14 to remind us of the 2 tens in 20. The product of 7 and 10 is 70, and 70 plus 20 equals 90. The sum of 90 and 8 is 98. The properties that allow us to use this "short-cut" method for multiplying 14 by 7 are the same as for 6×18, as illustrated in Section 5-8.

    In Example 2, we are actually multiplying $20 + 4$ by $10 + 3$, which is an application of the distributive property:

$$
\begin{array}{r}
24 \\
\times\ 13 \\
\hline
72 \\
24 \\
\hline
312
\end{array}
\qquad
\begin{aligned}
13 \times 24 &= 13 \times (20 + 4) \\
&= (13 \times 20) + (13 \times 4) \\
&= [(10 + 3) \times 20] + [(10 + 3) \times 4] \\
&= [(10 \times 20) + (3 \times 20)] + [(10 \times 4) + (3 \times 4)] \\
&= 200 + 60 + 40 + 12 \\
&= 312.
\end{aligned}
$$

Consider another algorithm for Example 2. This is sometimes called the "partial-products" method of multiplication.

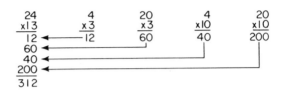

Exercises

1. Find each product. Use the familiar multiplication algorithm when applicable.

(a)   4
     ×2

(b)   40
     × 2

(c)   43
     × 2

(d)   40
     ×20

(e)   43
     ×21

2. The examples in Exercise 1 are arranged in a possible sequence of difficulty for teaching. Analyze each example and explain what skills a student must have developed to be able to work each example successfully.

3. Find each product. Use the familiar multiplication algorithm when applicable.

(a)   200
     ×   4

(b)   230
     ×   4

(c)   236
     ×   4

(d)   200
     × 40

(e)   230
     × 40

(f)   236
     × 40

(g)   236
     × 42

(h)   236
     × 78

4. The examples in Exercise 3 are arranged in a possible sequence of difficulty for teaching. Analyze each example and explain what skills a student must have developed to be able to work each example successfully.

5. Find each product, demonstrating the renaming that occurs when you use the partial-products method of multiplication.

(a)   38
      ×29

(b)   75
      ×14

(c)   298
     × 36

(d)   134
      ×218

6. Illustrate the use of a distributive property in determining each product in Exercise 5.

7. One of the older forms for multiplication is called the *lattice method*. Study the completed example in part (a) and then find the other products using the lattice method.

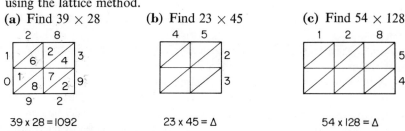

(a) Find 39 × 28

39 x 28 = 1092

(b) Find 23 × 45

23 x 45 = Δ

(c) Find 54 × 128

54 x 128 = Δ

5-10 Division of Whole Numbers

Division is related to multiplication in much the same way that subtraction is related to addition. In many of the modern school mathematics programs division of whole numbers is introduced as a series of repeated subtractions to improve the student's understanding of the usual division algorithm. In fact, modern desk calculators operate on this principle of repeated subtraction as the method for division. It is most important that children discover and understand the relationship between division and multiplication. In this chapter, we again find it necessary to extend our number system in order that we may provide solutions for certain division problems, such as $2 \div 3 = \square$.

The close relationship between multiplication and division may be observed in the following statements.

$$\text{If } 3 \times 6 = 18, \quad \text{then } 18 \div 6 = 3.$$
$$\text{If } 6 \times 3 = 18, \quad \text{then } 18 \div 3 = 6.$$
$$\text{If } 7 \times 8 = 56, \quad \text{then } 56 \div 8 = 7.$$
$$\text{If } 8 \times 7 = 56, \quad \text{then } 56 \div 7 = 8.$$

The commutative property for the multiplication of whole numbers allows us to state that $3 \times 6 = 6 \times 3$; therefore, if we have $3 \times 6 = 18$, we may change it to $6 \times 3 = 18$ and get both $18 \div 6 = 3$ and $18 \div 3 = 6$ from the original equation $3 \times 6 = 18$. This may also be expressed with a, b, and c as whole numbers ($b \neq 0$, $a \neq 0$):

If $a \times b = c$,   then $c \div b = a$ and $c \div a = b$.

The whole number n for the missing factor that makes $n \times 2 = 10$ a true statement is called the **quotient** $10 \div 2$ (read "ten divided by two"), and the operation of finding n is called **division.** The quotient $10 \div 2$ is another name for the number 5 in the set of whole numbers. Hence, $10 \div 2$ may be replaced by 5; that is, $10 \div 2 = 5$.

If $10 \div 2 = n$,   then $n \times 2 = 10$ and $n = 5$.

Division is a binary operation, since it combines a first number and a second number to produce a unique third number. Frequently we say that the first number is the **product,** the second is a **factor,** and the number to be determined is the **missing factor.** In the example $12 \div 3 = \square$, 12 is the product, 3 is a factor of 12, and the variable $\square$ stands for the missing factor that will produce 12 when used as a factor with 3.

Although we may relate the division operation to multiplication, it is important to recognize two kinds of division problems. The first is suggested by the question, "How many threes are in twelve?" This may be considered as a request to determine how many subsets of three elements each are contained in a given set of twelve elements.

This type of problem is sometimes referred to as a *measurement-type division* situation. The second type of division problem is that of separating the given set into a certain number of parts, or subsets. For example, we may be asked to divide twelve into fourths or to divide twelve into four equal parts. This type of problem is sometimes called a *partition-type division* situation.

Any quotient may be obtained by repeated subtraction. Consider the example $18 \div 6 = \square$. If the question, "How many sixes are in eighteen?" is asked, then the repeated-subtraction process produces the answer. The result of dividing 18 by 6 can be obtained using repeated subtraction, if we subtract 6 from 18 getting 12, then subtract 6 from 12 getting 6, and so on. We continue subtracting 6 until we obtain a difference of zero. The number of times that 6 has been subtracted is the result of 18 divided by 6. Hence, the quotient is 3.

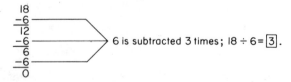

The following division example also illustrates the process of repeated subtraction.

$$156 \div 26 = \square$$

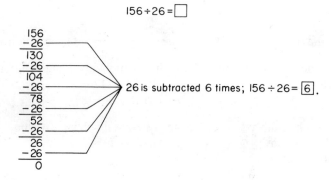

26 is subtracted 6 times; $156 \div 26 = \boxed{6}$.

If repeated subtraction does not produce a difference of zero, then the missing factor is not a whole number. Division of whole numbers may be represented on the number line by using a repeated-subtraction approach. Notice that in the case of $12 \div 4 = n$ we begin at the graph of 12 and count off to the left n jumps of 4, or n fours, and the coordinate of the terminal point is 0. Hence, three fours constitute twelve, and there are three fours in twelve; that is, $12 \div 4 = 3$ as shown on the following number line.

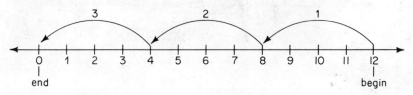

Division of whole numbers may also be represented by arrays. In the next picture, we see the result of placing 27 pegs in 3 columns, each containing the same number of pegs. How many such rows are there?

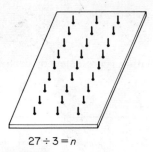

$$27 \div 3 = n$$

This again emphasizes the relation between division and multiplication. When we look at the array of 27 pegs, we see that there are several different ways in which we could express the problem. For example,

$$n \times 3 = 27, \qquad 27 \div 3 = n, \qquad \text{and} \qquad 27 \div n = 3.$$

Exercises

1. Find the quotient of 28 ÷ 7 by using the repeated-subtraction process.

2. Write each number sentence as an equation involving multiplication.
   (a) $16 \div 2 = n$ (b) $34 \div 17 = t$
   (c) $72 \div 18 = y$ (d) $a \div b = c$

3. Write each number sentence as an equation involving division.
   (a) $d \times 17 = 51$ (b) $g \times b = 39$
   (c) $26 \times 13 = n$ (d) $r \times s = t$

4. Find a whole number as a replacement for the variable such that the
   statement will be true.
   (a) $7 \times \square = 56$ (b) $56 \div \square = 7$
   (c) $\square \times 6 = 24$ (d) $\square = 24 \div 6$
   (e) $791 \div 7 = \square$ (f) $\square \times 7 = 791$

5. Write the division equation represented by each example:
   (a)   (b)

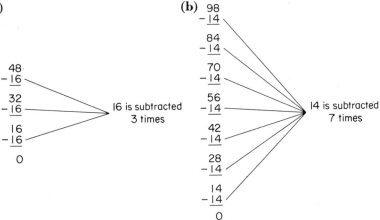

6. Write the division equation represented by each number line.
   (a)

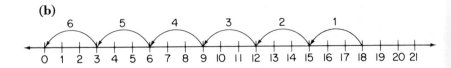

   (b)

7. Draw a number line to illustrate each equation.
   (a) $15 \div 3 = 5$ (b) $16 \div 2 = 8$
   (c) $18 \div 9 = 2$

8. Write the division equation represented by each array.
   (a)   (b)

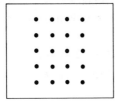

9. Draw an array to illustrate each equation.
   (a) $12 \div 6 = 2$ (b) $25 \div 5 = 5$
   (c) $8 \div 8 = 1$

10. Identify each problem situation as measurement-type division or as partition-type division.
    (a) Mary had 14 pieces of candy. She put all her candy into 2 bags so that she had the same number in each bag. How many pieces of candy were in each bag?
    (b) Bill had 48 baseball cards. He decided to give them away to his friends. He gave each friend 16 baseball cards until all his cards were gone. How many friends did he give baseball cards to?
    (c) Walter was having a birthday party. He bought 26 paper cups and gave away all the cups at the rate of 2 cups to each person attending his party. How many people attended his party?

5-11 Inverse Operations: Multiplication and Division

Consider the statements $3 \times 2 = 6$ and $6 \div 2 = 3$; that is, $(3 \times 2) \div 2 = 3$. Notice that multiplying with and dividing by 2 "undo" each other. That is we start by using 3 and 2 as factors to obtain 6, then divide by 2 and obtain 3. The two operations have the same net effect as the operation of multiplying by 1. Therefore, we state that using a given number as a factor in multiplication and dividing the product by that same number are **inverse operations.** Similarly, dividing by a given number and multiplying the quotient by the same number are inverse operations. For many years we have checked division problems by multiplication, that is, by an application of the property of inverse operations as shown in the next example.

$$\text{Check}$$

$$\begin{array}{r} 3 \\ 8\overline{)24} \\ 24 \\ \hline 0 \end{array} \qquad \begin{array}{r} 3 \\ \times\ 8 \\ \hline 24 \end{array}$$

In general, we say that multiplication is the inverse operation of division, and division is the inverse operation of multiplication.

Understanding that multiplication and division are inverse operations allows us to determine unknown division facts from our knowledge of multiplication facts, by finding missing factors. In other words, a division problem may always be thought of as a multiplication problem with the product and one factor known:

$$10 \div 2 = 5 \quad \text{because} \quad 5 \times 2 = 10,$$
$$16 \div 8 = 2 \quad \text{because} \quad 2 \times 8 = 16,$$
$$28 \div 7 = 4 \quad \text{because} \quad 4 \times 7 = 28,$$
$$45 \div 15 = 3 \quad \text{because} \quad 3 \times 15 = 45.$$

Exercises

1. Indicate whether each statement is true or false.
   (a) If $3 \times 4 = 12$, then $12 \div 4 = 3$ and $(3 \times 4) \div 4 = 3$.
   (b) If $63 \div 7 = 9$, then $9 \times 7 = 63$ and $(63 \div 7) \times 7 = 63$.
   (c) If $5 \times 15 = 75$, then $75 \div 15 = 5$ and $(5 \times 15) \div 75 = 5$.
   (d) If $98 \div 7 = 14$, then $14 \times 7 = 98$ and $(98 \div 7) \times 7 = 98$.

2. If you multiply 8 by a natural number n, what must you do to the product to obtain n?

3. If you divide a natural number n by 23, what must you do to the quotient to obtain n?

5-12 Properties of Division of Whole Numbers

Although multiplication and division are closely related operations, certain differences become obvious when we consider division with respect to some of the basic properties that hold for multiplication. Recall that one counter-example is sufficient evidence to state that a property does not hold in general.

Commutative property   Since $32 \div 4 \neq 4 \div 32$, division of whole numbers is not commutative.

Associative property   We may observe that in the following example the associative property does not hold for division of whole numbers:

$$18 \div (9 \div 3) \neq (18 \div 9) \div 3 \quad \text{because} \quad 18 \div 3 \neq 2 \div 3.$$

Identity element   One property that does hold in a limited way for division of whole numbers is the special property of operating with one. We have already agreed that $a \times 1 = 1 \times a = a$ for any whole number a. It is true that $a \div 1 = a$ although in general $1 \div a \neq a$. Thus we may think of 1 as a *right identity* that is not a left identity for division of whole numbers.

Closure and uniqueness properties If the set of whole numbers is to be closed with respect to division, there must exist a whole number for the quotient of every two whole numbers. Clearly, there is no whole number equal to $1 \div 2$, $2 \div 3$, or $5 \div 6$. Therefore, we conclude that quotients of whole numbers do not always exist within the set of whole numbers and that the set of whole numbers is not closed with respect to division. We note, however, that the quotient of two whole numbers, if it exists, is *unique*.

Distributive property of division over addition Since the operation of division is not commutative, we must consider two cases of distributing division over addition. In the first case we try to distribute from the *left:*

$$18 \div (6 + 2) \neq (18 \div 6) + (18 \div 2)   \text{since}   18 \div 8 \neq 3 + 9.$$

Hence, we see that the operation of division is not distributive from the left over the operation of addition. In the second case we try to distribute from the *right:*

$$(12 + 6) \div 3 = (12 \div 3) + (6 \div 3)   \text{because}   18 \div 3 = 4 + 2.$$

In more advanced texts it is proved that division is distributive from the right over the operation of addition.

Distributive property of division over subtraction We must again consider two cases since division is not commutative. In the first instance we see that division does not distribute over subtraction from the left:

$$12 \div (6 - 2) \neq (12 \div 6) - (12 \div 2)   \text{since}   12 \div 4 \neq 2 - 6.$$

In the second case we see that division is distributive over subtraction from the right:

$$(30 - 3) \div 3 = (30 \div 3) - (3 \div 3)   \text{since}   27 \div 3 = 10 - 1.$$

We assume that this property holds in general for the set of whole numbers.

Exercises

1. Give a counterexample to show that each of these properties does not hold for division of whole numbers.
   (a) Commutative   (b) Associative
   (c) Closure

2. Find each answer in the set of whole numbers. Indicate those examples for which no answer exists in the set of whole numbers.
   (a) $8 \div 2$ (b) $9 \div 27$
   (c) $14 \div (7 + 1)$ (d) $4 \times 6 \div 8$
   (e) $15 \div (10 \div 2)$ (f) $3 \times 8 \div 2 \div 2 \times 4$

3. Indicate whether each statement is true or false.

(a) $8 \div 2 \neq 2 \div 8$

(b) $24 \div (8 + 4) = (24 \div 8) + (24 \div 4)$

(c) $54 \div (18 \div 3) > (54 \div 18) \div 3$

(d) $(45 + 18) \div 9 = (45 \div 9) + (18 \div 9)$

(e) $14 \div 1 = 1 \div 14$

(f) $2 \div (1 + 1) = (1 + 1) \div 2$

(g) $(6 \div 1) \div 1 = 6 \div (1 \div 1)$

(h) $17 \times (1 + 1) = (17 \div 1) + (17 \div 1)$

5-13 Division by Zero

The mathematician considers division by zero to be *meaningless* in our number system. The statement $5 \div 0 = n$ is equivalent to the statement $n \times 0 = 5$. Since for every whole number n we have $n \times 0 = 0$, we cannot have $n \times 0 = 5$ for any whole number n. Therefore, to insure consistency in our number system, **division by zero is excluded or not defined.**

Notice that $0 \div 0 = n$ is equivalent to $n \times 0 = 0$, which is true for every whole number n. For example,

$$0 \div 0 = 1 \quad \text{because} \quad 1 \times 0 = 0,$$
$$0 \div 0 = 2 \quad \text{because} \quad 2 \times 0 = 0,$$
$$0 \div 0 = 3 \quad \text{because} \quad 3 \times 0 = 0.$$

To insure that quotients will represent *unique numbers*, we also consider $0 \div 0$ as meaningless and thus as not defined. Since division by zero has been excluded, *the denominator of a fractional number may never be zero or equivalent to zero.* The following is a summary of the operations with zero, where n represents any whole number.

$$n + 0 = 0 + n = n$$
$$n - 0 = n$$
$$0 - n \neq n \quad (n \neq 0)$$
$$0 \times n = n \times 0 = 0$$
$$0 \div n = 0 \quad (\text{provided } n \neq 0)$$
$$n \div 0 \quad \text{not defined}$$
$$0 \div 0 \quad \text{not defined}$$
$$0^0 \quad \text{not defined (recall from Chapter 3 that } a^0 = 1$$
$$\text{by definition, except when } a = 0)$$

Exercises

Indicate whether each statement is true or false for every natural number n. If false, tell why.

1. $0 \div n = 0$

2. $\dfrac{n \times 0}{n} + n = n$

3. $\dfrac{n \times 0}{0} = n$

4. $\dfrac{n - n}{n} = 0$

5. $\dfrac{n}{n - n} = 0$

6. $\dfrac{(n - n)}{n} \times \dfrac{n}{(n - n)} = 1$

5-14 Understanding the Division Algorithm

We mentioned previously that division of whole numbers is frequently introduced as a process of repeated subtraction. The purpose of this approach is to improve the student's understanding of the usual division algorithms. The student should be able, in fact, to discover short-cut methods of working division problems for himself, if he understands the repeated-subtraction process.

In the case of $36 \div 9$, the repeated-subtraction process looks like the following.

$$
\begin{array}{rl}
36 & \\
-\ 9 & \quad 1 \\
\hline
27 & \\
-\ 9 & \quad 1 \\
\hline
18 & \\
-\ 9 & \quad 1 \\
\hline
9 & \\
-\ 9 & \quad 1 \\
\hline
0 & \quad \overline{4}
\end{array}
$$

We see that 9 is subtracted 4 times from 36 before we get a remainder of 0. Hence, we conclude that $36 \div 9 = 4$. A student with a fair knowledge of the multiplication facts might discover a shorter method for working this same problem. Perhaps he would use the procedure below.

$$
\begin{array}{r|l}
9)\ \ 36 & \\
-18 & 2 \\
\hline
18 & \\
-18 & 2 \\
\hline
0 & \overline{4}
\end{array}
$$

Since this student knew that $9 \times 2 = 18$, he saw no need to make separate subtractions of 9 each time; hence, he reduced the number of subtractions.

With an example such as $176 \div 16$ this same student might apply his new process again in an effort to shorten the division procedure. Perhaps he would set the problem up as follows.

$$
\begin{array}{r|l}
16)\ \overline{176} & \\
-80 & 5 \\
\hline
96 & \\
-64 & 4 \\
\hline
32 & \\
-32 & 2 \\
\hline
0 & 11 \\
\end{array}
\qquad \text{or} \qquad
\begin{array}{r|l}
16)\ \overline{176} & \\
-160 & 10 \\
\hline
16 & \\
-16 & 1 \\
\hline
0 & 11 \\
\end{array}
$$

As the student's understanding improves, he will be able to refine this method of division until it resembles the commonly used algorithm. This is illustrated in the next examples.

Example 1
$$
\begin{array}{r|l}
238)\ \overline{5712} & \\
-2380 & 10 \\
\hline
3332 & \\
-2380 & 10 \\
\hline
952 & \\
-714 & 3 \\
\hline
238 & \\
-238 & 1 \\
\hline
0 & 24 \\
\end{array}
$$

Example 2
$$
\begin{array}{r|l}
238)\ \overline{5712} & \\
-4760 & 20 \\
\hline
952 & \\
-952 & 4 \\
\hline
0 & 24 \\
\end{array}
$$

Example 3
$$
\begin{array}{r}
24 \\
238)\ \overline{5712} \\
-4760 \\
\hline
952 \\
-952 \\
\hline
0 \\
\end{array}
$$

Example 4
$$
\begin{array}{r}
24 \\
238)\ \overline{5712} \\
-476 \\
\hline
952 \\
-952 \\
\hline
0 \\
\end{array}
$$

Example 1 illustrates a "guessing" method of working the problem using the repeated-subtraction approach. In Example 2 the student obviously knows his multiplication quite well, and the repeated-subtraction method very closely resembles the usual division algorithm, which is illustrated in Examples 3 and 4.

Example 3 points out that our division algorithm takes care of place value in the quotient. The 2 in the quotient actually represents 2 tens, because of its position; hence, the product of 2×238 is actually 20×238, that is, 4,760. Example 4 illustrates the streamlined version of our division algorithm, which is justified by our understanding of such processes as shown in Examples 1, 2, and 3.

The usual division algorithm, illustrated in Example 4, is highly recommended; however, the student should understand "why" it works, and the use of correct terminology should be emphasized. The "goes into" description of division should be discouraged, because it does not convey a precise mathematical idea.

We should also note that the quotient in each of the preceding division examples can be verified by the operation of multiplication, since division is the inverse of multiplication. Thus we see that $11 \times 16 = 176$ and $24 \times 238 = 5712$.

Two other methods sometimes employed to find quotients like $78 \div 6$ and $89 \div 4$ are illustrated in the following examples.

Example 5

$$
\begin{array}{r}
\underline{10 + 3} = 13 \\
6)\overline{78} = 6)\,\overline{60 + 18} \\
-(60 + 18) \\
\hline
0
\end{array}
$$

Example 6

$$
\left.\begin{array}{r} 3 \\ 10 \end{array}\right\} = 13
$$

$$
\begin{array}{r}
6)\quad 78 \\
-60 \\
\hline
18 \\
-18 \\
\hline
0
\end{array}
$$

Example 7

$$
\begin{array}{r}
\underline{100 + 20 + 1} = 121\,r\,2 \\
7)\overline{849} = 7)\,\overline{700 + 140 + 9} \\
-(700 + 140 + 7) \\
\hline
2 \quad \text{(remainder)}
\end{array}
$$

Example 8

$$
\left.\begin{array}{r} 1 \\ 20 \\ 100 \end{array}\right\} = 121\,r\,2
$$

$$
\begin{array}{r}
7)\quad 849 \\
-700 \\
\hline
149 \\
-140 \\
\hline
9 \\
-7 \\
\hline
2 \quad \text{(remainder)}
\end{array}
$$

In Examples 7 and 8, we should note that there are nonzero remainders. It was emphasized earlier that division was not possible for every pair of whole numbers. For example, $8 \div 5$ has no solution in the set of whole numbers. However, we can find an answer consisting of a *quotient* and a *remainder*, both of which are whole numbers. Thus we have $8 \div 5 = 1\,r\,3$.

In general, if a and b are whole numbers, $b \neq 0$, then there exist unique whole numbers q and r such that $a = b \times q + r$, where $0 \leq r < b$ (read "zero is less than or equal to r and r is less than b"). For example, $23 = 4 \times 5 + 3$ or $23 \div 4 = 5\,r\,3$, and $327 = 16 \times 20 + 7$ or $327 \div 16 = 20\,r\,7$.

Exercises

1. For each pair of numbers $a \div b$, find q and r such that $a = b \times q + r$, where $0 \leq r < b$.

(a) $23 \div 8$ (b) $85 \div 9$
(c) $392 \div 8$ (d) $408 \div 5$
(e) $426 \div 42$ (f) $923 \div 27$
(g) $51 \div 79$ (h) $95 \div 19$

2. Find each quotient by using a repeated-subtraction technique.
   (a) $15 \div 3$ (b) $76 \div 19$
   (c) $1,728 \div 216$

3. Verify each quotient obtained in Exercise 2 by the inverse operation.

4. Find each quotient by using expanded notation as shown in part (a).
   (a) $3\overline{)849} = 3\overline{)600 + 240 + 9}$ (b) $4\overline{)1272} = 4\overline{)\underline{} + \underline{} + \underline{}}$
   (c) $8\overline{)1529} = 8\overline{)\underline{} + \underline{} + \underline{}}$
   (d) $9\overline{)149} = 9\overline{)\underline{} + \underline{}}$

5-15 Summary of Properties for Multiplication and Division

The following table presents a summary of general properties for multiplication and division of whole numbers. The letters a, b, and c represent arbitrary whole numbers where $a \neq b$, and $b \neq c$.

Property	Multiplication	Division
Closure	$a \times b$ always produces a whole number for the product	$a \div b$ does not always produce a whole number for the quotient
Uniqueness	$a \times b$ is a unique whole number	$a \div b$, if it exists, is a unique whole number
Commutative	$a \times b = b \times a$	$a \div b \neq b \div a$
Associative	$a \times (b \times c) = (a \times b) \times c$	$a \div (b \div c) \neq (a \div b) \div c$ $(c \neq 1)$
Identity element	$a \times 1 = 1 \times a = a$	$a \div 1 = a$ $1 \div a \neq a$. $(a \neq 1)$ (right identity only)
Special property of zero	$a \times 0 = 0 \times a = 0$	$a \div 0$ not defined $0 \div a = 0$. $(a \neq 0)$ $0 \div 0$ not defined
Inverse operations	$a \times b = c$ if and only if $c \div b = a$	$a \div b = c$ if and only if $c \times b = a$
Distributive from left over $+$	$a \times (b + c) = (a \times b) + (a \times c)$	$a \div (b + c) \neq (a \div b) + (a \div c)$, $a \neq 0$
right over $+$	$(b + c) \times a = (b \times a) + (c \times a)$	$(b + c) \div a = (b \div a) + (c \div a)$, $a \neq 0$
left over $-$	$a \times (b - c) = (a \times b) - (a \times c)$	$a \div (b - c) \neq (a \div b) - (a \div c)$, $a \neq 0$
right over $-$	$(b - c) \times a = (b \times a) - (c \times a)$	$(b - c) \div a = (b \div a) - (c \div a)$, $a \neq 0$

5-16 Chapter Test

Indicate whether each statement is true or false. If false, tell why.

 1. The solution set for $0 \div 0 = \triangle$ is the set of whole numbers.

 2. If two factors are each multiplied by 5, their product is increased by a factor of 5.

 3. In the multiplication of 38 the algorithm is performed as $(20 + 7) \times 38$.

 4. This array illustrates $5 \times 2 = 10$.

 5. In general for sets A and B, $A \times B = B \times A$.

 6. The associative property permits the grouping of three numbers in any order for multiplication.

 7. Zero is the identity element for multiplication of whole numbers.

 8. If $n(G) \times n(H) = n(H)$ for sets G and H, then $n(G) = 1$.

 9. The set of numbers $\{5, 10, 15, 20, \ldots\}$ is closed under the operation of division.

10. For every natural number n, $\dfrac{n - n}{n + n} \times \dfrac{n + n}{n - n} = 1$.

Select the best possible answer.

11. Which of the following is a proper way to show $3 \times 5 = 15$?

      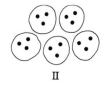   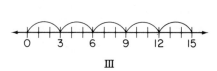

   I   II   III

   (a) I only   (b) II only
   (c) III only   (d) I and III only
   (e) I, II, and III

12. This number line shows an interpretation of $6 \div 2$. Which of the following describes the situation?

(a) Repeated subtraction   (b) Inverse operations
(c) Measurement-type division   (d) Partition-type division
(e) Commutative property

13. Which of the following represents an application of a commutative property?
    (a) $2 \times 2 = 2 + 2$ (b) $3 \times (4 + 5) = 12 + 15$
    (c) $3 + (4 \times 5) = 3 + (5 \times 4)$ (d) $2 \times 12 = 2 \times (2 \times 6)$
    (e) None of these

14. Which of the following represents an application of a distributive property?
    (a) $2 \times (8 + 8) = (4 \times 4) + (4 \times 4)$
    (b) $(4 \times 5) + (4 \times 3) = 4 \times (5 + 3)$
    (c) $6 \times (5 + 5) = (6 \times 6) + (6 \times 4)$
    (d) $3 \times (4 \times 0) = (3 \times 4) \times (3 \times 0)$
    (e) None of these

15. The distributive property of multiplication over subtraction can be used for which of the following?

    I. Discovery of new multiplication facts.
    II. Discovery of new subtraction facts.
    III. In finding the product 8×79.

    (a) I only   (b) II only
    (c) III only   (d) I and II only
    (e) I and III only

16. Which of the following products could not be established by repeated addition?
    (a) 1×1 (b) 8×0
    (c) 3×1 (d) 10×10
    (e) 126×295

17. Which property or properties does one make use of in showing that $30 \times 40 = (3 \times 4) \times (10 \times 10)$?

    I. Commutative.
    II. Associative.
    III. Distributive.

    (a) I only   (b) II only
    (c) I and II only   (d) II and III only
    (e) I and III only

18. If $r \div s = t$, and r, s, and t are whole numbers ($t \neq 0$), what will be the effect on t if a whole number n is added to r and s?
    (a) t will remain the same.
    (b) t will increase.

(c) t will decrease.

(d) The effect cannot be predicted in general.

(e) None of these.

19. If $15 \div 3 = 5$, then $5 \times 3 = 15$ illustrates which of these general patterns?

(a) If $\square \div \triangle = \diamondsuit$, then $\square \div \diamondsuit = \triangle$.

(b) If $\square \div \triangle = \diamondsuit$, then $\triangle \div \diamondsuit = \square$.

(c) If $\square \div \triangle = \diamondsuit$, then $\square \times \triangle = \diamondsuit$.

(d) If $\square \div \triangle = \diamondsuit$, then $\diamondsuit \times \triangle = \square$.

(e) If $\square \div \triangle = \diamondsuit$, then $\triangle \div \square = \diamondsuit$.

20. In the example shown below, the number 1284 is best explained as being the product of:

(a) 600×214

(b) 60×214

(c) 6×214

(d) 6×214 without a zero added

(e) None of these

$$\begin{array}{r} 214 \\ \times\,364 \\ \hline 856 \\ 1284 \\ 642 \\ \hline 77896 \end{array}$$

chapter 6

Elementary
Number Theory

Some of the most fascinating aspects of the study of mathematics may be introduced by a study of number theory. The opportunities for discovery and exploration are manyfold. These explorations often result in the real thrill of finding a pattern, making a conjecture, and testing it successfully. The basic tools required for such explorations are neither sophisticated nor highly abstract. In this chapter we will explore a few of the elementary ideas of number theory.

6-1 Multiples and Divisibility

We can generate the set of whole number **multiples** of 2 by using 2 as a factor with each whole number beginning with zero. Thus $2 \times 0 = 0$, $2 \times 1 = 2, 2 \times 2 = 4, 2 \times 3 = 6$, and so on. From this, the set of multiples of 2 is $\{0, 2, 4, 6, 8, 10, 12, 14, 16, 18, 20, 22, \ldots\}$. A careful observation of this set of numbers reveals one of the simple rules of divisibility. We note that the ones digit for each numeral is always 0, 2, 4, 6, or 8. Thus, only whole numbers whose numerals end in 0, 2, 4, 6, or 8 are **divisible** by 2.

In other words the multiples of 2 (numbers divisible by 2) are even numbers.

Similar observations are possible for the multiples of 5 and 10. The multiples of 5 are {0, 5, 10, 15, 20, 25, 30, 35, . . .}. What pattern is revealed by the ones digits in this set? Is it true that a multiple of 5 (number divisible by 5) is always an even number? Is it true that a number is divisible by 5 if its numeral ends in a 0 or a 5?

What pattern do you observe in this set of multiples of 10?

$$\{0, 10, 20, 30, 40, 50, 60, 70, . . .\}.$$

How can you tell if a number is a multiple of 10? Are all multiples of 10 divisible by 10? You should also note that all multiples of 10 are divisible by 2. Can you explain why?

Next we shall consider the set of multiples of 3. The following chart may help focus attention on a pattern relative to divisibility by 3.

Factors	Product	Sum of digit numbers in product
$3 \times 0 =$	0	0
$3 \times 1 =$	3	3
$3 \times 2 =$	6	6
$3 \times 3 =$	9	9
$3 \times 4 =$	12	$1 + 2 = 3$
$3 \times 5 =$	15	$1 + 5 = 6$
$3 \times 6 =$	18	$1 + 8 = 9$
$3 \times 7 =$	21	$2 + 1 = 3$
$3 \times 8 =$	24	$2 + 4 = 6$
$3 \times 9 =$	27	$2 + 7 = 9$
$3 \times 10 =$	30	$3 + 0 = 3$
$3 \times 11 =$	33	$3 + 3 = 6$
$3 \times 12 =$	36	$3 + 6 = 9$
$3 \times 13 =$	39	$3 + 9 = 12; 1 + 2 = 3$
$3 \times 14 =$	42	$4 + 2 = 6$
$3 \times 15 =$	45	$4 + 5 = 9$
$3 \times 16 =$	48	$4 + 8 = 12; 1 + 2 = 3$
$3 \times 17 =$	51	$5 + 1 = 6$
$3 \times 18 =$	54	$5 + 4 = 9$
$3 \times 19 =$	57	$5 + 7 = 12; 1 + 2 = 3$
$3 \times 20 =$	60	$6 + 0 = 6$

For each multiple of 3 in the above chart, is it true that the sum of the digit numbers is divisible by 3? Test this idea on some other multiples of 3. Test this idea on some numbers that are not multiples of 3. Is it true that a number is not a multiple of 3 if the sum of its digit numbers is not divisible by 3? Many other interesting divisibility rules exist. We will explore a few of them in the exercises.

Exercises

1. By inspection, indicate which numbers are divisible by 2.
   (a) 27   (b) 60
   (c) 35   (d) 84
   (e) 101   (f) 516
   (g) 278,190   (h) 6,783
   (i) 1,103   (j) 0

2. By inspection, indicate which numbers in Exercise 1 are divisible by:
   (a) 5   (b) 10
   (c) 3

3. List the set of multiples of:
   (a) 4   (b) 6
   (c) 8   (d) 9

4. Jim proposed that "a number is divisible by 4 if and only if the last two digits of its numeral represent a number that is divisible by 4." Test Jim's rule on these numbers.
   (a) 972   (b) 497
   (c) 113,528   (d) 386,045
   (e) 777,980

5. Use Jim's rule and list:
   (a) Five numbers that are divisible by 4.
   (b) Five numbers that are not divisible by 4.

6. (a) Test Jim's divisibility by 4 rule on the numbers you listed in Exercise 5(a). In Exercise 5(b).
   (b) Did the rule work in both cases?
   (c) Try to explain why the rule works.

7. Make up a rule for divisibility by 8. (Hint: Think about the divisibility rules for 2 and 4.)

8. Is there any need to test for divisibility by 8 if a number is not divisible by 2?

9. Is it true that a number is divisible by 6 if it is divisible by both 2 and 3? Explain your answer.

10. Is there any need to test for divisibility by 6 if a number is not divisible by 2?

11. Marty made the following chart and proposed that "a number is divisible by 9 if and only if the sum of the digit numbers in its numeral is divisible by 9." Study this chart and then try the rule on some other numbers. Does Marty's rule appear to always work?

Factors	Product	Sum of digit numbers in product
$9 \times 0 =$	0	0
$9 \times 1 =$	9	9
$9 \times 2 =$	18	$1 + 8 = 9$
$9 \times 3 =$	27	$2 + 7 = 9$
$9 \times 4 =$	36	$3 + 6 = 9$
$9 \times 5 =$	45	$4 + 5 = 9$
$\vdots$	$\vdots$	$\vdots$
$9 \times 11 =$	99	$9 + 9 = 18; 1 + 8 = 9$
$9 \times 12 =$	108	$1 + 0 + 8 = 9$
$9 \times 13 =$	117	$1 + 1 + 7 = 9$
$\vdots$	$\vdots$	$\vdots$
$9 \times 111 =$	999	$9 + 9 + 9 = 27; 2 + 7 = 9$

12. Use Marty's rule and test each number for divisibility by 9.
    (a) 234   (b) 701
    (c) 5,112   (d) 42,714
    (e) 930,018

13. Is there any need to test for divisibility by 9:
    (a) If a number is not divisible by 3?
    (b) If a number is not divisible by 6?

14. Indicate whether each statement is true or false. If false, tell why.
    (a) If a number is divisible by 2 and 6, then it is divisible by 12.
    (b) If a number is divisible by 2 and 5, then it is divisible by 10.
    (c) If a number is divisible by 9, then it is divisible by 3.
    (d) If a number is divisible by 10, then it is divisible by 20.

6-2 Prime and Composite Numbers

Any whole number that is a factor of a given whole number is called a **divisor** of that number. For example, $3 \times 4 = 12$. Here 3 and 4 are called factors (divisors) of 12. We should note that 3 and 4 are not the only factors (divisors) of 12. We could also have $2 \times 6 = 12$ with 2 and 6 as factors, or we could have $1 \times 12 = 12$, where 1 and 12 are factors. For our purposes, the word "factor" is defined as a whole-number factor (that is, divisor) of a number. In the case of 12 we have determined *all possible factors:* 1, 2, 3, 4, 6, and 12.

In the case of the number 3 there are only two factors, 1 and 3. Similarly, each of the numbers 5, 7, 11, 13, and 17 have only two factors, 1 and the number itself. We define a **prime number** as any whole number greater than 1 that has only itself and 1 as factors. Thus 2, 3, 5, 7, 11, 13, 17 are the first seven prime numbers.

We can find all the prime numbers less than 100 by listing all the whole numbers from 2 to 99 (inclusive) and then crossing out all multiples of each succeeding prime; see the array.

2	3	~~4~~	5	~~6~~	7	~~8~~	~~9~~	~~10~~
11	~~12~~	13	~~14~~	~~15~~	~~16~~	17	~~18~~	19	~~20~~
~~21~~	~~22~~	23	~~24~~	~~25~~	~~26~~	~~27~~	~~28~~	29	~~30~~
31	~~32~~	~~33~~	~~34~~	~~35~~	~~36~~	37	~~38~~	~~39~~	~~40~~
41	~~42~~	43	~~44~~	~~45~~	~~46~~	47	~~48~~	~~49~~	~~50~~
~~51~~	~~52~~	53	~~54~~	~~55~~	~~56~~	~~57~~	~~58~~	59	~~60~~
61	~~62~~	~~63~~	~~64~~	~~65~~	~~66~~	67	~~68~~	~~69~~	~~70~~
71	~~72~~	73	~~74~~	~~75~~	~~76~~	~~77~~	~~78~~	79	~~80~~
~~81~~	~~82~~	83	~~84~~	~~85~~	~~86~~	~~87~~	~~88~~	89	~~90~~
~~91~~	~~92~~	~~93~~	~~94~~	~~95~~	~~96~~	97	~~98~~	~~99~~

Since 2 is a prime number, cross out all numbers that have 2 as a factor, except 2; that is, 4, 6, 8, 10, 12, ... Since 3 is a prime number, cross out all numbers that have 3 as a factor, except 3; that is, 6, 9, 12, 15, 18, ... Next, cross out all numbers that have 5 as a factor, except 5; that is, 10, 15, 20, 25, 30, ... Do the same with all numbers that have 7 as a factor, except 7; that is, 14, 21, 28, 35, 42, ...

The next prime number is 11 and 11×11 is greater than 99. Thus all the multiples of 11 that are less than 100 have already been crossed out. Similarly, the multiples of all prime numbers greater than 11 have been crossed out already, and we may stop testing.

The twenty-five remaining numbers are the prime numbers less than 100 and may be designated as set P:

$$P = \{2, 3, 5, 7, 11, 13, 17, 19, 23, 29, 31, 37, 41, 43, 47,$$
$$53, 59, 61, 67, 71, 73, 79, 83, 89, 97\}.$$

This process for determining prime numbers is called the *Sieve of Eratosthenes*, after the Greek mathematician who originated the idea.

Each of the numbers that was crossed out of the table may be expressed as a product of prime numbers. These whole numbers are called **composite numbers.** For example, 4 may be expressed as 2×2, 6 may be expressed as 2×3, 8 may be expressed as $2 \times 2 \times 2$, 10 may be expressed as 2×5, and so forth.

The number 1 is a special number and is neither prime nor composite. It is not included in the set of prime numbers partly because its only whole-number factors are 1. The discussion in Section 6-3 will provide another reason for excluding 1 from the set of prime numbers. The number 0 also has special properties and is not included in the set of prime numbers. Thus, the whole numbers may be classified in four disjoint subsets: 0, 1, prime numbers, and composite numbers.

How do we determine whether a number is prime? Since a prime number has only itself and 1 as divisors, a systematic approach to this problem would be to test in turn all possible divisors of the number in question: if the number has a divisor other than itself and 1, it is a composite number. Actually, we need to try only the prime divisors of the number, because the composite divisors are composed of smaller divisors that will already have been tried. For example, a number that is not divisible by 2 cannot be divisible by 4 or any other multiple of 2.

Is 29 a prime number? We try to divide 29 by 2, then by 3, then by 5, then by 7, then by 11, and so on. Do we continue to test prime divisors until the divisor is greater than 29? In this particular case, we could stop testing with the divisor 7, since $7 \times 7 = 49$ and $49 > 29$. If 29 were divided by a number equal to or greater than 7, the quotient would be a number less than 7. Since we have already tested for divisors less than 7 and determined that none exists, we conclude that 29 is a prime number.

Is 101 a prime number? The answer is yes, because 101 is not divisible by the prime numbers 2, 3, 5, and 7. We need not try other prime divisors, since $11^2 = 121$ and $121 > 101$. Note that 101 divided by 11 produces a quotient of 9 (remainder 2). When the quotient is less than the divisor we may stop testing divisors.

The following display of the pairs of factors of each of the first ten whole numbers as stories of a house may help you to see more clearly the classification of the whole numbers into four disjoint subsets: 0, 1, prime numbers, composite numbers.

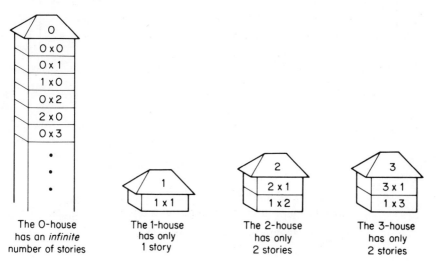

The 0-house
has an *infinite*
number of stories

The 1-house
has only
1 story

The 2-house
has only
2 stories

The 3-house
has only
2 stories

A careful observation will lead you to the conclusion that there is only one house (the 0-house) with an infinite number of stories. Thus 0 is in a

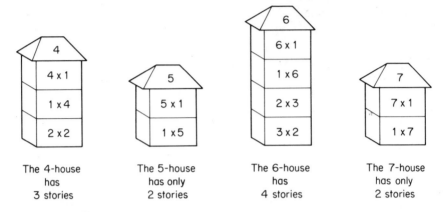

The 4-house has 3 stories   The 5-house has only 2 stories   The 6-house has 4 stories   The 7-house has only 2 stories

class by itself. Similarly you should note that there is only one house (the 1-house) with a single story. Thus 1 is in a class by itself. Additional study of the pictures should lead you to the observation that several houses (2-house, 3-house, 5-house, 7-house) have two and only two stories each. Such houses

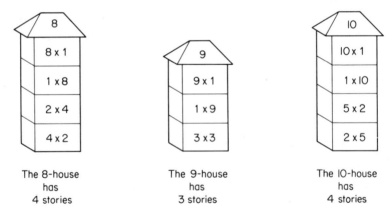

The 8-house has 4 stories   The 9-house has 3 stories   The 10-house has 4 stories

name *prime numbers*. The remaining houses (4-house, 6-house, 8-house, 9-house, 10-house) have more than two stories each and they name the *composite numbers*.

In summary,

> 0-house has infinite number of stories; neither a prime nor a composite number.
>
> 1-house has only one story; neither a prime nor a composite number.

2-, 3-, 5-, and 7-houses each have two and only
two stories; prime numbers.

4-, 6-, 8-, 9-, and 10-houses each have more than
two stories; composite numbers.

Exercises

1. List the prime numbers between 100 and 200.

2. Indicate whether each of the following is a composite number or a prime
   number.*
   (a) 41   (b) 313
   (c) 115   (d) 823

3. Is zero a prime number? A composite number?

4. Is one a prime number? A composite number?

5. What is the union of the set of prime numbers and the set of:
   (a) Composite numbers?   (b) Whole numbers?

6. What is the intersection of the set of prime numbers and the set of:
   (a) Composite numbers?   (b) Whole numbers?
   (c) Odd numbers less than 30?   (d) Even numbers?

7. To determine whether or not each number is prime, what is the largest
   prime divisor that must be checked or tested?
   (a) 37   (b) 113
   (c) 201   (d) 329
   (e) 659

8. In the illustration using stories of houses to indicate the number of possible
   pairs of factors for a number, which numbers had 4-story houses? Is
   there anything special about these numbers? Name five other numbers
   that would have 4-story houses.

6-3 Fundamental Theorem of Arithmetic

Every composite number can be expressed as a product of prime numbers
that is unique except for the order of the factors. This statement is called
the **Fundamental Theorem of Arithmetic** and is also often referred to as
the **Unique Factorization Theorem.** Consider the following examples.

Example 1 $18 = 2 \times 9 = 2 \times 3 \times 3$,

$18 = 3 \times 6 = 3 \times 2 \times 3$.

Example 2 $24 = 4 \times 6 = 2 \times 2 \times 2 \times 3$,

$24 = 3 \times 8 = 3 \times 2 \times 4 = 3 \times 2 \times 2 \times 2$.

Notice that the two ways of factoring in each of the examples result in the same product of prime numbers, but the orders of the factors are different. Application of the commutative and associative properties could change the order of the factors so that the two expressions would be identical.

If 1 were called a prime number, the Fundamental Theorem of Arithmetic would have to be rephrased. The number 18, for example, could then be factored as a product of prime numbers in each of these ways:

$$2 \times 3 \times 3 \times 1 = 18,$$

$$2 \times 3 \times 3 \times 1 \times 1 = 18,$$

$$2 \times 3 \times 3 \times 1 \times 1 \times 1 = 18.$$

The expressions would not be identical except for the order of the factors. Therefore, to enable us to keep the Fundamental Theorem of Arithmetic in a simple form (and also for other related reasons) we define the set of prime numbers so as to exclude the number 1.

Factor trees are sometimes used to display the factorization of a number. For example, 36 is shown on these two factor trees. The factorizations may appear to be different, but the prime factorization results in the same prime factors except for the order.

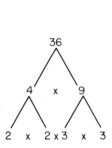
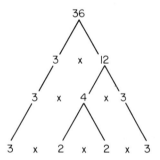

Exercises

1. Express each composite number as a product of prime numbers.
   (a) 42   (b) 99
   (c) 148   (d) 375

2. Develop an algorithm that can be used to determine the prime factors of 1,592.

3. (a) List all factors of 64.
   (b) List the factors of 64 that are composite numbers.
   (c) List the factors of 64 that are prime numbers.

4. If possible, express 64 as a product of :
   (a) Two equal factors.   (b) Two composite numbers.
   (c) Two prime numbers.   (d) A prime and a composite number.

5. Represent 18 in all possible ways as a product of counting numbers greater than 1.

6. Draw a factor tree to display the prime factorization of each number.
   (a) 48   (b) 36
   (c) 78   (d) 80

7. Express each even number between 2 and 26 as a sum of two prime numbers. Can you do this for all other even numbers?

6-4 Least Common Multiple

Consider the natural numbers 2 and 3. The set of multiples of 2 is

$$\{2, 4, 6, 8, 10, 12, 14, 16, 18, 20, \ldots\},$$

and the set of multiples of 3 is

$$\{3, 6, 9, 12, 15, 18, 21, \ldots\}.$$

Observe that some of the multiples of 2 are also multiples of 3. For example, 6, 12, and 18 are represented in the intersection of the two sets. Hence, we call 6, 12, 18, . . . the *common multiples* of 2 and 3.

If we let M denote the set of common multiples of 2 and 3,

$$M = \{6, 12, 18, 24, 30, 36, \ldots\}.$$

Notice that there is no largest common multiple of 2 and 3 in set M. By continually adding 6 to the preceding common multiple of 2 and 3, we may always find a larger common multiple. There is, however, a smallest element in set M. This element is the number 6. Since 6 is the smallest of the common multiples of 2 and 3, we call 6 the **least common multiple** (L.C.M.) of 2 and 3. Notice that only natural number multiples are considered and that the least common multiple is the common multiple that is a factor of each of the common multiples; that is, every common multiple is also a multiple of the least common multiple. Thus, zero multiples and negative-integral multiples are not considered; and in the preceding illustration, the least common multiple 6 is a factor of each of the common multiples in the set $\{6, 12, 18, 24, 30, 36, \ldots\}$.

Example 1   Find the least common multiple of 3 and 4 by using the intersection of sets.

Let A equal the set of multiples of 3,

$$A = \{3, 6, 9, 12, 15, 18, 21, 24, 27, \ldots\},$$

and B equal the set of multiples of 4,

$$B = \{4, 8, 12, 16, 20, 24, 28, 32, \ldots\}.$$

The least number represented in the set resulting from the intersection of A and B is the least common multiple. Since

$$A \cap B = \{12, 24, 36, \ldots\},$$

the least common multiple of 3 and 4 is 12.

Example 2   Find the least common multiple of 3 and 4 by using a number line.

Observe that on the number line 12 is the least common multiple of 3 and 4, since the graph of 12 is the first point greater than 0 on the number line that is common to the set of multiples of 3 and the set of multiples of 4.

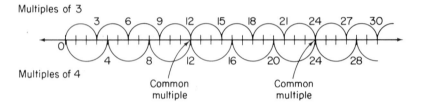

Multiples of 3

Multiples of 4

Common multiple   Common multiple

Example 3   Find the least common multiple of 6, 8, and 9 by using prime factors.

We may find the least common multiple of 6, 8, and 9 by separating each number into a product of prime factors:

$$6 = 2 \times 3,$$
$$8 = 2 \times 2 \times 2,$$
$$9 = 3 \times 3.$$

The least common multiple is the product of the different prime factors, each factor being used the *greatest number of times it occurs as a factor of any one number.* Thus the L.C.M. of 6, 8, and 9 is $2 \times 2 \times 2 \times 3 \times 3$; that is, 72.

Exercises

1. Use the intersection of sets of multiples to find the least common multiple (L.C.M.) of:
   (a) 6 and 8   (b) 9 and 15
   (c) 12 and 32   (d) 48 and 64

2. Use a number line to find the least common multiple of:
   (a) 2 and 3   (b) 5 and 7
   (c) 2 and 5   (d) 3 and 7

3. What observation can you make about the least common multiple of two prime numbers?

4. What is the least common multiple of:
   (a) 1 and 6?   (b) 1 and 10?
   (c) 1 and n, where $n \in N$?

5. What is the least common multiple of:
   (a) 6 and 25?   (b) 4 and 21?
   (c) 9 and 10?

6. Numbers that have no common factors other than 1 are said to be "*relatively prime*" with respect to each other. Which pairs of numbers in Exercise 5 are relatively prime?

7. List four pairs of relatively prime numbers.

8. What observation can you make about the least common multiple of two relatively prime numbers?

9. Use prime factorization to find the least common multiple of:
   (a) 4, 18, and 20   (b) 2, 5, and 11
   (c) 32 and 396   (d) 239 and 785

10. (a) Will the product of two natural numbers always produce a common multiple of the two numbers?
    (b) Under what conditions will the product of two natural numbers be the least common multiple?

6-5 Greatest Common Factor

The **greatest common factor** (G.C.F.), also called the **greatest common divisor** (G.C.D.), of a set of natural numbers is the largest natural number that is a factor of all numbers in the set. For example, the greatest common factor of 18 and 24 is the largest number that is a factor of both 18 and 24. Observe that the set of factors of 18 is

$$\{1, 2, 3, 6, 9, 18\},$$

and the set of factors of 24 is

$$\{1, 2, 3, 4, 6, 8, 12, 24\}.$$

The *common factors* are those found in the intersection of the two sets, namely $\{1, 2, 3, 6\}$, and the greatest common factor is 6. Hence, the G.C.F. of 18 and 24 is 6. Notice that the greatest common factor of two numbers has each of the common factors as a factor.

Example 1   Find the greatest common factor of 180 and 390 by using the intersection of sets.

Let D equal the set of factors of 180,

$$D = \{1, 2, 3, 4, 5, 6, 9, 10, 12, 15, 18, 20, 30, 36, 45, 60, 90, 180\},$$

and E equal the set of factors of 390,

$$E = \{1, 2, 3, 5, 6, 10, 13, 15, 26, 30, 39, 65, 78, 130, 195, 390\}.$$

The largest number represented in the set resulting from the intersection of D and E is the greatest common factor. Since

$$D \cap E = \{1, 2, 3, 5, 6, 10, 15, 30\},$$

the greatest common factor of 180 and 390 is 30.

Example 2   Find the greatest common factor of 180 and 390 by prime factorization.

We may find the G.C.F. of 180 and 390 if we separate each number into a product of prime factors and then form a product of the different prime factors, each factor being used the same number of times as the *least number of times it occurs as a factor of any one of the given numbers.* Thus,

$$180 = 2 \times 90 = 2 \times 2 \times 45 = 2 \times 2 \times 5 \times 9 = 2 \times 2 \times 3 \times 3 \times 5,$$

$$390 = 2 \times 195 = 2 \times 5 \times 39 = 2 \times 3 \times 5 \times 13,$$

and $2 \times 3 \times 5$ is the product of prime factors common to the prime factors of 180 and 390. Since 2, 3, and 5 are the only prime factors common to the prime factors of 180 and 390, the G.C.F. of 180 and 390 is $2 \times 3 \times 5 = 30$.

The process of finding the G.C.F. by prime factorization may be applied to three or more numbers as well as to two numbers.

Exercises

1. Use the intersection of sets of factors to find the greatest common factor (G.C.F.) of:
   (a) 12 and 16   (b) 32 and 48
   (c) 64 and 72   (d) 256 and 87

2. Use prime factorization to find the G.C.F. of:
   (a) 29 and 13   (b) 3 and 5
   (c) 17 and 31   (d) 53 and 97

3. What observation can you make about the G.C.F. of two prime numbers?

4. What is the greatest common factor of:
   (a) 1 and 6?   (b) 1 and 25?
   (c) 1 and n, $n \in N$?

5. What is the G.C.F. of :
   (a) 6 and 25?   (b) 4 and 21?
   (c) 9 and 10?

6. Which pairs of numbers in Exercise 5 are relatively prime?

7. What observation can you make about the G.C.F. of two relatively prime numbers?

8. Use prime factorization to find the G.C.F. of:
   (a) 8, 12, and 30   (b) 36, 54, and 72
   (c) 19, 31, and 5   (d) 27, 8, and 125

9. Find the L.C.M. of:
   (a) 8 and 12   (b) 10 and 15
   (c) 12 and 45   (d) 258 and 570

10. Find the G.C.F. of each pair of numbers given in Exercise 9.

11. Is it true that the product of the "L.C.M. of 8 and 12" and the "G.C.F. of 8 and 12" is equal to the product of 8 and 12?

12. Does the statement in Exercise 11 hold true for
    (a) 10 and 15?   (b) 12 and 45?

13. What is the greatest common factor of 18 and 18?

14. What is the smallest common factor of 9 and 6?

15. Is it possible for the G.C.F. of a pair of natural numbers to be the same as the L.C.M. of the two numbers?

16. Can the L.C.M. of a pair of natural numbers ever be less than the G.C.F. of the same numbers?

17. Can the L.C.M. of a pair of natural numbers ever be greater than the G.C.F. of the same numbers?

6-6 Perfect, Abundant, and Deficient Numbers

We have seen that the natural numbers may be classified into subsets in several ways. Two of these classification systems are depicted in the following figure.

Classification system *A*   Classification system *B*

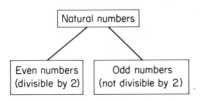

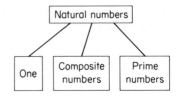

There are other ways to classify special subsets of the natural numbers. For example, a number greater than 1 is called a **perfect number** if it is equal to the sum of its proper divisors. The *proper divisors* of a number consist of all the divisors of the number, except the number itself. Consider the number 6. Its proper divisors are 1, 2, and 3. Since $1 + 2 + 3 = 6$, we say that 6 is a perfect number.

Is 12 a perfect number? The proper divisors of 12 are 1, 2, 3, 4, and 6. Since $1 + 2 + 3 + 4 + 6 \neq 12$, we say that 12 is *not* a perfect number.

Each natural number greater than 1 that is not a perfect number may be classified as an *abundant number* or as a *deficient number*.

A number may be classified as an **abundant number** if the sum of its proper divisors is greater than the number. In the case of 12 we note that $1 + 2 + 3 + 4 + 6 > 12$, and we call 12 an abundant number.

A number may be classified as a **deficient number** if the sum of its proper divisors is less than the number. Consider the number 4. The sum of the proper divisors 1 and 2 is less than 4, that is, $1 + 2 < 4$. Thus we say that 4 is a deficient number.

Exercises

1. List the proper divisors of each number.
   (a) 10   (b) 16
   (c) 24   (d) 28
   (e) 36   (f) 45
   (g) 93   (h) 117
   (i) 248   (j) 374

2. Classify each number in Exercise 1 as perfect, abundant, or deficient.

3. (a) List five prime numbers.
   (b) Classify each number in part (a) as perfect, abundant, or deficient.
   (c) Is it true that all prime numbers are deficient numbers?

4. (a) List the numbers 2 through 15.
   (b) Classify each number in part (a) as perfect, abundant, or deficient.
   (c) Which type of number occurs most frequently in part (b)?

5. (a) Two perfect numbers have been identified (6 and 28). Find one or more additional perfect numbers.

   (b) Several abundant numbers have been identified. Find five additional abundant numbers.

   (c) Several deficient numbers have been identified. Find five additional deficient numbers.

6. Two numbers are said to be **amicable** if each number equals the sum of the proper divisors of the other. Show that 220 and 284 are amicable numbers. Try to find another pair of amicable numbers. (Hint: One member of such a pair is 1,184.)

7. Two consecutive prime numbers whose difference is 2 are called **prime twins**. Find five examples of prime twins.

8. Three consecutive prime numbers that differ by 2, such as 3, 5, and 7, are called **prime triplets**. Can you find one or more sets of prime triplets?

9. In 1742, Christian Goldbach conjectured that every even number greater than or equal to 6 can be represented as the sum of two prime numbers. This conjecture has never been proved to be correct or incorrect. One counterexample is all that is needed to prove that the conjecture is incorrect. Test this conjecture on ten numbers.

10. Goldbach also conjectured that every odd number greater than or equal to 9 may be represented as the sum of three odd prime numbers. This conjecture has likewise never been proved or disproved. Test this conjecture on ten numbers.

6-7 Chapter Test

Indicate whether each statement is true or false. If false, tell why.

1. The multiples of 3 are 3^0, 3^1, 3^2, 3^3, ...

2. A number is divisible by 12 if it is divisible by both 6 and 2.

3. 257 is a prime number.

4. The L.C.M. of two numbers is greater than the G.C.F. of the same two numbers.

5. Any whole number is either a prime or a composite number.

6. The number 36 can be expressed as a product of a prime and a composite number.

7. 496 is a perfect number.

8. All prime numbers are odd numbers.

9. The numbers 36 and 81 are relatively prime.

10. The product of two prime numbers is the least common multiple of the two numbers.

Select the best possible answer.

11. The greatest common factor of two different prime numbers is:
    - (a) Nonexistent
    - (b) A prime number
    - (c) A composite number
    - (d) An odd number
    - (e) An even number

12. If n represents an odd number, the next consecutive odd number can be represented by:
    - (a) $2n$
    - (b) n^2
    - (c) $n + 1$
    - (d) $2 + n$
    - (e) $4 + n$

13. Observe this pattern:

$$1 = 1^2$$
$$1 + 3 = 4 = 2^2$$
$$1 + 3 + 5 = 9 = 3^2$$
$$1 + 3 + 5 + 7 = 16 = 4^2$$
$$\vdots$$

Which of the following conjectures may be drawn from this pattern?
    - (a) The sum of any set of odd numbers is a perfect square.
    - (b) Between any two odd numbers, there exists an even number.
    - (c) The sum of a series of n odd numbers is equal to n^2 where $n \neq 0$.
    - (d) The sum of a series of the first n odd numbers is equal to n^2, where $n \neq 0$.
    - (e) No conjecture is possible from the pattern.

14. The G.C.F. of 48 and 72 is:
    - (a) 144
    - (b) 24
    - (c) 18
    - (d) 8
    - (e) None of these

15. Which numbers are divisible by 9?
    I. 345,681   II. 27,061,542   III. 992,041,099
    - (a) I only
    - (b) II only
    - (c) III only
    - (d) I and II only
    - (e) II and III only

16. Which number is an abundant number?
    - (a) 6
    - (b) 16
    - (c) 27
    - (d) 29
    - (e) 30

17. Which number in Exercise 16 is a perfect number?

18. What is the largest prime divisor that must be tested in order to determine
    if 117 is a prime number?
    (a) 3   (b) 7
    (c) 11   (d) 37
    (e) 117

19. Which of the following is a set of prime triplets?
    (a) {3, 3, 3}   (b) {3, 3^2, 3^3}
    (c) {1, 2, 3}   (d) {17, 19, 21}
    (e) {7, 5, 3}

20. The proper divisors of 48 are:
    (a) 1 and 48
    (b) 2, 4, 6, 8, 12, and 24
    (c) 1, 2, 4, 6, 8, 12, and 24
    (d) 1, 2, 3, 4, 6, 8, 12, 16, and 24
    (e) 1, 2, 3, 4, 6, 8, 12, 16, 24, and 48

chapter 7

Integers: Operations and Properties

In our previous study, we discovered that the set of whole numbers is not closed under the operation of subtraction. In other words, there is at least one (actually many) subtraction equation involving whole numbers which has no solution in the set of whole numbers. For example, in $3 - 5 = n$, there is no whole number n such that $n + 5 = 3$. In order to provide solutions for equations like $3 - 5 = n$, we will extend our study of numbers to the set of integers.

7-1 The Set of Integers

Recall that the set N of natural numbers (also called the counting numbers) is the set $\{1, 2, 3, 4, \ldots\}$, and the set W of whole numbers is the set $\{0, 1, 2, 3, \ldots\}$. We observed earlier that the set of whole numbers may be thought of as the union of the set consisting of zero with the set of natural numbers; that is, $\{0\} \cup \{1, 2, 3, 4, \ldots\} = \{0, 1, 2, 3, \ldots\}$.

We have also noted that the set of whole numbers is not closed for subtraction, and that we need to create a set of numbers that is closed under subtraction so that each equation of the form $a - b = \triangle$, where a and b are members of W, will have a unique solution.

To do this we construct the set of **integers** consisting of the natural numbers, their negatives, and the number zero. The set of integers may be represented by

$$I = \{\ldots, {}^-4, {}^-3, {}^-2, {}^-1, 0, 1, 2, 3, 4, \ldots\}.$$

The numbers in the subset $\{1, 2, 3, 4, \ldots\}$ may be called **positive integers** and the numbers in the subset $\{{}^-1, {}^-2, {}^-3, {}^-4, \ldots\}$ may be called **negative integers.** The negative integers are read "negative one, negative two, negative three, negative four," and so on. Notice that zero is an integer that is neither positive nor negative. The set of integers may be pictured as on the next number line.

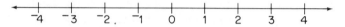

In many arithmetic programs, the existence of the negative integers has not been recognized. Children in the lower elementary grades may have been taught that a larger number cannot be subtracted from a smaller number and that in subtraction problems the larger number always "goes on top." It is important that a child's concept of number not be so restricted, even in the primary grades. When a restriction is necessary, it should be explained in the form "We do not yet have a number to represent . . ." and not as an indication of problems that can never be solved.

In modern programs, children in the primary and intermediate grades should develop intuitive notions of negative numbers through the use of a number line and other activities. The notions, suggesting a need for numbers not in the set of whole numbers, should be encouraged in order to provide a basis for the future introduction of negative numbers.

A look at the following diagram may clarify the extensions used to develop our number system up to this point. The first set of numbers that we men-

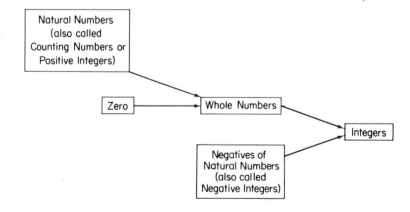

tioned was the set of natural numbers. By including zero with the set of natural numbers, we extended our system to the set of whole numbers. Finally, by introducing the negative integers we extended our system to the set of integers.

We consider the whole numbers greater than zero as positive integers; we consider the opposites (with respect to addition) of the positive integers as the negative integers. As noted, the integer zero is neither positive nor negative. The set of numbers formed by the union of the positive integers and zero is sometimes called the **nonnegative integers.** In other words, the whole numbers may also be referred to as the nonnegative integers. Some authors refer to the integers as *signed numbers* and use the + symbol with the natural numbers to indicate that they are positive integers. Then, it is easy to lead children to discover that for each positive integer to the right of zero on the number line, there is an *opposite number* represented to the left of zero, and vice versa. This is illustrated in the next diagram where semi-circles denote pairing of opposites on the number line.

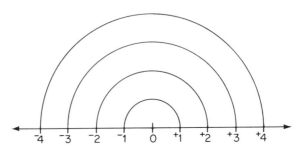

We should notice that the + and the − symbols have dual uses. When the + symbol is used to denote the operation of addition, it is read "plus"; however, when this same + symbol is used to denote that the number is a positive integer, it is read "positive." The − symbol, when used to denote subtraction, is read "minus"; however, when this same − symbol is used to indicate that the number is negative, it is read "negative." Some authors suggest that the + symbol used to denote a positive integer and the − symbol used to denote a negative integer should be written in the raised position we have used here for ⁺3, ⁺4, etc., read "positive three, positive four," etc.

5 + 3	This expression is read "five *plus* three."
9 − 3	This expression is read "nine *minus* three."
⁺5, ⁺4, ⁺3	The + symbols indicate that these are *positive* integers and would be represented to the right of zero on the number line (read "positive five, positive four, positive three").

$^-6, ^-5, ^-4$ The $-$ symbols indicate that these are *negative* integers and would be represented to the left of zero on the number line (read "negative six, negative five, negative four").

$^+6 + ^-2 = ^+4$ This equation is read "*positive* six *plus negative* two is equal to *positive* four."

$^-5 - ^-7 = ^+2$ This equation is read "*negative* five *minus negative* seven is equal to *positive* two."

Exercises

Read each of the statements 1 through 9, translating into words.

1. $^+4 + ^+9 = ^+13$

2. $^+7 + ^-3 = ^+4$

3. $^-6 + ^+8 = ^+2$

4. $^-5 + ^-2 = ^-7$

5. $0 + ^-14 = ^-14$

6. $^+6 + 0 = ^+6$

7. $^+7 - ^+5 = ^+2$

8. $^+10 - ^-9 = ^+19$

9. $^-3 - ^+5 = ^-8$

10. When possible, find each solution in the set of whole numbers.
    (a) $11 - 7 = \triangle$
    (b) $7 - 11 = \triangle$
    (c) $15 - 15 = \triangle$
    (d) $4 - 9 = \triangle$
    (e) $9 - 4 = \triangle$
    (f) $9 - 17 = \triangle$ ~6

11. Draw a number line to show the set of:
    (a) Nonnegative integers
    (b) Negative integers
    (c) Positive integers
    (d) Integers

12. Write the numeral for the opposite of:
    (a) $^+2$
    (b) $^-5$
    (c) $^-9$
    (d) $^+27$ – 27

13. Indicate whether each statement is true or false. If false, tell why.
    (a) The union of the positive integers with the negative integers forms the set of integers.
    (b) The set of natural numbers is equal to the set of positive integers.
    (c) Every natural number has an opposite in the set of integers.
    (d) The set of integers may be represented as $\{^-4, ^-3, ^-2, ^-1, 0, 1, 2, 3, 4\}$.

7-2 Additive Inverse

The number line may be used to picture the set of integers and to show that each positive integer may be paired with its opposite, which is called a negative integer. In other words, the positive integers may be placed in one-to-one correspondence with the negative integers. The + symbol is often omitted when we write the numeral for a positive integer; that is, $^{+}1$ is 1, $^{+}2$ is 2, $^{+}3$ is 3, $^{+}4$ is 4, $^{+}5$ is 5, and so on.

Additive inverse is a term used to convey the notion of an opposite with respect to addition. The opposite of 6 is $^{-}6$ and the opposite of $^{-}6$ is 6. Since $6 + {}^{-}6 = 0$ and $^{-}6 + 6 = 0$, where 0 is the identity element for addition, $^{-}6$ is called the additive inverse of 6, and 6 is called the additive inverse of $^{-}6$. *The sum of an integer and its additive inverse is always zero.* The *additive inverse* of a number is often called **the negative** of the number. Thus, $^{-}3$ is the negative of 3, and 3 is the negative of $^{-}3$. In general, the negative of a positive number is a negative number; the negative of a negative number is a positive number; the negative of zero is zero.

A number line is an excellent model for illustrating operations with signed numbers. Arrows may be used to represent directed line segments for positive and negative numbers. Notice that in the following number-line illustrations the arrow points to the *right* if it represents a *positive* number and to the *left* if it represents a *negative* number.

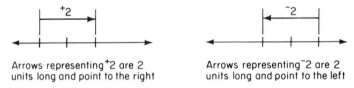

Arrows representing $^{+}2$ are 2   Arrows representing $^{-}2$ are 2
units long and point to the right   units long and point to the left

To picture the sum of two signed numbers on the number line, we draw the first arrow with its tail (initial point) at the graph of 0 and the second arrow with its tail at the head (terminal point) of the first arrow. The coordinate of the point representing the head of the second arrow is the sum of the two signed numbers.

The equation $4 + {}^{-}4 = 0$ may be pictured on the next number line by using arrows (directed line segments) for 4 and $^{-}4$. Notice that 4 is represented by a line segment of 4 units' length directed to the right from the graph of 0, and $^{-}4$ is represented by a line segment of 4 units' length directed to the left from the graph of 4. The coordinate of the other endpoint of the directed line segment representing $^{-}4$ is 0. Thus, $^{-}4$ is the number that yields 0 when added to 4; that is, $^{-}4$ is the additive inverse, or negative, of 4.

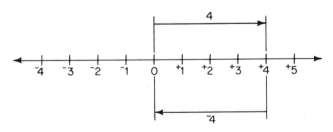

We may similarly show that 4 is the additive inverse, or negative, of $^-4$; that is, $^-4 + 4 = 0$.

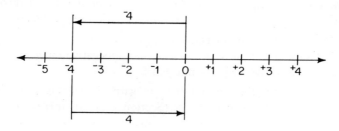

Every integer has an additive inverse, since each positive integer has a negative integer as its additive inverse, each negative integer has a positive integer as its additive inverse, and zero is its own additive inverse. In general, for every n where n is an integer, there exists a negative of n such that

$$n + {}^-n = {}^-n + n = 0.$$

Exercises

1. Write the numeral for the additive inverse of each integer.
   (a) $^-3$
   (b) 5
   (c) 17
   (d) $^-26$
   (e) 0
   (f) b
   (g) ^-b
   (h) $^-(a + b)$
   (i) $a - b$
   (j) $a \cdot b$

2. Complete each statement to make a true number sentence.
   (a) $7 + {}^-7 = \underline{\;0\;}$
   (b) $^-4 + \underline{\quad} = 0$
   (c) $\underline{\quad} + 0 = 0$
   (d) $\underline{\quad} + {}^-18 = 0$
   (e) $n + \underline{\quad} = 0, n \in I$

3. Draw a number line picture to show:
   (a) $2 + {}^-2 = 0$
   (b) $^-6 + 6 = 0$

4. What number does n represent if:
   (a) $n + {}^-8 = 0$?
   (b) $n + 8 = 0$?
   (c) $n + (5 \times 6) = 0$?
   (d) $n + (4 + 9) = 0$?

5. (a) What is the meaning of $^-(^-5)$?
   (b) What is $^-(^-5)$ equal to?

6. (a) If $n \in I$, what is the meaning of $^-(^-n)$?
   (b) What is $^-(^-n)$ equal to?

7-3 Addition of Integers

Intuitive notions of the addition of integers may be developed through the use of a number line. There are three types of addition problems:

(1) The sum of two nonnegative integers.

(2) The sum of a negative integer and a nonnegative integer.

(3) The sum of two negative integers.

The representation of the sum of two nonnegative integers is the same as that for the sum of two whole numbers (Section 4-1). In this section we will concentrate on the other two types of sums of integers.

We first use a number line to determine the sum $4 + {}^-3$. The 4 is represented by a line segment of 4 units' length directed to the right from the graph of 0, and $^-3$ is represented by a line segment of 3 units' length directed to the left from the graph of 4. The coordinate of the other endpoint of the directed line segment representing $^-3$ is 1. Thus, $4 + {}^-3 = 1$ as shown on the following number line.

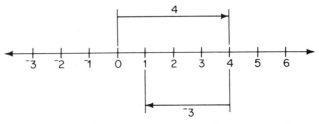

The sum $2 + {}^-3$ may be represented as on the next number line. Thus, we see that $2 + {}^-3 = {}^-1$.

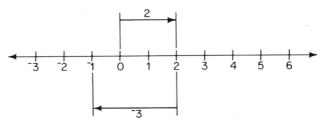

Consider the sum $^-2 + {}^-3$. In this case the line segments are both directed to the left. Observe on the next number line that $^-2 + {}^-3 = {}^-5$.

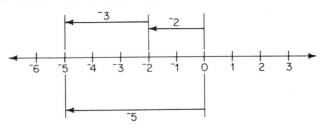

After considering many examples of the three types of sums of integers, such as those shown on the number lines, the student should be able to discover some generalizations (rules) for the addition of integers.

We will assume that the basic properties of addition of whole numbers may be extended to the addition of integers. These properties may be summarized as follows:

Closure and uniqueness properties   If a and b are integers, then

$$a + b \text{ is a unique integer.}$$

Commutative property   If a and b are integers, then

$$a + b = b + a.$$

Associative property   If a, b, and c are integers, then

$$a + (b + c) = (a + b) + c.$$

Identity element   If a is an integer, then

$$a + 0 = 0 + a = a.$$

We have already discovered in our study of the integers a new property concerning the existence and uniqueness of the additive inverse of any integer. This property may be stated as follows.

Additive inverse property   If a is an integer, then

$$a + {}^-a = {}^-a + a = 0.$$

These properties may be used to illustrate some of the reasoning involved in the formulation of generalizations or rules for the different types of addition problems involving integers. Study these examples.

Example 1   Consider the sum ${}^-2 + {}^-4 = \triangle$.

${}^-2 + {}^-4 = \triangle$	
$({}^-2 + {}^-4) + 4 = \triangle + 4$	If $a = b$, then $a + n = b + n$, where a, b, and n are integers.
${}^-2 + ({}^-4 + 4) = \triangle + 4$	Associative property.
${}^-2 + 0 = \triangle + 4$	Additive inverse property.
${}^-2 = \triangle + 4$	Additive identity property.
${}^-2 + 2 = \triangle + 4 + 2$	If $a = b$, then $a + n = b + n$, where a, b, and n are integers.
$0 = \triangle + 4 + 2$	Additive inverse property.
$0 = \triangle + (4 + 2)$	Associative property.

Thus we conclude that $\triangle$ is the additive inverse of $4 + 2$; that is ${}^-(4 + 2)$. Since $\triangle = {}^-2 + {}^-4$ in the beginning, we conclude that $\triangle = {}^-2 + {}^-4 = {}^-(4 + 2) = {}^-6$.

From Example 1, we make the conjecture that $^-a + {}^-b = {}^-(a + b)$, where a and b are whole numbers. For instance, $^-5 + {}^-3 = {}^-(5 + 3) = {}^-8$ and $^-8 + {}^-9 = {}^-(8 + 9) = {}^-17$.

Example 2   Consider the sum $9 + {}^-4 = \triangle$.

$9 + {}^-4 = (5 + 4) + {}^-4$	Renamed 9 as $5 + 4$.
$(5 + 4) + {}^-4 = 5 + (4 + {}^-4)$	Associative property.
$5 + (4 + {}^-4) = 5 + 0$	Additive inverse property.
$5 + 0 = 5$	Additive identity property.

Thus $9 + {}^-4 = 5$.

From Example 2, we make the conjecture that $a + {}^-b = {}^-b + a = a - b$ if $a > b$ and both a and b are whole numbers. For instance, $7 + {}^-3 = {}^-3 + 7 = 7 - 3 = 4$ and $^-9 + 15 = 15 + {}^-9 = 15 - 9 = 6$.

Example 3   Consider the sum $^-7 + 5 = \triangle$.

$^-7 + 5 = ({}^-2 + {}^-5) + 5$	Renamed $^-7$ as $^-2 + {}^-5$.
$({}^-2 + {}^-5) + 5 = {}^-2 + ({}^-5 + 5)$	Associative property.
$^-2 + ({}^-5 + 5) = {}^-2 + 0$	Additive inverse property.
$^-2 + 0 = {}^-2$	Additive identity property.

Thus $^-7 + 5 = {}^-2$.

From Example 3, we make the conjecture that $a + {}^-b = {}^-b + a = {}^-(b - a)$ if $b > a$ and both a and b are whole numbers. For instance, $^-8 + 3 = 3 + {}^-8 = {}^-(8 - 3) = {}^-5$ and $6 + {}^-12 = {}^-12 + 6 = {}^-(12 - 6) = {}^-6$.

Each of the sums in Examples 1, 2, and 3 may be verified on number lines as shown previously in this section.

Exercises

1. Use a number line to determine each sum.
   - (a) $5 + {}^-4$
   - (b) $^-1 + {}^-3$
   - (c) $^-6 + {}^+5$
   - (d) $^-2 + 2$

2. Find each sum. Use a number line if you need help.
   - (a) $^-3 + {}^-6$
   - (b) $^-1 + {}^-7$
   - (c) $^-4 + {}^-7$
   - (d) $^-5 + {}^-5$
   - (e) $6 + {}^-2$
   - (f) $8 + {}^-5$
   - (g) $^-7 + 9$
   - (h) $^-1 + 8$
   - (i) $^-4 + 2$
   - (j) $^-6 + 3$
   - (k) $5 + {}^-8$
   - (l) $9 + {}^-14$
   - (m) $8 + 0$
   - (n) $^-9 + 0$
   - (o) $0 + {}^-7$
   - (p) $^-5 + 5$

3. Complete this addition table.

+	$^-5$	$^-4$	$^-3$	$^-2$	$^-1$	0	1	2	3	4	5
$^-5$											
$^-4$											
$^-3$											
$^-2$											
$^-1$											
0											
1											
2											
3											
4											
5											

4. Make a list of some of the patterns you can find in the addition table for Exercise 3.

5. Assume that $a + {}^-b = a - b$ if $a > b$ and both a and b are whole numbers. What does $a + {}^-b$ equal when $a \geq b$?

6. Assume that $a + {}^-b = {}^-(b - a)$ if $b > a$ and both a and b are whole numbers. What does $a + {}^-b$ equal when $b \geq a$?

7. Find each sum. Show each step and justify it. (See Examples 1, 2, and 3 in this section.)

(a) $^-8 + 7$ (b) $^-4 + {}^-9$

(c) $5 + {}^-2$

7-4 Subtraction of Integers

The importance of clearly understanding the relationship between the operations of addition and subtraction becomes evident when we need to subtract integers. In Section 4-8 we studied the relationship between addition and subtraction of whole numbers. We will assume that this relationship may be extended to the addition and subtraction of integers.

If $a + b = c$, then $c - b = a$ and $c - a = b$, where a, b, and c are integers. For example, if $8 + {}^-3 = 5$, then $5 - {}^-3 = 8$ and $5 - 8 = {}^-3$. These equations may be shown on number lines. The next number line represents $8 + {}^-3 = 5$.

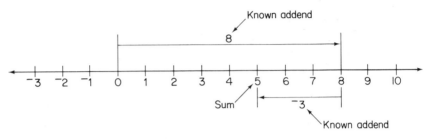

The additive point of view may be used to interpret $5 - {}^-3 = \triangle$ and to show it on the number line. In other words, we think of the number that must be used as an addend with ${}^-3$ to produce a sum of 5. The next number line shows $\triangle + {}^-3 = 5$. We should note that the missing addend is 8. Thus $8 + {}^-3 = 5$ and $5 - {}^-3 = 8$.

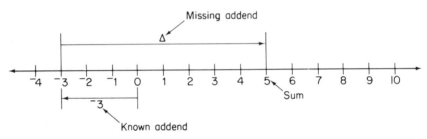

An additive interpretation of $5 - 8 = \square$ directs us to find the number that must be used as an addend with 8 to produce a sum of 5; that is, $\square + 8 = 5$. The next number line shows that the missing addend is ${}^-3$ and that ${}^-3 + 8 = 5$. We may also conclude that $5 - 8 = {}^-3$.

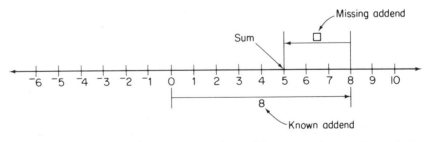

Using the number line in connection with our understanding of the relationship between addition and subtraction, we can develop an interesting pattern as shown in the following table. Is it true that "subtracting an integer is equivalent to adding its additive inverse"?

Subtraction	*Interpretation of subtraction*
$5 - 8 = {}^-3$	$5 + {}^-8 = {}^-3$
$5 - 14 = {}^-9$	$5 + {}^-14 = {}^-9$
$7 - 3 = 4$	$7 + {}^-3 = 4$
$6 - {}^-2 = 8$	$6 + 2 = 8$
$10 - {}^-3 = 13$	$10 + 3 = 13$
$8 - {}^-1 = 9$	$8 + 1 = 9$
${}^-2 - 3 = {}^-5$	${}^-2 + {}^-3 = {}^-5$
${}^-4 - 6 = {}^-10$	${}^-4 + {}^-6 = {}^-10$
${}^-9 - 5 = {}^-14$	${}^-9 + {}^-5 = {}^-14$
${}^-3 - {}^-6 = 3$	${}^-3 + 6 = 3$
${}^-5 - {}^-4 = {}^-1$	${}^-5 + 4 = {}^-1$
${}^-12 - {}^-7 = {}^-5$	${}^-12 + 7 = {}^-5$

A study of the pairs of examples given in this table should lead us to the general statement

$$a - b = a + {}^-b$$

where a and b are integers.

Recall (Section 7-2) that signed numbers may be represented by arrows (directed line segments) when the number line is used to picture the addition of two integers. For example, $3 + 4$ is represented on the following number line.

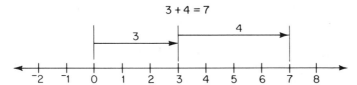

The number line also may be used to picture the subtraction of signed numbers. The minus sign, designating the operation of subtraction, is a direction-changing signal when we use arrows to illustrate the subtraction of signed numbers. For example, when 3 and 4 are added, 4 is represented by an arrow of 4 units' length pointing to the right; however, in order to subtract 4 from 3 we add $^-4$ to 3 and represent $^-4$ by an arrow of 4 units' length directed to the left as shown on the next number line.

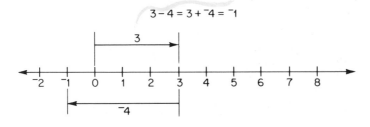

Consider the examples $2 - 5$ and $2 - {}^-5$ pictured on the next two number lines respectively.

$$2 - {}^-5 = 2 + {}^+5 = 7$$

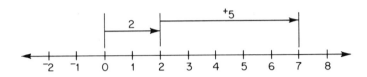

$$2 - 5 = 2 + {}^-5 = {}^-3$$

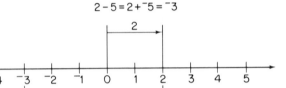

Exercises

1. For each addition equation, write two related subtraction equations.
   (a) $9 + 7 = 16$ (b) $6 + {}^-9 = {}^-3$
   (c) ${}^-4 + {}^-1 = {}^-5$

2. Show each of the three equations in Exercise 1(a) on a number line.

3. Find the additive inverse of each integer.
   (a) ${}^-6$ (b) x
   (c) 4 (d) ${}^-1$
   (e) $a + b$ (f) $4 + {}^-3$
   (g) ${}^-13 + {}^-4$ (h) ${}^-a + 2$

4. Find each difference.
   (a) $7 - 9$ (b) ${}^-3 - 10$
   (c) ${}^-8 - {}^-6$ (d) $14 - {}^-6$
   (e) $(6 + {}^-2) - {}^-11$ (f) $({}^-2 - {}^-5) - {}^-9$

5. Copy and complete each statement correctly.
   (a) ${}^-5 - 4 = {}^-9$ and ${}^-5 + \underline{} = {}^-9$
   (b) $12 - {}^-1 = 13$ and $12 + \underline{} = 13$
   (c) ${}^-7 - 8 = {}^-15$ and ${}^-7 + \underline{} = {}^-15$
   (d) ${}^-6 - {}^-5 = {}^-1$ and ${}^-6 + \underline{} = {}^-1$
   (e) $x - y = z$ and $x + \underline{} = z$, where x, y, and z are integers

6. Use a number line to show each part of Exercise 5(d) and t o verify the correctness of the statement.

7. Tell whether or not each set of numbers is closed with respect to subtraction. If the set is not closed, give one counterexample.
   (a) Integers   (b) Positive integers
   (c) Negative integers   (d) Nonnegative integers
   (e) Even integers   (f) Odd integers

8. Does the commutative property hold for the operation of subtraction on the set of integers? Give a numerical example to illustrate your answer.

9. Does the associative property hold for the operation of subtraction on the set of integers? Give a numerical example to illustrate your answer.

10. Find each difference.
    (a) $^-3 - 0$ (b) $4 - 0$
    (c) $0 - 7$ (d) $0 - {}^-6$

11. For every integer n, is it true that
    (a) $n - 0 = n$?   (b) $0 - n = n$?
    (c) $n - 0 = 0 - n = n$?

12. Indicate whether each statement is true or false. If false, tell why.
    (a) Zero is the identity element for subtraction of integers.
    (b) Zero is a right only identity for subtraction of integers.
    (c) Zero is a left only identity for subtraction of integers.
    (d) One is the identity element for subtraction of integers.
    (e) There is no identity element for subtraction of integers.

7-5 Multiplication of Integers

Intuitive notions of the multiplication of integers may be developed through the use of a multiplication table. There are three types of multiplication problems involving integers:

   (1) The product of two nonnegative integers.
   (2) The product of a negative integer and a nonnegative integer.
   (3) The product of two negative integers.

Since we have already worked with the first case, when we dealt with multiplication of whole numbers (Section 5-1), the following discussion will concentrate on the other two possibilities for the multiplication of integers. Some of the blocks in the following table have been filled in as a result of our understanding of the multiplication of whole numbers.

×	⁻3	⁻2	⁻1	0	1	2	3
⁻3							
⁻2							
⁻1							
0				0	0	0	0
1				0	1	2	3
2				0	2	4	6
3				0	3	6	9

In order to complete the table, we will look for patterns that exist and assume that these patterns are reliable. We observe in the right-hand column that each number is three less than the number represented below it. Applying this notion to the empty blocks in this column, we have

3
⁻9
⁻6
⁻3
0
3
6
9

In a like manner, the corresponding row would become

3	⁻9	⁻6	⁻3	0	3	6	9

If we extend this pattern to the remainder of the empty blocks, the table may be completed as follows.

×	-3	-2	-1	0	1	2	3
-3	9	6	3	0	-3	-6	-9
-2	6	4	2	0	-2	-4	-6
-1	3	2	1	0	-1	-2	-3
0	0	0	0	0	0	0	0
1	-3	-2	-1	0	1	2	3
2	-6	-4	-2	0	2	4	6
3	-9	-6	-3	0	3	6	9

By studying the completed table carefully we should be able to discover a pattern in the multiplication of the following:

(1) A nonnegative integer and a nonnegative integer.

(2) A positive integer and a negative integer or zero and a negative integer.

(3) A negative integer and a negative integer.

We will assume that the basic properties of multiplication of whole numbers may be extended to the multiplication of integers. These properties may be summarized as follows:

Closure and uniqueness properties   If a and b are integers, then

$$a \times b \text{ is a unique integer.}$$

Commutative property   If a and b are integers, then

$$a \times b = b \times a.$$

Associative property   If a, b, and c are integers, then

$$a \times (b \times c) = (a \times b) \times c.$$

Identity element   If a is an integer, then

$$a \times 1 = 1 \times a = a.$$

Special property of zero   If a is an integer, then

$$a \times 0 = 0 \times a = 0.$$

Distributive properties   If a, b, and c are integers, then

$$a \times (b + c) = (a \times b) + (a \times c);$$
$$(b + c) \times a = (b \times a) + (c \times a);$$

$$a \times (b - c) = (a \times b) - (a \times c); \text{ and}$$
$$(b - c) \times a = (b \times a) - (c \times a).$$

These properties may be used to illustrate some of the reasoning involved in the formulation of generalizations or rules for the different types of multiplication problems involving integers. Study these examples.

Example 1 To establish the product $^-2 \times 4$, we consider the following.

$0 = 0 \times 4$	Special property of 0 in multiplication.
$0 \times 4 = (2 + {}^-2) \times 4$	Additive inverse property.
$(2 + {}^-2) \times 4 = (2 \times 4) + ({}^-2 \times 4)$	Distributive property for multiplication over addition from the right.
$(2 \times 4) + ({}^-2 \times 4) = 8 + ({}^-2 \times 4)$	Multiplication facts for 2×4.
$8 + ({}^-2 \times 4) = 8 + {}^-8$	For $^-2 \times 4$ to be the additive inverse of 8, it must be another name for $^-8$.

From the above discussion, we conclude that $^-2 \times 4 = {}^-(2 \times 4) = {}^-8$ since $^-2 \times 4$ must be the additive inverse of (2×4) or 8. Thus we make the conjecture that $^-a \times b = {}^-(a \times b)$, where a and b are whole numbers. For instance, $^-5 \times 3 = {}^-(5 \times 3) = {}^-15$ and $^-4 \times 7 = {}^-(4 \times 7) = {}^-28$.

Since multiplication is assumed to be commutative, it may also be stated that $^-a \times b = b \times {}^-a = {}^-(a \times b)$.

Example 2 To establish the product $^-3 \times {}^-2$, we consider the following.

$0 = 0 \times {}^-2$	Special property of zero in multiplication.
$0 \times {}^-2 = (3 + {}^-3) \times {}^-2$	Additive inverse property.
$(3 + {}^-3) \times {}^-2 = (3 \times {}^-2) + ({}^-3 \times {}^-2)$	Distributive property of multiplication over addition from right.
$(3 \times {}^-2) + ({}^-3 \times {}^-2) = {}^-6 + ({}^-3 \times {}^-2)$	Illustrated in Example 1 $[a \times {}^-b = {}^-(a \times b)]$.
$^-6 + ({}^-3 \times {}^-2) = {}^-6 + 6$	For $({}^-3 \times {}^-2)$ to be the additive inverse of $^-6$, it must be another name for 6.

From the above discussion, we conclude that $^-3 \times {}^-2 = 3 \times 2 = 6$, since $^-3 \times {}^-2$ must be the additive inverse of $3 \times {}^-2$, or $^-6$. Thus we make the conjecture that $^-a \times {}^-b = a \times b$, where a and b are integers. For instance, $^-2 \times {}^-5 = 2 \times 5 = 10$, and $^-6 \times {}^-3 = 6 \times 3 = 18$.

Each of the products in Examples 1 and 2 may be verified from a multiplication table like the one given earlier in this section.

Exercises

1. Use the pattern revealed in each of these columns from a multiplication table and find the missing products.

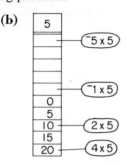
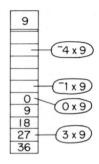

2. Draw a number line to show each multiplication equation as repeated addition.
   (a) $2 \times {}^-6 = \triangle$ (b) $4 \times {}^-1 = \triangle$
   (c) $7 \times {}^-3 = \triangle$

3. Find each product.
   (a) $3 \times {}^-7$ (b) $4 \times {}^-6$
   (c) $^-2 \times 8$ (d) $^-1 \times 9$
   (e) $^-4 \times {}^-5$ (f) $^-3 \times {}^-9$
   (g) $^-5 \times {}^-1$ (h) $^-6 \times {}^-7$
   (i) $^-3 \times 0$ (j) $0 \times {}^-12$
   (k) 7×0 (l) $^-4 \times {}^-9 \times 0$
   (m) $^-4 \times {}^-5 \times {}^-7$ (n) $^-9 \times {}^-6 \times 10$
   (o) $19 \times 12 \times {}^-6$ (p) $^-1 \times {}^-1 \times {}^-1$

4. Make a multiplication table for $\{^-5, {}^-4, {}^-3, {}^-2, {}^-1, 0, 1, 2, 3, 4, 5\}$.

5. Make a list of some of the patterns you can find in the multiplication table for Exercise 4.

6. From the set of integers, find the solution set for each equation.
   (a) $12 \times {}^-12 = \square$ (b) $^-7 \times \square = 0$
   (c) $\square \times 8 = {}^-56$ (d) $^-9 \times \square = 81$
   (e) $\square \times {}^-13 = 13$ (f) $75 \times \square = {}^-75$

7-6 Division of Integers

The inverse nature of the multiplication and division operations on whole numbers was established in Chapter 5. We will assume that this relationship may be extended to the multiplication and division of integers, i.e., if $a \times b = c$, then $c \div b = a$ and $c \div a = b$, where a, b, and c are integers $(b \neq 0, a \neq 0)$. For example if $^-4 \times 7 = ^-28$, then $^-28 \div 7 = ^-4$ and $^-28 \div ^-4 = 7$.

We may think of division as finding a missing factor in a multiplication equation with the product and one factor given. Thus $^-28 \div ^-4 = \triangle$ may be interpreted as $\triangle \times ^-4 = ^-28$, and $^-28 \div 7 = \square$ may be interpreted as $\square \times 7 = ^-28$. This inverse relationship may also be observed in the following statements.

$$^-15 \div 3 = ^-5 \quad \text{because} \quad ^-5 \times 3 = ^-15.$$
$$^-18 \div 6 = ^-3 \quad \text{because} \quad ^-3 \times 6 = ^-18.$$
$$^-21 \div ^-3 = 7 \quad \text{because} \quad 7 \times ^-3 = ^-21.$$
$$^-12 \div ^-4 = 3 \quad \text{because} \quad 3 \times ^-4 = ^-12.$$
$$16 \div ^-8 = ^-2 \quad \text{because} \quad ^-2 \times ^-8 = 16.$$
$$24 \div ^-6 = ^-4 \quad \text{because} \quad ^-4 \times ^-6 = 24.$$

If we study the above examples and others like them, we should be able to discover a pattern for:

(1) A negative integer divided by a positive integer.
(2) A negative integer divided by a negative integer.
(3) A nonnegative integer divided by a negative integer.

The pattern for division of a nonnegative integer by a positive integer is identical to that for division of whole numbers. We should also note that it is unnecessary to learn a set of rules for division of integers. By thinking of each division equation as a multiplication equation with a missing factor, we can apply the rules for multiplication of integers.

Exercises

1. For each multiplication equation, write two related division equations.
   (a) $^-6 \times 9 = ^-54$ (b) $^-7 \times ^-5 = 35$
   (c) $4 \times ^-8 = ^-32$

2. Give an example to show that each of these properties does not hold for division of integers.
   (a) Commutative   (b) Associative
   (c) Closure

3. Indicate whether each statement is true or false.
    (a) If $^-3 \times {}^-4 = 12$, then $12 \div {}^-4 = {}^-3$ and $({}^-3 \times {}^-4) \div {}^-4 = {}^-3$.
    (b) If $63 \div {}^-7 = {}^-9$, then $^-9 \times {}^-7 = 63$ and $(63 \div {}^-7) \times {}^-7 = 63$.
    (c) If $5 \times {}^-15 = {}^-75$, then $^-75 \div {}^-15 = 5$ and $(5 \times {}^-15) \div 75 = 5$.
    (d) If $^-98 \div 7 = {}^-14$, then $^-14 \times 7 = {}^-98$ and $({}^-98 \div 7) \times {}^-7 = {}^-98$.

4. If you multiply $^-6$ by an integer j, what must you do to the product to obtain j?

5. If you divide an integer j by $^-4$, what must you do to the quotient to obtain j?

6. Write each division equation as a multiplication equation with a missing factor. Then find the quotient.
    (a) $52 \div {}^-13 = \triangle$ (b) $^-63 \div 9 = \triangle$
    (c) $^-72 \div {}^-12 = \triangle$

7. Find each quotient.
    (a) $^-16 \div 4$ (b) $^-9 \div {}^-9$
    (c) $13 \div {}^-13$ (d) $0 \div {}^-7$
    (e) $^-19 \div {}^-19$ (f) $^-20 \div 2$
    (g) $36 \div 4$ (h) $40 \div {}^-5$
    (i) $^-54 \div 6$

8. Find each quotient, if it exists in the set of integers.
    (a) $16 \div 1$ (b) $^-5 \div 1$
    (c) $^-14 \div 1$ (d) $3 \div {}^-1$
    (e) $^-6 \div {}^-1$ (f) $0 \div {}^-1$
    (g) $1 \div 4$ (h) $1 \div {}^-8$
    (i) $1 \div {}^-2$

9. For every integer j, is it true that:
    (a) $j \div 1 = j$?   (b) $j \div {}^-1 = j$?
    (c) $1 \div j = j$?   (d) $j \div 1 = 1 \div j = j$?

10. Indicate whether each statement is true or false. If false, tell why.
    (a) One is the identity element for division of integers.
    (b) One is a right only identity for division of integers.
    (c) One is a left only identity for division of integers.
    (d) Negative one is the identity element for division of integers.
    (e) Zero is the identity element for division of integers.
    (f) There is no identity element for division of integers.

11. Recall for the whole numbers that an equation like $4 \div 0 = n$ had no meaning because there was no whole number n such that $n \times 0 = 5$.

(a) Does an integer j exist such that $3 \div 0 = j$ and $j \times 0 = 3$?

(b) Is division by zero excluded or not defined for the set of integers?

12. Indicate whether each statement is true or false.

(a) $(14 + 8) \div {}^-2 = (14 \div {}^-2) + (8 \div {}^-2)$

(b) $({}^-12 + 6) \div 3 = ({}^-12 \div 3) + (6 \div 3)$

(c) $(16 - {}^-16) \div {}^-4 = (16 \div {}^-4) - ({}^-16 \div {}^-4)$

(d) $({}^-15 - 10) \div 5 = ({}^-15 \div 5) - (10 \div 5)$

(e) $24 \div ({}^-2 + 6) = (24 \div {}^-2) + (24 \div 6)$

(f) ${}^-18 \div (6 - 3) = ({}^-18 \div 6) - ({}^-18 \div 3)$

13. What property or counterexample of a property is illustrated by:

(a) Exercises 12(a) and 12(b)?   (b) Exercises 12(c) and 12(d)?

(c) Exercises 12(e) and 12(f)?

7-7 Summary of Properties for Operations on Integers

The following table presents a summary of the general properties for addition and subtraction of integers. The letters a, b, and c represent arbitrary integers.

Property	Addition	Subtraction
Closure	$a + b$ always produces an integer for the sum.	$a - b$ always produces an integer for the difference.
Uniqueness	$a + b$ is a unique integer.	$a - b$ is a unique integer.
Commutative	$a + b = b + a$.	$a - b \neq b - a$. $(a \neq b)$
Associative	$a + (b + c) = (a + b) + c$.	$a - (b - c) \neq (a - b) - c$. $(c \neq 0)$
Identity element	$a + 0 = a$. $0 + a = a$.	$a - 0 = a$. $0 - a \neq a$. (right identity only, $a \neq 0$)
Inverse operations	$a + b = c$ if and only if $c - b = a$.	$a - b = c$ if and only if $c + b = a$.
Inverse element	For every a, there exists an integer ${}^-a$ such that $a + {}^-a = {}^-a + a = 0$.	

The next table is a summary of the general properties for multiplication and division of integers. The letters a, b, and c represent arbitrary integers.

Property	Multiplication	Division
Closure	$a \times b$ always produces an integer for the product.	$a \div b$ does *not* always produce an integer for the quotient.
Uniqueness	$a \times b$ is a unique integer.	$a \div b$, if it exists, is a unique integer.
Commutative	$a \times b = b \times a$.	$a \div b \neq b \div a.$ $(a \neq b)$
Associative	$a \times (b \times c) = (a \times b) \times c$.	$a \div (b \div c) \neq (a \div b) \div c.$ $(c \neq 1)$
Identity element	$a \times 1 = 1 \times a = a$.	$a \div 1 = a.$ $1 \div a \neq a.$ (right identity only, $a \neq 1$)
Special property of zero	$a \times 0 = 0 \times a = 0$.	$a \div 0$ not defined. $0 \div a = 0$ $(a \neq 0).$ $0 \div 0$ not defined.
Inverse operations	$a \times b = c$ if and only if $c \div b = a$.	$a \div b = c$ if and only if $c \times b = a$.
Distributive from left over $+$ right over $+$ left over $-$ right over $-$	$a \times (b + c) = (a \times b) + (a \times c)$. $(b + c) \times a = (b \times a) + (c \times a)$. $a \times (b - c) = (a \times b) - (a \times c)$. $(b - c) \times a = (b \times a) - (c \times a)$.	$a \div (b + c) \neq (a \div b) + (a \div c).$ $(a \neq 0)$ $(b + c) \div a = (b \div a) + (c \div a).$ $(a \neq 0)$ $a \div (b - c) \neq (a \div b) - (a \div c).$ $(a \neq 0)$ $(b - c) \div a = (b \div a) - (c \div a).$ $(a \neq 0)$

7-8 Chapter Test

Indicate whether each statement is true or false. If false, tell why.

1. The negative of $^-3$ is 3.

2. Zero is its own additive inverse.

3. The set of integers is closed under the operation of division.

4. An addition sentence related to $6 - {}^-8 = 14$ is $6 + 8 = 14$.

5. The set of integers consists of the natural numbers and their opposites.

6. The sum of an integer and its additive inverse is 0.

7. The statement $^-4 + {}^-6 = {}^-10$ should be read "minus four plus minus six is equal to minus ten."

8. The inverse element for multiplication of integers is one.

9. The sum of two negative integers is a negative integer.

10. The product of two negative integers is a negative integer.

Select the best possible answer.

11. Which of these represents the set of nonnegative integers?
    (a) $\{1, 2, 3, 4, 5, 6, 7, 8, 9\}$ (b) $\{0, 1, 2, 3, 4, 5, 6, 7, 8, 9\}$
    (c) $\{1, 2, 3, \ldots\}$ (d) $\{0, 1, 2, 3, \ldots\}$
    (e) None of these

12. Why do we need to extend our systems of numbers to include the integers?
    (a) To provide solutions for all division problems involving whole numbers.
    (b) To provide solutions for all subtraction problems involving whole numbers.
    (c) To provide solutions for all multiplication problems involving whole numbers.
    (d) To provide solutions for all addition problems involving whole numbers.
    (e) To provide solutions for all problems involving both multiplication and addition of whole numbers.

13. Which of these is a true statement?
    (a) $6 + {}^-2 = 8$ (b) $6 + {}^-2 = {}^-8$
    (c) $6 + {}^-2 = {}^-4$ (d) $6 + {}^-2 = 4$
    (e) $6 + {}^-2$ has no solution

14. Which of these is a true statement?
    (a) $^-7 \times {}^-8 = {}^-56$ (b) $^-7 \times 8 = {}^-56$
    (c) $7 \times 8 = {}^-56$ (d) $^-56 \div {}^-8 = {}^-7$
    (e) $^-56 \div 7 = 8$

15. Which set of numbers is closed with respect to subtraction?
    (a) Whole numbers   (b) Negative integers
    (c) Positive integers   (d) Nonnegative integers
    (e) Even integers

chapter 8

Rational Numbers: Addition and Subtraction

Quotients $c \div b$, where c and b are whole numbers and $b \neq 0$, have been defined as $c \div b = a$, where $a \times b = c$. We have mentioned that the quotient a does not always exist in the set of whole numbers. For example, if $c = 4$ and $b = 5$, then there is no whole number a such that $a \times 5 = 4$. Therefore, we must construct an extension of our number system, if each equation of the form $a \times b = c$ is to have a unique solution when $b \neq 0$. In other words, we must try to develop a set of numbers that is closed under the operation of division by any number other than zero.

8-1 The Set of Rational Numbers

The set of **rational numbers** consists of all numbers that may be expressed in the form $\frac{a}{b}$, where the *numerator* a and the *denominator* b are integers and $b \neq 0$. This set of numbers is constructed in order that the operation of division (except division by zero) may always be possible. Note that $\frac{a}{b}$ means $a \div b$.

We should observe that the set of natural numbers, the set of whole numbers, and the set of integers may be considered as subsets of the set of rational numbers. For example, the natural number 6 may be expressed in the form $\frac{6}{1}$ and thus is a rational number, and, for any natural number n, $n = \frac{n}{1}$. The integer $^{-}7$ may be expressed as $\frac{-7}{1}$ and thus is a rational number, and, for any integer k, $k = \frac{k}{1}$. We should also note that $\frac{k}{k} = 1$ for any integer $k \neq 0$.

The set of rational numbers consists of two major subsets:

(1) The set of *nonnegative rational numbers,* such as

$$\frac{0}{2}, \frac{4}{9}, \frac{1}{2}, \frac{3}{5}, \frac{7}{7}, \frac{9}{1}, \frac{11}{4}.$$

(2) The set of *negative rational numbers,* such as

$$\frac{-4}{3}, \frac{-7}{8}, \frac{-8}{1}, \frac{-1}{10}, \frac{5}{-9}, \frac{-15}{3}, \frac{11}{-12}.$$

The nonnegative rational numbers consist of zero and all the positive rational numbers. On the number line these numbers would generally be graphed at zero or to the right of zero. The negative rational numbers would be graphed to the left of zero.

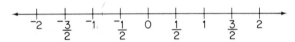

Exercises

1. State whether or not each numeral represents a rational number.
   (a) 5   (b) 0

   (c) $\frac{6}{4}$ (d) $\frac{2}{0}$

   (e) $\frac{0}{9}$ (f) $\frac{-1}{7}$

   (g) $\frac{-3}{-8}$ (h) $^{-}12$

   (i) $\frac{3}{11}$ (j) $\frac{11}{3}$

(k) $\dfrac{9}{-10}$ (l) $\dfrac{27}{4}$

(m) $6 \div 3$ (n) $5 \div 13$

(o) $0 \div 3$ (p) $12 \div 0$ ∞

2. Indicate whether each statement is true or false. If false, tell why.

 (a) The set of odd whole numbers is a subset of the set of rational numbers.

 (b) The set of nonnegative integers is a subset of the set of rational numbers.

 (c) For any integer a, $\dfrac{a}{a} = 1$. $a \neq 0$

 (d) The union of the set of positive rational numbers with the set of negative rational numbers is the set of rational numbers.

8-2 Fractional Numbers

The set of rational numbers also includes many numbers that are not natural numbers, whole numbers, or integers. Some examples of these are

$$\frac{1}{2}, \frac{2}{3}, \frac{3}{4}, \frac{9}{7}, \frac{10}{6}, \frac{76}{18}, \frac{29}{35}.$$

We have probably been calling them fractional numbers rather than rational numbers. This is acceptable, since a **fractional number** is a number that may be expressed in the form $\dfrac{c}{d}$, where c and d represent whole numbers (except $d \neq 0$). Thus, all fractional numbers are rational numbers.

It should be emphasized, however, that *not all* rational numbers are fractional numbers. The definition of a fractional number specifies that the numerator and denominator are whole numbers (denominator $\neq 0$) and some rational numbers do not satisfy this definition. For example,

$$\frac{-1}{2}, \frac{7}{-8}, \quad \text{and} \quad \frac{-5}{-9}$$

are rational numbers that are not fractional numbers.

A fractional number is a number that is represented by a "fractional numeral" (also referred to as a "fraction"). In the past, "fraction" has been used to refer to both the number and the numeral representing the concept. To avoid confusion, we shall use the word *fraction* only when referring to the numeral. We shall use the term *fractional number* to discuss the idea of number.

The term **like fractions** refers to fractions that have identical numerals representing denominators. Likeness is a property of numerals and not a property of numbers. For example, $\frac{5}{9}$ and $\frac{7}{9}$ are like fractions, while $\frac{5}{9}$ and $\frac{10}{18}$ are unlike fractions (even though they represent the same number).

There are several techniques for presenting the idea of a fractional number. One technique is to show parts of a geometric region that has been subdivided into parts of the same size and shape. For example, this rectangular region has been subdivided into three parts of the same size and shape. One of the parts is shaded. Thus we write the fraction $\frac{1}{3}$ to mean that one of the three equivalent parts is shaded.

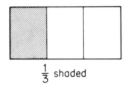

$\frac{1}{3}$ shaded

Some of the other geometric regions which may be used to illustrate fractional numbers are shown in the following pictures.

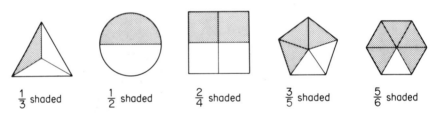

$\frac{1}{3}$ shaded $\frac{1}{2}$ shaded $\frac{2}{4}$ shaded $\frac{3}{5}$ shaded $\frac{5}{6}$ shaded

The number line may also be used to illustrate the meaning of a fractional number. For example, the unit segment from 0 to 1 may be subdivided into 3 parts of equal length. Each part is then $\frac{1}{3}$ of the unit segment as shown on this number line.

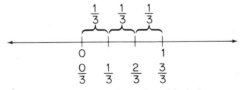

We may develop a set of like fractions for thirds by extending the number line and marking each additional unit segment into 3 parts of equal length.

It should be observed from this number line that, in some instances, two numbers are graphed at the same point. In other words, we have two names for the same number where $0 = \frac{0}{3}$, $1 = \frac{3}{3}$, $2 = \frac{6}{3}$, $3 = \frac{9}{3}$, and $4 = \frac{12}{3}$.

Another technique for presenting fractional numbers is to consider subsets of a set of objects. For example, in this set of 6 stars, there is one white star and we say that $\frac{1}{6}$ of the stars are white. What fractional number of the stars are not white?

$\frac{1}{6}$ white $\frac{\triangle}{\square}$ not white

In this set of 8 dots, we have 4 black dots and we say that $\frac{4}{8}$ of the dots are black. What fractional number of the dots are not black?

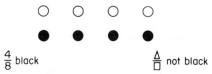

$\frac{4}{8}$ black $\frac{\triangle}{\square}$ not black

It is also possible to think of these 8 dots in another way. Suppose we form two equivalent subsets. We then have 1 set of black dots out of 2 sets of dots, or $\frac{1}{2}$. What fractional number of the *sets* of dots is not black?

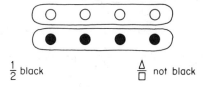

$\frac{1}{2}$ black $\frac{\triangle}{\square}$ not black

We should observe from the two previous illustrations with the set of 8 dots that we again have two names for the same number. Thus $\frac{4}{8} = \frac{1}{2}$.

Exercises

1. Indicate whether the fractions of each pair are like fractions or unlike fractions.

(a) $\dfrac{3}{4}$ and $\dfrac{3}{5}$ (b) $\dfrac{2}{9}$ and $\dfrac{6}{9}$

(c) $\dfrac{1}{2}$ and $\dfrac{3}{2}$ (d) $\dfrac{4}{8}$ and $\dfrac{7}{14}$

2. State whether or not each rational number is a fractional number.

(a) $\dfrac{9}{4}$ (b) $\dfrac{2}{-5}$

(c) $\dfrac{0}{2}$ (d) $\dfrac{0}{-6}$

(e) $\dfrac{-5}{-5}$ (f) $\dfrac{-8}{1}$

(g) 15 (h) $\dfrac{23}{2}$

3. Draw a geometric region to show each of the following.

(a) $\dfrac{2}{3}$ (b) $\dfrac{1}{6}$

(c) $\dfrac{8}{8}$ (d) $\dfrac{0}{5}$

4. For each fractional number, draw a number line and graph the number.

(a) $\dfrac{1}{2}$ (b) $\dfrac{3}{4}$

(c) $\dfrac{2}{5}$ (d) $\dfrac{3}{3}$

5. Draw a number line and mark it to show that $0 = \dfrac{0}{2}$, $1 = \dfrac{2}{2}$, $2 = \dfrac{4}{2}$, and $3 = \dfrac{6}{2}$.

6. For each number, make a set of dots and shade properly to illustrate:

(a) $\dfrac{1}{3}$ (b) $\dfrac{5}{6}$

(c) $\dfrac{4}{7}$ (d) $\dfrac{4}{4}$

7. Show a set of dots in two ways to illustrate:

(a) $\dfrac{1}{3} = \dfrac{2}{6}$ (b) $\dfrac{5}{10} = \dfrac{1}{2}$

(c) $\dfrac{1}{4} = \dfrac{2}{8}$ (d) $\dfrac{1}{3} = \dfrac{3}{9}$

8-3 Equality and Inequality of Rational Numbers

Consider the possible relations between any two fractional numbers $\frac{a}{b}$ and $\frac{c}{d}$; either $\frac{a}{b}$ is equal to $\frac{c}{d}$ $\left(\text{that is, } \frac{a}{b} = \frac{c}{d}\right)$, or $\frac{a}{b}$ is not equal to $\frac{c}{d}$ $\left(\text{that is, } \frac{a}{b} \neq \frac{c}{d}\right)$. Study the following examples of relations between two fractional numbers.

Example 1	$\dfrac{1}{2} = \dfrac{27}{54}$	**Example 2** $\dfrac{1}{3} = \dfrac{2}{6}$

$$1 \times 54 = 2 \times 27 \qquad\qquad 1 \times 6 = 3 \times 2$$
$$54 = 54 \qquad\qquad\qquad 6 = 6$$

Example 3	$\dfrac{2}{3} \neq \dfrac{3}{4}$	**Example 4** $\dfrac{1}{2} \neq \dfrac{9}{16}$

$$2 \times 4 \neq 3 \times 3 \qquad\qquad 1 \times 16 \neq 2 \times 9$$
$$8 \neq 9 \qquad\qquad\qquad 16 \neq 18$$

In Example 1, we observe that $\frac{1}{2} = \frac{27}{54}$ and $1 \times 54 = 2 \times 27$, whereas, in Example 3, $\frac{2}{3} \neq \frac{3}{4}$ and $2 \times 4 \neq 3 \times 3$. In general, we say that $\frac{a}{b} = \frac{c}{d}$ if and only if $a \times d = b \times c$, and it follows that $\frac{a}{b} \neq \frac{c}{d}$ if and only if $a \times d \neq b \times c$. We should note that the products $a \times d$ and $b \times c$ are products of whole numbers.

If $\frac{a}{b} \neq \frac{c}{d}$, we may go one step further and state that $\frac{a}{b} > \frac{c}{d}$ or $\frac{a}{b} < \frac{c}{d}$. Thus, we say that the set of fractional numbers is **ordered**; that is, if we have two fractional numbers, one is either less than, greater than, or equal to the other. Consider the following examples.

Example 5	$\dfrac{2}{3} > \dfrac{7}{11}$	**Example 6** $\dfrac{2}{5} < \dfrac{5}{9}$

$$2 \times 11 > 3 \times 7 \qquad\qquad 2 \times 9 < 5 \times 5$$
$$22 > 21 \qquad\qquad\qquad 18 < 25$$

In Example 5, we observe that $\frac{2}{3} > \frac{7}{11}$ and $2 \times 11 > 3 \times 7$, whereas, in Example 6, $\frac{2}{5} < \frac{5}{9}$ and $2 \times 9 < 5 \times 5$. In general for the fractional numbers $\frac{a}{b}$ and $\frac{c}{d}$, we make the definition

$$\frac{a}{b} > \frac{c}{d}, \quad \text{if and only if } a \times d > b \times c,$$

and it follows that

$$\frac{a}{b} < \frac{c}{d}, \quad \text{if and only if } a \times d < b \times c.$$

Similar relations may be considered for two rational numbers $\frac{m}{n}$ and $\frac{r}{s}$.

Either $\frac{m}{n} = \frac{r}{s}$, or $\frac{m}{n} > \frac{r}{s}$, or $\frac{m}{n} < \frac{r}{s}$. Thus, the set of rational numbers is also ordered.

Consider these examples of relations between two rational numbers.

Example 7

$$\frac{^-3}{4} = \frac{^-9}{12}$$

$$^-3 \times 12 = 4 \times {}^-9$$

$$^-36 = {}^-36$$

Example 8

$$\frac{^-1}{5} < \frac{^-1}{6}$$

$$^-1 \times 6 < 5 \times {}^-1$$

$$^-6 < {}^-5$$

Example 9

$$\frac{^-4}{9} > \frac{^-2}{3}$$

$$^-4 \times 3 > 9 \times {}^-2$$

$$^-12 > {}^-18$$

Example 10

$$\frac{1}{2} > \frac{^-1}{2} \left(\text{Note: } \frac{^-1}{2} = \frac{1}{^-2}\right)$$

$$1 \times 2 > 2 \times {}^-1$$

$2 > {}^-2$. However when $\frac{1}{^-2}$ is used as

a replacement for $\frac{^-1}{2}$

$$\frac{1}{2} \ngtr \frac{1}{^-2}$$

$$1 \times {}^-2 \ngtr 2 \times 1$$

$$^-2 \ngtr 2$$

In Example 7, we observe that $\frac{^-3}{4} = \frac{^-9}{12}$ and $^-3 \times 12 = 4 \times {}^-9$. In Example 8, $\frac{^-1}{5} < \frac{^-1}{6}$ and $^-1 \times 6 < 5 \times {}^-1$, whereas in Example 9, $\frac{^-4}{9} > \frac{^-2}{3}$ and $^-4 \times 3 > 9 \times {}^-2$. In Example 10, we observe a special problem. When we have $\frac{1}{2} > \frac{^-1}{2}$ and $1 \times 2 > 2 \times {}^-1$, the relation seems to be correctly stated; however, when we use $\frac{1}{^-2}$ as a replacement for $\frac{^-1}{2}$, $\frac{1}{2} \ngtr \frac{1}{^-2}$ since $1 \times {}^-2 \ngtr 2 \times 1$. An investigation of other similar examples would lead to the observation that in general the denominator must be greater than 0 in order for the relation to be tested using the technique displayed in Examples 7, 8, 9, and 10.

In general for the rational numbers $\dfrac{e}{f}$ and $\dfrac{g}{h}$, when $f > 0$ and $h > 0$, we have

$$\frac{e}{f} = \frac{g}{h} \qquad \text{if and only if } e \times h = f \times g,$$

$$\frac{e}{f} > \frac{g}{h} \qquad \text{if and only if } e \times h > f \times g, \text{ and}$$

$$\frac{e}{f} < \frac{g}{h} \qquad \text{if and only if } e \times h < f \times g.$$

Exercises

1. Indicate whether each statement is true or false.

   (a) $\dfrac{9}{39} = \dfrac{6}{26}$

   (b) $\dfrac{-31}{79} = \dfrac{-113}{317}$

   (c) $\dfrac{-12}{25} > \dfrac{-11}{23}$

   (d) $\dfrac{213}{789} < \dfrac{97}{353}$

   (e) $\dfrac{17}{41} \neq \dfrac{35}{83}$

   (f) $\dfrac{67}{-5} > \dfrac{141}{-11}$

2. Use one of the symbols, $=$ or $\neq$, to indicate the relation between the rational numbers of each pair.

   (a) $\dfrac{7}{8} \bigcirc \dfrac{21}{24}$

   (b) $\dfrac{8}{15} \bigcirc \dfrac{7}{13}$

   (c) $\dfrac{29}{-8} \bigcirc \dfrac{177}{-51}$

   (d) $\dfrac{9}{14} \bigcirc \dfrac{5}{8}$

3. Use one of the symbols, $=$, $>$, or $<$, to indicate the relation between the rational numbers of each pair.

   (a) $\dfrac{6}{7} \bigcirc \dfrac{5}{8}$

   (b) $\dfrac{18}{19} \bigcirc \dfrac{23}{24}$

   (c) $\dfrac{-5}{4} \bigcirc \dfrac{-7}{6}$

   (d) $\dfrac{2}{3} \bigcirc \dfrac{42}{63}$

4. Indicate whether or not each set is ordered.
   (a) Natural numbers   (b) Whole numbers
   (c) Integers   (d) Fractional numbers
   (e) Rational numbers   (f) Nonnegative rational numbers

8-4 Numerals—the Names for Numbers

A number may have several different names, that is, numerals. Consider the abstract number concept called "threeness." Each of the following is a name for the number three: 3, III, $1 + 1 + 1$, 3×1, and $7 - 4$. Notice that $\frac{3}{1}, \frac{6}{2}, \frac{9}{3}, \frac{12}{4}$, and $\frac{15}{5}$ are also names for the number three. In Section 8-3 we stated that the rational numbers $\frac{a}{b}$ and $\frac{c}{d}$ are equal if and only if $a \times d = b \times c$. This definition leads us to a very useful relation of equality,

$$\frac{a}{b} = \frac{a \times k}{b \times k} \qquad \text{for any integer } k \neq 0,$$

since $a \times b \times k = b \times a \times k$. Thus, $\frac{a \times k}{b \times k}$ represents the same number as $\frac{a}{b}$, and we may find as many names for a rational number as we like. For example,

$$\frac{2}{3} = \frac{2 \times 2}{3 \times 2} = \frac{4}{6},$$

$$\frac{2}{3} = \frac{2 \times 3}{3 \times 3} = \frac{6}{9},$$

$$\frac{2}{3} = \frac{2 \times {}^-4}{3 \times {}^-4} = \frac{{}^-8}{{}^-12},$$

$$\frac{2}{3} = \frac{2 \times 19}{3 \times 19} = \frac{38}{57}.$$

The equality $\frac{a}{b} = \frac{a \times k}{b \times k}$ may be interpreted as meaning that the numerator a and the denominator b of any rational number may both be multiplied by the same integer k, where $k \neq 0$, without any change in the rational number represented. Thus we may express any rational number as a fraction with a positive integer as denominator. For example,

$$\frac{{}^-8}{{}^-12} = \frac{{}^-8 \times {}^-1}{{}^-12 \times {}^-1} = \frac{8}{12}, \qquad \frac{3}{{}^-4} = \frac{3 \times {}^-1}{{}^-4 \times {}^-1} = \frac{{}^-3}{4},$$

and so on. This allows us to apply the general statements on order relation in Section 8-3.

Two fractions that name the same rational number are called **equivalent fractions**. Thus, $\frac{2}{3}, \frac{4}{6}, \frac{6}{9}, \frac{{}^-8}{{}^-12}$, and $\frac{38}{57}$ are called equivalent fractions.

The relation $\frac{a \times k}{b \times k} = \frac{a}{b}$ allows us to express rational numbers in **lower terms** (that is, with smaller numbers for numerator and denominator) by finding common factors k in both the numerator and denominator. For example,

$$\frac{6}{9} = \frac{2 \times 3}{3 \times 3} = \frac{2}{3},$$

$$\frac{48}{64} = \frac{3 \times 16}{4 \times 16} = \frac{3}{4},$$

$$\frac{29}{58} = \frac{1 \times 29}{2 \times 29} = \frac{1}{2},$$

$$\frac{75}{25} = \frac{3 \times 25}{1 \times 25} = \frac{3}{1}.$$

The equality $\dfrac{a \times k}{b \times k} = \dfrac{a}{b}$ means that both the numerator $a \times k$ and the denominator $b \times k$ of any rational number may be divided by the same integer k, where $k \neq 0$, without changing the rational number represented. Consider the following examples.

Example 1 $\dfrac{18}{27} = \dfrac{18 \div 3}{27 \div 3} = \dfrac{6}{9}$ **Example 2** $\dfrac{18}{27} = \dfrac{18 \div 9}{27 \div 9} = \dfrac{2}{3}$

Example 3 $\dfrac{7}{13} = \dfrac{7 \div 1}{13 \div 1} = \dfrac{7}{13}$ **Example 4** $\dfrac{8}{17} = \dfrac{8 \div 1}{17 \div 1} = \dfrac{8}{17}$

In Example 1 the rational number $\dfrac{18}{27}$ has been expressed in *lower terms*.

It is important to note that in Example 2, $\dfrac{18}{27}$ has been expressed in **lowest terms**. A rational number is said to be in lowest terms (also called **simplest form**) when it cannot be reduced further; that is, when the greatest common factor of its numerator and denominator is 1. In Example 3 the rational number $\dfrac{7}{13}$ is in simplest form, since the only common factor of 7 and 13 is 1; thus $\dfrac{7}{13}$ cannot be reduced further. The same situation applies to $\dfrac{8}{17}$ in Example 4.

The greatest common factor has an important application in the renaming of a rational number. If we want to express $\dfrac{39}{429}$ in simplest form, we must find the largest common divisor of the numerator and the denominator. The G.C.F. of 39 and 429 is the divisor we need. The factors of 39 are the set $\{1, 3, 13, 39\}$ and the factors of 429 are the set $\{1, 3, 11, 13, 33, 39, 143, 429\}$. Thus the set of the common factors of 39 and 429 is $\{1, 3, 13, 39\}$, and the G.C.F. of 39 and 429 is 39. Hence, we have

$$\frac{39}{429} = \frac{39 \div 39}{429 \div 39} = \frac{1}{11},$$

where $\frac{1}{11}$ is the simplest form of the rational number $\frac{39}{429}$.

We can show equivalent fractions using the "parts of a region" technique, provided that we use the same size and shape for all the regions involved in the particular comparison. For example, to show $\frac{1}{2} = \frac{2}{4}$, we have

$\frac{1}{2}$ shaded $\frac{2}{4}$ shaded

Number lines may also be used to illustrate equivalent fractions, provided that the unit segment on each number line is the same length. These number lines should help us to see that $\frac{2}{3} = \frac{4}{6}$.

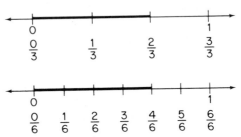

An illustration of equivalent fractions using sets was given in Section 8-3.

Exercises

1. Use the "parts of a region technique" to illustrate:

   (a) $\frac{1}{2} = \frac{3}{6}$ (b) $\frac{3}{4} = \frac{6}{8}$

   (c) $\frac{1}{3} = \frac{3}{9}$

2. Use the "number line technique" to illustrate:

   (a) $\frac{1}{3} = \frac{4}{12}$ (b) $\frac{1}{4} = \frac{2}{8}$

   (c) $\frac{4}{5} = \frac{8}{10}$

3. Use the "subsets of a set technique" to illustrate:

(a) $\dfrac{1}{2} = \dfrac{2}{4}$ (b) $\dfrac{1}{3} = \dfrac{2}{6}$

(c) $\dfrac{3}{4} = \dfrac{6}{8}$

4. Give three different names for each rational number.

(a) 7 (b) $\dfrac{1}{5}$

(c) $\dfrac{-2}{3}$ (d) $\dfrac{0}{4}$

(e) $\dfrac{-9}{-7}$ (f) $\dfrac{125}{75}$

(g) $\dfrac{26}{-4}$ (h) $\dfrac{183}{792}$

(i) $\dfrac{0}{-225}$

5. Replace the variable with an integer to make each statement true.

(a) $\dfrac{6}{7} = \dfrac{6 \times \square}{7 \times \square} = \dfrac{-54}{-63}$ (b) $\dfrac{2}{3} = \dfrac{\square}{78}$

(c) $\dfrac{5}{12} = \dfrac{\square}{144}$ (d) $\dfrac{\square}{6} = \dfrac{-20}{24}$

(e) $\dfrac{0}{3} = \dfrac{\square}{39}$ (f) $\dfrac{8}{9} = \dfrac{\square}{9}$

6. Indicate whether or not each pair of fractions may be called equivalent fractions.

(a) $\dfrac{6}{9}$ and $\dfrac{38}{57}$ (b) $\dfrac{25}{75}$ and $\dfrac{9}{29}$

(c) $\dfrac{0}{19}$ and $\dfrac{0}{-47}$ (d) $\dfrac{-8}{-12}$ and $\dfrac{14}{21}$

(e) $\dfrac{2}{8}$ and $\dfrac{1}{3}$ (f) $\dfrac{117}{13}$ and $\dfrac{863}{96}$

7. Express each rational number in simplest form.

(a) $\dfrac{36}{45}$ (b) $\dfrac{51}{85}$

(c) $\dfrac{-12}{-28}$ (d) $\dfrac{97}{101}$

(e) $\dfrac{0}{78}$ (f) $\dfrac{165}{225}$

8. Find the greatest common factor of the numerator and denominator of each rational number.

(a) $\dfrac{18}{24}$ (b) $\dfrac{-29}{66}$

(c) $\dfrac{72}{135}$ (d) $\dfrac{496}{368}$

(e) $\dfrac{-224}{-784}$ (f) $\dfrac{63}{-119}$

8-5 Addition of Rational Numbers

If two rational numbers are to be added, they are usually represented by like fractions whose denominators are the least common denominator. Finding the least common denominator is a problem of finding the least common multiple of the numbers represented by the denominators. Recall (Section 6-4) that the least common multiple of two or more numbers is the smallest number contained in the sets of multiples of the numbers under consideration.

We may use geometric regions to develop an understanding of the addition of nonnegative rational numbers. For example, $\dfrac{1}{5} + \dfrac{2}{5}$ may be shown as follows:

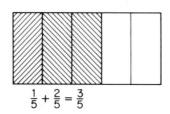

$$\frac{1}{5} + \frac{2}{5} = \frac{3}{5}$$

We observe that $\dfrac{1}{5}$ of the region is shaded in one direction and $\dfrac{2}{5}$ of the region is shaded in a different manner. This gives us a total of $\dfrac{3}{5}$ of the region that is shaded in some manner.

Study these illustrations of addition of nonnegative rational numbers using regions.

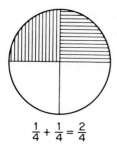

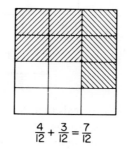

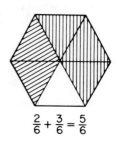

$$\frac{1}{4} + \frac{1}{4} = \frac{2}{4} \qquad\qquad \frac{4}{12} + \frac{3}{12} = \frac{7}{12} \qquad\qquad \frac{2}{6} + \frac{3}{6} = \frac{5}{6}$$

The number line is also helpful in developing an understanding of the addition of rational numbers. To find the sum of $\frac{1}{7}$ and $\frac{3}{7}$ we find a point on the number line that represents $\frac{1}{7}$. We then move from the graph of $\frac{1}{7}$ to the right to a point that represents $\frac{3}{7}$ more. The sum of $\frac{1}{7}$ and $\frac{3}{7}$ may be determined as $\frac{4}{7}$ from the number line; $\frac{1}{7} + \frac{3}{7} = \frac{4}{7}$.

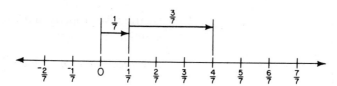

To add $\frac{2}{5}$ and $\frac{4}{5}$ on the number line, we operate in a similar manner as shown in the next figure; $\frac{2}{5} + \frac{4}{5} = \frac{6}{5}$.

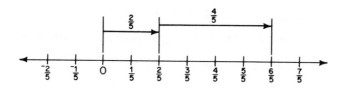

The sum of $\frac{3}{6}$ and $\frac{-2}{6}$ is represented on the next number line; $\frac{3}{6} + \frac{-2}{6} = \frac{1}{6}$.

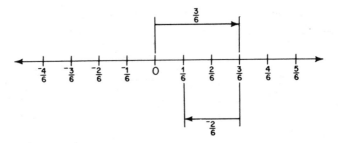

A study of these illustrations reveals that the sum of two rational numbers with the same denominator is the rational number represented by the fraction whose numerator is the sum of the numerators and whose denominator is the common denominator. In general, we write

$$\frac{a}{b} + \frac{c}{b} = \frac{a+c}{b}.$$

Observe that the sum of the numerators, $a + c$, is the sum of two integers.

We may justify this algorithm for the addition of rational numbers $\frac{a}{b}$ and $\frac{c}{b}$ from our knowledge of the operations and properties for integers.

$\frac{a}{b} + \frac{c}{b} = (a \div b) + (c \div b)$ Interpretation of $\frac{m}{n}$ as $m \div n$.

$(a \div b) + (c \div b) = (a + c) \div b$ Distributive property of division over addition from right.

$(a + c) \div b = \frac{a+c}{b}$ Interpretation of $m \div n$ as $\frac{m}{n}$.

Now let us consider the more difficult case of adding rational numbers represented by unlike fractions. For example, add $\frac{1}{2}$ and $\frac{1}{3}$. We cannot add these rational numbers as we did previously, because they have different denominators. We must find the least common multiple of the denominators and then find equivalent fractions with the least common multiple as the denominator for $\frac{1}{2}$ and $\frac{1}{3}$. The L.C.M. of 2 and 3 is 6.

Recall from Section 8-4 that we may develop a set of equivalent fractions for a rational number $\frac{a}{b}$ from the equality relation $\frac{a}{b} = \frac{a \times k}{b \times k}$ for any integer $k \neq 0$. Thus we have

$$\frac{1}{2} = \frac{1 \times 3}{2 \times 3} = \frac{3}{6}, \text{ and}$$

$$\frac{1}{3} = \frac{1 \times 2}{3 \times 2} = \frac{2}{6}.$$

Since $\dfrac{1}{2} = \dfrac{3}{6}$ and $\dfrac{1}{3} = \dfrac{2}{6}$, it follows that

$$\frac{1}{2} + \frac{1}{3} = \frac{3}{6} + \frac{2}{6} = \frac{5}{6}.$$

If we let $\dfrac{a}{b}$ and $\dfrac{c}{d}$ represent any two rational numbers, we may find the sum of $\dfrac{a}{b}$ and $\dfrac{c}{d}$ as follows.

$\dfrac{a}{b} = \dfrac{a \times d}{b \times d} = \dfrac{ad}{bd}$ From the equality relation for rational numbers.

and $\dfrac{c}{d} = \dfrac{c \times b}{d \times b} = \dfrac{cb}{db}$

$\dfrac{cb}{db} = \dfrac{bc}{bd}$ Commutative property for multiplication of integers. Recall that b, c, and d are integers, $b \neq 0$, $d \neq 0$.

$\dfrac{a}{b} + \dfrac{c}{d} = \dfrac{ad}{bd} + \dfrac{bc}{bd} = \dfrac{ad + bc}{bd}$ Definition of addition of rational numbers represented by like fractions.

In general, for rational numbers $\dfrac{a}{b}$ and $\dfrac{c}{d}$, we define addition as

$$\frac{a}{b} + \frac{c}{d} = \frac{ad + bc}{bd}.$$

Note that the numerator may be represented as an integer $ad + bc$, where ad and bc are integers, and the denominator may be represented as an integer bd, where b and d are integers different from zero. Thus, the sum of any two rational numbers is a rational number. This definition of the addition of rational numbers includes the addition of an integer and a rational number, since $\dfrac{a}{b} + \dfrac{c}{d}$ may represent $\dfrac{a}{1} + \dfrac{c}{d}$, where $\dfrac{a}{1}$ is an integer. For example,

$$6 + \frac{2}{3} = \frac{6}{1} + \frac{2}{3} = \frac{18 + 2}{3} = \frac{20}{3} = 6\frac{2}{3}.$$

Thus the sum of an integer and a rational number such as $6 + \dfrac{2}{3}$ is often written $6\dfrac{2}{3}$, and the numeral is called a **mixed numeral**, since the numeral includes the representation of both an integer and a fractional number. All mixed numerals represent rational numbers.

Exercises

1. Use shadings of a geometric region to represent each sum.

    (a) $\dfrac{1}{3} + \dfrac{1}{3}$　　　　　　**(b)** $\dfrac{2}{4} + \dfrac{2}{4}$　　　　　　**(c)** $\dfrac{3}{10} + \dfrac{4}{10}$

2. Illustrate each sum on a number line.

    (a) $\dfrac{4}{7} + \dfrac{1}{7}$　　　　　　**(b)** $\dfrac{5}{8} + \dfrac{5}{8}$　　　　　　**(c)** $\dfrac{5}{6} + \dfrac{^-4}{6}$

3. Find each sum.

    (a) $\dfrac{3}{7} + \dfrac{2}{7}$　　　　　　**(b)** $\dfrac{1}{8} + \dfrac{4}{8}$　　　　　　**(c)** $\dfrac{4}{7} + \dfrac{9}{7}$

    (d) $\dfrac{6}{13} + \dfrac{^-5}{13}$　　　　　**(e)** $\dfrac{^-1}{4} + \dfrac{^-2}{4}$　　　　　**(f)** $\dfrac{3}{17} + \dfrac{5}{17} + \dfrac{6}{17}$

4. What is the lowest common denominator of:

    (a) $\dfrac{5}{6}$ and $\dfrac{7}{8}$?　　　**(b)** $\dfrac{7}{12}$ and $\dfrac{2}{3}$?　　　**(c)** $\dfrac{8}{15}$ and $\dfrac{4}{9}$?

5. Find each sum.

    (a) $\dfrac{5}{6} + \dfrac{7}{8}$　　　　　　**(b)** $\dfrac{7}{12} + \dfrac{2}{3}$　　　　　　**(c)** $\dfrac{8}{15} + \dfrac{4}{9}$

    (d) $\dfrac{2}{3} + \dfrac{4}{7}$　　　　　　**(e)** $\dfrac{4}{9} + \dfrac{5}{6}$　　　　　　**(f)** $\dfrac{11}{5} + \dfrac{10}{3}$

    (g) $\dfrac{^-13}{16} + \dfrac{7}{12}$　　　**(h)** $7 + \dfrac{5}{9}$　　　**(i)** $\dfrac{7}{10} + \dfrac{4}{15}$

6. Write a mixed numeral for each of the following.

    (a) $5 + \dfrac{1}{4}$　　　　　　**(b)** $25 + \dfrac{2}{5}$　　　　　　**(c)** $^-3 + \dfrac{^-4}{7}$

8-6 Properties of Addition of Rational Numbers

The following properties of addition for rational numbers are illustrated and assumed. In more advanced courses, these properties may be proved using our definitions and the properties of addition and multiplication of integers.

Closure and uniqueness properties　The property of *closure* applied to the addition of rational numbers means that the sum of any two rational numbers is also a rational number. The property of *uniqueness* indicates that there is only one possible sum when two rational numbers are added. For example, $\frac{1}{5} + \frac{2}{5} = \frac{3}{5}$ illustrates that the addends and the sum are elements of the set of rational numbers. We should also note that $\frac{3}{5}$ is the only possible sum when $\frac{1}{5}$ and $\frac{2}{5}$ are added; that is, the number $\frac{3}{5}$ is the unique sum of $\frac{1}{5}$ and $\frac{2}{5}$. Notice that the number $\frac{3}{5}$ may also be expressed as $\frac{6}{10}, \frac{9}{15}, \frac{12}{20}$, and so on.

In general, for the rational numbers $\frac{a}{b}$ and $\frac{c}{d}$, as explained in Section 8-5,

$$\frac{a}{b} + \frac{c}{d} = \frac{ad + bc}{bd},$$

where the rational number $\frac{ad + bc}{bd}$ is the unique sum of $\frac{a}{b}$ and $\frac{c}{d}$.

Commutative property   The order of the addition of two rational numbers does not affect the sum. This may be illustrated by the case of $\frac{2}{7}$ and $\frac{4}{7}$. Since

$$\frac{2}{7} + \frac{4}{7} = \frac{6}{7} \quad \text{and} \quad \frac{4}{7} + \frac{2}{7} = \frac{6}{7},$$

we have

$$\frac{2}{7} + \frac{4}{7} = \frac{4}{7} + \frac{2}{7}.$$

Addition of rational numbers is *commutative* and, for any two rational numbers $\frac{a}{b}$ and $\frac{c}{d}$,

$$\frac{a}{b} + \frac{c}{d} = \frac{c}{d} + \frac{a}{b}.$$

Associative property   Three or more rational numbers may be grouped differently for addition, without changing their order, and the sum will not be affected. For example,

$$\frac{2}{9} + \frac{4}{9} + \frac{5}{9}$$

may be grouped for addition as

$$\left(\frac{2}{9} + \frac{4}{9}\right) + \frac{5}{9} \quad \text{or} \quad \frac{2}{9} + \left(\frac{4}{9} + \frac{5}{9}\right).$$

Since

$$\left(\frac{2}{9} + \frac{4}{9}\right) + \frac{5}{9} = \frac{6}{9} + \frac{5}{9} = \frac{11}{9}$$

and

$$\frac{2}{9} + \left(\frac{4}{9} + \frac{5}{9}\right) = \frac{2}{9} + \frac{9}{9} = \frac{11}{9},$$

we have

$$\left(\frac{2}{9} + \frac{4}{9}\right) + \frac{5}{9} = \frac{2}{9} + \left(\frac{4}{9} + \frac{5}{9}\right).$$

In general, for the rational numbers $\frac{a}{b}, \frac{c}{d},$ and $\frac{e}{f},$

$$\left(\frac{a}{b} + \frac{c}{d}\right) + \frac{e}{f} = \frac{a}{b} + \left(\frac{c}{d} + \frac{e}{f}\right),$$

and we say that the addition of rational numbers is *associative*.

Identity element   The sum of any rational number and the number zero in either order is always the original rational number. For example,

$$\frac{1}{2} + 0 = 0 + \frac{1}{2} = \frac{1}{2}, \quad \frac{3}{4} + 0 = 0 + \frac{3}{4} = \frac{3}{4}, \quad \frac{^-7}{9} + 0 = 0 + \frac{^-7}{9} = \frac{^-7}{9},$$

and so on. Thus, $\frac{a}{b} + 0 = 0 + \frac{a}{b} = \frac{a}{b}$ for any rational number $\frac{a}{b},$ and 0 is called the *identity element for addition* of rational numbers (also referred to as the *additive identity*).

Additive inverse property   Two rational numbers are called *additive inverses* of each other if their sum is zero. For example $\frac{1}{2}$ and $\frac{^-1}{2}$ are additive inverses since $\frac{1}{2} + \frac{^-1}{2} = \frac{1 + ^-1}{2} = \frac{0}{2} = 0.$

Every rational number $\frac{a}{b}$ has an additive inverse $\frac{^-a}{b},$ since the positive rational numbers may be placed in one-to-one correspondence with the negative rational numbers and since $0 + 0 = 0.$ In general, for every rational number $\frac{a}{b},$ there exists a negative of $\frac{a}{b}$ such that

$$\frac{a}{b} + \frac{^-a}{b} = \frac{^-a}{b} + \frac{a}{b} = 0.$$

Exercises

1. Find each sum.

    (a) $\frac{6}{7} + \frac{3}{91}$

    (b) $\frac{3}{91} + \frac{6}{7}$

2. What may be concluded from a comparison of the two answers in Exercise 1?

3. What property of addition for rational numbers is suggested in Exercises 1 and 2?

4. (a) Does $\dfrac{5}{6} + \dfrac{7}{8}$ equal a rational number?

   (b) How many different rational numbers can you find for the sum of $\dfrac{5}{6}$ and $\dfrac{7}{8}$?

   (c) What properties of addition for rational numbers are suggested by your answers to parts (a) and (b)?

5. What is the additive identity for the rational number system?

6. Find each sum.

   (a) $\left(\dfrac{1}{2} + \dfrac{1}{3}\right) + \dfrac{1}{4}$

   (b) $\dfrac{1}{2} + \left(\dfrac{1}{3} + \dfrac{1}{4}\right)$

7. What may be concluded from a comparison of the two answers in Exercise 6?

8. What property of addition for rational numbers is suggested in Exercises 6 and 7?

9. Compute the following.

   (a) $\dfrac{2}{3} + 1$

   (b) $\dfrac{3}{4} + 0$

   (c) $\dfrac{9}{17} + \dfrac{-9}{17}$

   (d) $\left(\dfrac{5}{17} + \dfrac{4}{17}\right) + \dfrac{7}{17}$

   (e) $\dfrac{7}{17} + \left(\dfrac{5}{17} + \dfrac{4}{17}\right)$

   (f) $\dfrac{7}{9} + \dfrac{0}{9}$

10. What property of addition of rational numbers, if any, is employed in
    (a) Exercise 9(a)?   (b) Exercise 9(b)?
    (c) Exercise 9(c)?   (d) Exercise 9(d)?
    (e) Exercise 9(e)?   (f) Exercise 9(f)?

11. What property of addition of rational numbers is illustrated by a comparison of Exercise 9(d) and Exercise 9(e)?

12. State the additive inverse of each rational number.

    (a) $\dfrac{-5}{8}$

    (b) $\dfrac{2}{9}$

    (c) $\dfrac{0}{4}$

    (d) $\dfrac{-11}{9}$

    (e) $\dfrac{-10}{-10}$

    (f) $\dfrac{23}{19}$

8-7 Subtraction of Rational Numbers

The negative of the rational number $\frac{3}{4}$ may be expressed in any one of these ways:

$$-\frac{3}{4}, \frac{^-3}{4}, \frac{3}{^-4}.$$

The form $\frac{^-3}{4}$ is generally preferred for the sake of consistency and convenience.

Two rational numbers are called *additive inverses* of each other if their sum is zero. In the example

$$\frac{3}{4} + \frac{^-3}{4} = \frac{3 + {}^-3}{4} = \frac{0}{4} = 0,$$

$\frac{^-3}{4}$ is the additive inverse of $\frac{3}{4}$, and $\frac{3}{4}$ is the additive inverse of $\frac{^-3}{4}$. Recall (Section 7-2) that the additive inverse of a number is also called *the negative of* the number. Thus $\frac{^-3}{4}$ is the negative of $\frac{3}{4}$, and $\frac{3}{4}$ is the negative of $\frac{^-3}{4}$.

Every rational number $\frac{a}{b}$ has an additive inverse $\frac{^-a}{b}$. Thus, as in the case of integers, we may define subtraction of rational numbers in terms of an equivalent addition statement. In general, for rational numbers $\frac{a}{b}$ and $\frac{c}{d}$,

$$\frac{a}{b} - \frac{c}{d} = \frac{a}{b} + \frac{^-c}{d},$$

and we recall (Section 8-5) that

$$\frac{a}{b} + \frac{^-c}{d} = \frac{ad + b(^-c)}{bd}.$$

Note that the products ad, $b(^-c)$, and bd are products of integers.

Perhaps the following examples will clarify the definition of subtraction of rational numbers.

Example 1 Find the difference $\frac{7}{9} - \frac{3}{6}$.

$$\frac{7}{9} - \frac{3}{6} = \frac{7}{9} + \frac{^-3}{6} \qquad \left(\text{Note that } \frac{^-3}{6} \text{ is the additive inverse of } \frac{3}{6}.\right)$$

$$= \frac{(7 \times 6) + (9 \times {}^-3)}{9 \times 6}$$

$$= \frac{42 + {}^-27}{54}$$

$$= \frac{15}{54}.$$

If the simplified form of $\frac{15}{54}$ is desired, we may divide both the numerator

15 and the denominator 54 by 3; thus,

$$\frac{15}{54} = \frac{15 \div 3}{54 \div 3} = \frac{5}{18}.$$

Example 2   Find the difference $\frac{2}{3} - \frac{{}^-3}{4}$.

$$\frac{2}{3} - \frac{{}^-3}{4} = \frac{2}{3} + \frac{3}{4} \qquad \left(\text{Note that } \frac{3}{4} \text{ is the additive inverse of } \frac{{}^-3}{4}. \right)$$

$$= \frac{(2 \times 4) + (3 \times 3)}{3 \times 4}$$

$$= \frac{8 + 9}{12}$$

$$= \frac{17}{12}.$$

Example 3   Find the difference $\frac{{}^-4}{5} - \frac{3}{7}$.

$$\frac{{}^-4}{5} - \frac{3}{7} = \frac{{}^-4}{5} + \frac{{}^-3}{7} \qquad \left(\text{Note that } \frac{{}^-3}{7} \text{ is the additive inverse of } \frac{3}{7}. \right)$$

$$= \frac{({}^-4 \times 7) + (5 \times {}^-3)}{5 \times 7}$$

$$= \frac{{}^-28 + {}^-15}{35}$$

$$= \frac{{}^-43}{35}.$$

Example 4   Find the difference $\frac{{}^-1}{2} - \frac{{}^-3}{11}$.

$$\frac{{}^-1}{2} - \frac{{}^-3}{11} = \frac{{}^-1}{2} + \frac{3}{11} \qquad \left(\text{Note that } \frac{3}{11} \text{ is the additive inverse of } \frac{{}^-3}{11}. \right)$$

$$= \frac{({}^-1 \times 11) + (2 \times 3)}{2 \times 11}$$

$$= \frac{-11 + 6}{22}$$

$$= \frac{-5}{22}.$$

We may also use a number line to represent the subtraction of rational numbers. The difference $\frac{1}{3} - \frac{2}{3}$ is pictured on the next number line.

$$\frac{1}{3} - \frac{2}{3} = \frac{1}{3} + \frac{-2}{3} = \frac{1 + {}^{-}2}{3} = \frac{-1}{3}.$$

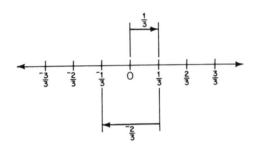

In general, the difference between two rational numbers with the same denominator is

$$\frac{a}{b} - \frac{c}{b} = \frac{a}{b} + \frac{-c}{b} = \frac{a + {}^{-}c}{b}.$$

Exercises

1. Represent each difference on a number line.

   (a) $\dfrac{6}{7} - \dfrac{2}{7}$ (b) $\dfrac{-2}{5} - \dfrac{-4}{5}$

   (c) $\dfrac{9}{4} - \dfrac{5}{4}$ (d) $\dfrac{-1}{3} - \dfrac{-1}{3}$

2. Find each difference.

   (a) $\dfrac{12}{7} - \dfrac{9}{7}$ (b) $\dfrac{11}{13} - \dfrac{5}{13}$

   (c) $\dfrac{7}{9} - \dfrac{7}{9}$ (d) $\dfrac{3}{6} - \dfrac{4}{6}$

   (e) $\dfrac{4}{5} - \dfrac{1}{5}$ (f) $\dfrac{5}{8} - \dfrac{-3}{8}$

   (g) $\dfrac{-12}{7} - \dfrac{9}{7}$ (h) $\dfrac{-6}{10} - \dfrac{-3}{10}$

3. Find each difference.

(a) $\dfrac{4}{12} - \dfrac{1}{3}$ (b) $0 - \dfrac{3}{8}$

(c) $\dfrac{13}{4} - 0$ (d) $\dfrac{^-12}{9} - \dfrac{^-4}{3}$

(e) $\dfrac{2}{15} - \dfrac{4}{5}$ (f) $\begin{array}{r} \dfrac{6}{13} \\ -\dfrac{2}{7} \\ \hline \end{array}$

(g) $\begin{array}{r} \dfrac{9}{10} \\ -\dfrac{14}{15} \\ \hline \end{array}$ (h) $\begin{array}{r} \dfrac{1}{12} \\ -\dfrac{3}{32} \\ \hline \end{array}$

8-8 Properties of Subtraction of Rational Numbers

Recall (Section 4-10) that a basic property of mathematics does not hold for a given operation on the elements of a set of numbers if it leads us to a false conclusion in at least one case.

Closure and uniqueness properties The difference between any two given rational numbers is always a *unique* rational number. Thus we may say that the set of rational numbers is *closed* with respect to the operation of subtraction. For any rational numbers $\dfrac{a}{b}$ and $\dfrac{c}{d}$,

$$\frac{a}{b} - \frac{c}{d} = \frac{a}{b} + \frac{^-c}{d} = \frac{ad + b(^-c)}{bd} = \frac{ad + {^-bc}}{bd} = \frac{ad - bc}{bd},$$

where $\dfrac{ad - bc}{bd}$ is the unique rational number called the difference $\dfrac{a}{b} - \dfrac{c}{d}$.

Observe that ad and bc are integers and, consequently, the subtraction of rational numbers behaves like the subtraction of integers.

Commutative property Since the commutative property does not hold for the subtraction of integers, and since rational numbers with common denominators behave like their numerators, we may assume that the subtraction of rational numbers is not commutative. For example,

$$\frac{3}{4} - \frac{1}{4} \neq \frac{1}{4} - \frac{3}{4}$$

because

$$\frac{3-1}{4} \neq \frac{1-3}{4}; \quad \text{that is,} \quad \frac{2}{4} \neq \frac{^-2}{4}.$$

Associative property   Since the associative property does not hold for the subtraction of integers, we may assume that it does not hold for the subtraction of rational numbers. For example,

$$\left(\frac{7}{8}-\frac{5}{8}\right)-\frac{1}{8}\neq\frac{7}{8}-\left(\frac{5}{8}-\frac{1}{8}\right)$$

because

$$\left(\frac{7-5}{8}\right)-\frac{1}{8}\neq\frac{7}{8}-\left(\frac{5-1}{8}\right);$$

that is,

$$\frac{2}{8}-\frac{1}{8}\neq\frac{7}{8}-\frac{4}{8}\quad\text{and}\quad\frac{1}{8}\neq\frac{3}{8}.$$

Identity element   Zero subtracted from any rational number yields the given rational number as the difference. For example, $\frac{2}{3}-0=\frac{2}{3}$; however, since subtraction of rational numbers is not commutative, $0-\frac{2}{3}\neq\frac{2}{3}$. In general, $\frac{a}{b}-0=\frac{a}{b}$ for each rational number, but $0-\frac{a}{b}\neq\frac{a}{b}$. Thus, 0 is a *right identity* that is not a left identity for the subtraction of rational numbers.

Inverse operations   The inverse relationship of addition and subtraction may be extended to the set of rational numbers. For example,

$$\frac{3}{4}+\frac{1}{4}=\frac{3+1}{4}=\frac{4}{4},$$

and

$$\frac{4}{4}-\frac{1}{4}=\frac{4+{}^-1}{4}=\frac{3}{4}.$$

Thus the operation of adding $\frac{1}{4}$ to $\frac{3}{4}$ is undone by subtracting $\frac{1}{4}$ from the sum $\frac{4}{4}$. In general, for the rational numbers $\frac{a}{b}$ and $\frac{c}{d}$,

$$\frac{a}{b}+\frac{c}{d}=\frac{ad+bc}{bd}$$

if and only if

$$\frac{ad+bc}{bd}-\frac{c}{d}=\frac{a}{b}.$$

Similarly,

$$\frac{a}{b}-\frac{c}{d}=\frac{ad-bc}{bd}$$

if and only if

$$\frac{ad-bc}{bd}+\frac{c}{d}=\frac{a}{b}.$$

Exercises

1. (a) What is n if $n = \dfrac{13}{14} - \dfrac{8}{14}$?

   (b) What is n if $n = \dfrac{8}{14} - \dfrac{13}{14}$?

   (c) Does $\dfrac{13}{14} - \dfrac{8}{14} = \dfrac{8}{14} - \dfrac{13}{14}$?

   (d) Does the commutative property hold for subtraction of rational numbers?

2. (a) What is n if $n = \left(\dfrac{9}{11} - \dfrac{5}{11}\right) - \dfrac{3}{11}$?

   (b) What is n if $n = \dfrac{9}{11} - \left(\dfrac{5}{11} - \dfrac{3}{11}\right)$?

   (c) Does $\left(\dfrac{9}{11} - \dfrac{5}{11}\right) - \dfrac{3}{11} = \dfrac{9}{11} - \left(\dfrac{5}{11} - \dfrac{3}{11}\right)$?

   (d) Does the associative property hold for subtraction of rational numbers?

3. Find each difference.

   (a) $\dfrac{7}{8} - 0$

   (b) $\dfrac{6}{5} - \dfrac{0}{5}$

   (c) $\dfrac{3}{7} - \dfrac{0}{8}$

   (d) $\dfrac{-4}{9} - 0$

4. What do you observe in each part of Exercise 3 about the difference $\dfrac{a}{b} - 0$ for any rational number $\dfrac{a}{b}$?

5. Find each difference.

   (a) $\dfrac{9}{13} - \dfrac{9}{13}$

   (b) $\dfrac{7}{8} - \dfrac{7}{8}$

   (c) $\dfrac{-2}{3} - \dfrac{-2}{3}$

   (d) $\dfrac{24}{8} - \dfrac{12}{4}$

6. What do you observe in each part of Exercise 5 about the difference $\dfrac{a}{b} - \dfrac{a}{b}$ for any rational number $\dfrac{a}{b}$?

7. Indicate whether each statement is true or false.

   (a) $\dfrac{5}{16} - 2$ is a rational number

(b) $\left(\dfrac{2}{3} - \dfrac{1}{4}\right) - \dfrac{1}{8} > \dfrac{2}{3} - \left(\dfrac{1}{4} - \dfrac{1}{8}\right)$

(c) $\dfrac{5}{13} - \dfrac{13}{5} = \dfrac{13}{5} - \dfrac{5}{13}$

(d) $0 - \dfrac{3}{7} = 0$

8. Tell whether or not each set of numbers is closed with respect to subtraction. If the set is not closed, give one counterexample.
   (a) Fractional numbers
   (b) Negative rational numbers
   (c) Positive rational numbers
   (d) Rational numbers
   (e) Nonnegative rational numbers

8-9 Mixed Numerals and Computation

In Section 8-5 we learned that numerals such as $6\dfrac{2}{3}$, $2\dfrac{5}{8}$, and $^-5\dfrac{7}{4}$ are called *mixed numerals*, and that all such numerals represent rational numbers. We have developed computational procedures for adding and subtracting rational numbers. Now we may apply these same computational procedures to rational numbers represented by mixed numerals, if we change the mixed numerals to fractional form.

The definition for the addition of two rational numbers,

$$\frac{a}{b} + \frac{c}{d} = \frac{ad + bc}{bd},$$

provides us with a procedure for working with mixed numerals such as $2\dfrac{1}{3}$. The numeral $2\dfrac{1}{3}$ (read "two and one-third") actually represents the sum $2 + \dfrac{1}{3}$. Since $\dfrac{2}{1}$ is another name for the integer 2,

$$2 + \frac{1}{3} = \frac{2}{1} + \frac{1}{3} = \frac{6 + 1}{3} = \frac{7}{3}.$$

Using a similar procedure, we may represent any mixed numeral in the form $\dfrac{a}{b}$.

A second procedure for addition of rational numbers represented by mixed numerals is to regroup, using the commutative and associative properties, in order to work with the integers and rational numbers separately. Consider these examples illustrating several procedures for computation with rational numbers represented by mixed numerals. Explanations for each step are given in Examples 1(a) and 1(b). Try to provide the necessary explanations for the steps in the other examples.

Example 1   Find the sum $1\frac{1}{3} + 2\frac{1}{4}$.

(a) $1\frac{1}{3} + 2\frac{1}{4} = \left(1 + \frac{1}{3}\right) + \left(2 + \frac{1}{4}\right)$ Definition of mixed numeral.

$= \left(\frac{1}{1} + \frac{1}{3}\right) + \left(\frac{2}{1} + \frac{1}{4}\right)$ Rename integer a as $\frac{a}{1}$.

$= \frac{3+1}{3} + \frac{8+1}{4}$ Definition of addition of rational numbers.

$= \frac{4}{3} + \frac{9}{4}$ Addition of integers in each numerator.

$= \frac{16 + 27}{12}$ Definition of addition of rational numbers.

$= \frac{43}{12}$ Addition of integers in numerator.

$= \frac{36 + 7}{12}$ Rename integer 43 as $36 + 7$.

$= \frac{36}{12} + \frac{7}{12}$ Definition of addition of rational numbers with common denominators.

$= 3 + \frac{7}{12}$ Rename $\frac{36}{12}$ as 3.

$= 3\frac{7}{12}$.   Definition of mixed numeral.

(b) $1\frac{1}{3} + 2\frac{1}{4} = \left(1 + \frac{1}{3}\right) + \left(2 + \frac{1}{4}\right)$ Definition of mixed numeral.

$= (1 + 2) + \left(\frac{1}{3} + \frac{1}{4}\right)$ Commutative and associative properties for addition of rational numbers.

$= 3 + \frac{4 + 3}{12}$ Definitions of addition of rational numbers.

$= 3 + \frac{7}{12}$ Addition of integers in numerator.

$= 3\frac{7}{12}$.   Definition of a mixed numeral.

(c) $1\dfrac{1}{3} = 1\dfrac{4}{12}$

$+\,2\dfrac{1}{4} = 2\dfrac{3}{12}$

$\qquad\qquad 3\dfrac{7}{12}.$

The steps outlined in parts (a) and (b) of this example provide the basis for the algorithm presented to the left.

Example 2   Find the sum $6\dfrac{5}{7} + 2\dfrac{1}{2}.$

(a) $6\dfrac{5}{7} + 2\dfrac{1}{2} = \left(6 + \dfrac{5}{7}\right) + \left(2 + \dfrac{1}{2}\right)$

$\qquad\qquad = \left(\dfrac{6}{1} + \dfrac{5}{7}\right) + \left(\dfrac{2}{1} + \dfrac{1}{2}\right)$

$\qquad\qquad = \dfrac{42 + 5}{7} + \dfrac{4 + 1}{2}$

$\qquad\qquad = \dfrac{47}{7} + \dfrac{5}{2}$

$\qquad\qquad = \dfrac{94 + 35}{14}$

$\qquad\qquad = \dfrac{129}{14}$

$\qquad\qquad = \dfrac{126 + 3}{14}$

$\qquad\qquad = \dfrac{126}{14} + \dfrac{3}{14}$

$\qquad\qquad = 9 + \dfrac{3}{14}$

$\qquad\qquad = 9\dfrac{3}{14}.$

(b) $6\dfrac{5}{7} + 2\dfrac{1}{2} = \left(6 + \dfrac{5}{7}\right) + \left(2 + \dfrac{1}{2}\right)$

$\qquad\qquad = (6 + 2) + \left(\dfrac{5}{7} + \dfrac{1}{2}\right)$

$\qquad\qquad = 8 + \dfrac{10 + 7}{14}$

$\qquad\qquad = 8 + \dfrac{17}{14}$

$$= 8 + \frac{14 + 3}{14}$$

$$= 8 + \frac{14}{14} + \frac{3}{14}$$

$$= 8 + 1 + \frac{3}{14}$$

$$= 9 + \frac{3}{14}$$

$$= 9 \frac{3}{14}.$$

(c) $6 \frac{5}{7} = 6 \frac{10}{14}$

$+ 2 \frac{1}{2} = 2 \frac{7}{14}$

$8 \frac{17}{14} = 9 \frac{3}{14}.$

Example 3   Find the difference $5 \frac{1}{2} - 3 \frac{1}{3}.$

(a) $5 \frac{1}{2} - 3 \frac{1}{3} = \left(5 + \frac{1}{2}\right) - \left(3 + \frac{1}{3}\right)$

$$= \left(\frac{5}{1} + \frac{1}{2}\right) - \left(\frac{3}{1} + \frac{1}{3}\right)$$

$$= \frac{10 + 1}{2} - \frac{9 + 1}{3}$$

$$= \frac{11}{2} - \frac{10}{3}$$

$$= \frac{33 - 20}{6}$$

$$= \frac{13}{6}$$

$$= \frac{12 + 1}{6}$$

$$= \frac{12}{6} + \frac{1}{6}$$

$$= 2 + \frac{1}{6}$$

$$= 2 \frac{1}{6}.$$

(b) $5\frac{1}{2} - 3\frac{1}{3} = \left(5 + \frac{1}{2}\right) - \left(3 + \frac{1}{3}\right)$

$$= (5 - 3) + \left(\frac{1}{2} - \frac{1}{3}\right)$$

$$= 2 + \left(\frac{3 - 2}{6}\right)$$

$$= 2 + \frac{1}{6}$$

$$= 2\frac{1}{6}.$$

(c) $5\frac{1}{2} = 5\frac{3}{6}$

$-\ 3\frac{1}{3} = 3\frac{2}{6}$

$$\rule{2cm}{0.4pt}$$

$$2\frac{1}{6}.$$

Example 4   Find the difference $6\frac{1}{3} - 3\frac{2}{3}$.

(a) $6\frac{1}{3} - 3\frac{2}{3} = \left(6 + \frac{1}{3}\right) - \left(3 + \frac{2}{3}\right)$

$$= \left(\frac{6}{1} + \frac{1}{3}\right) - \left(\frac{3}{1} + \frac{2}{3}\right)$$

$$= \frac{18 + 1}{3} - \frac{9 + 2}{3}$$

$$= \frac{19}{3} - \frac{11}{3}$$

$$= \frac{19 - 11}{3}$$

$$= \frac{8}{3}$$

$$= \frac{6 + 2}{3}$$

$$= \frac{6}{3} + \frac{2}{3}$$

$$= 2 + \frac{2}{3}$$

$$= 2\frac{2}{3}.$$

(b) $6\frac{1}{3} = 6 + \frac{1}{3} = 5 + 1 + \frac{1}{3} = 5 + \frac{3}{3} + \frac{1}{3} = 5 + \frac{4}{3}$

$-3\frac{2}{3} = 3 + \frac{2}{3} = 3 \qquad + \frac{2}{3} = 3 \qquad + \frac{2}{3} = 3 + \frac{2}{3}$

$$2 + \frac{2}{3} = 2\frac{2}{3}.$$

Exercises

1. Express each number in the form $\frac{a}{b}$ where a and b are integers, $b \neq 0$.

(a) $3\frac{2}{3}$ (b) $6\frac{1}{4}$ (c) $2\frac{5}{8}$

(d) $1\frac{3}{5}$ (e) $5\frac{1}{6}$ (f) $8\frac{6}{7}$

(g) $7\frac{1}{2}$ (h) $19\frac{5}{6}$ (i) $14\frac{7}{8}$

2. Find each sum.

(a) $3\frac{2}{3} + 6\frac{1}{4}$ (b) $2\frac{5}{8} + 1\frac{3}{5}$ (c) $5\frac{1}{6} + 8\frac{6}{7}$

(d) $7\frac{1}{2} + 19\frac{5}{6}$ (e) $1\frac{5}{8} + {}^-3\frac{2}{8}$ (f) ${}^-4\frac{1}{3} + {}^-2\frac{1}{6}$

3. Find each difference.

(a) $3\frac{2}{3} - 6\frac{1}{4}$ (b) $2\frac{5}{8} - 1\frac{3}{5}$ (c) $5\frac{1}{6} - {}^-8\frac{6}{7}$

(d) $7\frac{1}{2} - 7\frac{8}{16}$ (e) $13 - \frac{14}{15}$ (f) $1\frac{1}{8} - 2\frac{2}{7}$

8-10 Chapter Test

Indicate whether each statement is true or false. If false, tell why.

1. All rational numbers are fractional numbers.
2. Two fractions that name the same number are called like fractions.
3. A rational number may always be expressed in lower terms if both the numerator and denominator are composite numbers.
4. Every rational number has an additive inverse.
5. $\frac{6}{9}$ and $\frac{38}{57}$ are equivalent fractions.

6. Zero is the smallest rational number.

7. ⁻5 is a fractional number.

8. The integers are a subset of the rational numbers.

9. $\dfrac{-12}{25} > \dfrac{-11}{23}$.

10. The set of rational numbers is ordered.

Select the best possible answer.

11. A common denominator larger than the least common denominator is used in addition involving unlike fractions. Will the sum be such that it may be reduced to lower terms?
   (a) Yes, always   (b) No, never
   (c) Sometimes   (d) Impossible to determine

12. How many rational numbers belong to the set of rational numbers between $\dfrac{1}{5}$ and $\dfrac{3}{5}$?

   (a) 0   (b) 1
   (c) 2   (d) 5
   (e) Infinitely many

13. The process of finding the least common multiple is used in the addition of two fractional numbers when:
   (a) The denominators are different numbers
   (b) The fractions are expressed in lowest terms
   (c) The denominators are prime numbers
   (d) The denominators are composite numbers
   (e) None of these

14. The justification of the algorithm for addition of rational numbers is most dependent upon an understanding of:
   (a) Multiplication facts
   (b) Commutative properties for addition and multiplication of integers
   (c) Addition and subtraction facts involving 0
   (d) Closure property
   (e) Right distributive property of division over addition for integers

15. To add fractional numbers:
   (a) Find the sum of the numerators and divide by the sum of the denominators
   (b) Find the product of the numerators and divide by the product of the denominators
   (c) Find the sum of the numerators and divide by the product of the denominators

(**d**) Find a common denominator and multiply the numerators

(**e**) None of the above applies

16. The name for a number such as $2\frac{1}{4}$ is:

(**a**) A mixed number   (**b**) A mixed numeral

(**c**) A proper fraction   (**d**) An improper fraction

(**e**) None of these

17. If the pattern suggested by $\frac{1}{3} + \frac{1}{5} = \frac{8}{15}$, $\frac{1}{4} + \frac{1}{6} = \frac{10}{24}$, $\frac{1}{3} + \frac{1}{6} = \frac{9}{18}$ is applied to $\frac{1}{a} + \frac{1}{b}$, the sum is:

(**a**) $\dfrac{ab}{a+b}$ (**b**) $\dfrac{a+b}{ab}$

(**c**) $\dfrac{a+a}{bb}$ (**d**) $\dfrac{b+b}{aa}$

(**e**) None of these

18. The definition of equality of two fractional numbers $\frac{a}{c}$ and $\frac{d}{b}$ is:

(**a**) $a \times d = c \times b$ (**b**) $a \times c = d \times b$

(**c**) $a \times b = c \times d$ (**d**) $a \div c = d \div b$

(**e**) None of these

19. A fractional number may be defined as the quotient of:

(**a**) Any two whole numbers

(**b**) Any two integers

(**c**) Any two numbers

(**d**) A whole number and a natural number

(**e**) A natural number and a whole number

20. $3\frac{1}{4}$ means:

(**a**) $3 \times \dfrac{1}{4}$ (**b**) $3 + \dfrac{1}{4}$

(**c**) $3 - \dfrac{1}{4}$ (**d**) $3 \div \dfrac{1}{4}$

(**e**) None of these

chapter 9

Rational Numbers: Multiplication and Division

In the last chapter, we extended our study of numbers to the set of rational numbers. This was done for the purpose of developing a set of numbers that is closed under the operation of division by any number in the set except zero. In this chapter, the operations of multiplication and division of rational numbers are defined, and some of the basic properties are presented as assumptions.

9-1 Multiplication of Rational Numbers

We learned previously for whole numbers that in some instances multiplication may be thought of as repeated addition. The same is true for multiplication of rational numbers. There are some instances when multiplication may be interpreted as repeated addition. For example, $3 \times \frac{1}{2}$ may be illustrated on the number line as $\frac{1}{2} + \frac{1}{2} + \frac{1}{2}$.

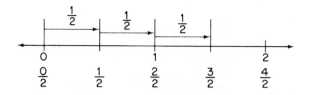

We see that $3 \times \dfrac{1}{2} = \dfrac{3}{2}$, or $1\dfrac{1}{2}$.

We may show $4 \times \dfrac{1}{3}$ with geometric regions as follows:

If we join the shaded parts of each region as shown below, we have $4 \times \dfrac{1}{3} = \dfrac{4}{3}$, or $1\dfrac{1}{3}$.

Suppose we have the problem $\dfrac{1}{2} \times \dfrac{3}{5}$. The repeated addition approach using regions or the number line is not adequate for this situation. We may use another approach with a region to convey the idea $\dfrac{1}{2} \times \dfrac{3}{5}$. First we construct a unit region (either rectangular or square).

Unit region

Next we divide the unit region horizontally into halves and vertically into fifths and shade as follows to show $\dfrac{1}{2} \times \dfrac{3}{5} = \dfrac{3}{10}$.

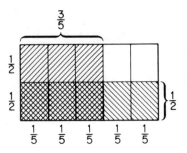

We should observe from the shading that $\frac{3}{5}$ is shaded in one direction while $\frac{1}{2}$ is shaded in another direction. Thus $\frac{1}{2} \times \frac{3}{5}$ $\left(\text{interpreted as } \frac{1}{2} \text{ of } \frac{3}{5} \right)$ is the part of the unit region that is shaded in both directions; namely, 3 out of 10 blocks in the unit region.

We may use a similar procedure to show $\frac{1}{4} \times \frac{2}{3} = \frac{2}{12}$ as follows.

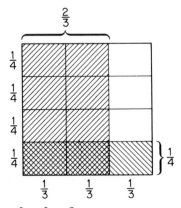

To show the product $\frac{3}{4} \times \frac{3}{2} = \frac{9}{8}$ we need more than one unit region. Since $\frac{3}{2}$ is greater than 1 and less than 2, we will use 2 unit regions as shown in the following diagram.

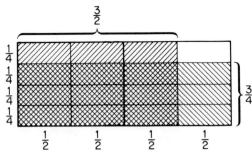

We should observe from the shading that $\frac{3}{2}$ is shaded in one direction while $\frac{3}{4}$ is shaded in another direction. Thus $\frac{3}{4} \times \frac{3}{2}$ is the part that is shaded in both directions; namely, 9 blocks such that there are 8 blocks in the unit region. Special note should be made of the fact that reference is always made to the number of subdivisions in the unit region, even though more than one unit region may be used. Thus, we may have 9 blocks shaded as compared to only 8 blocks in the unit region.

In the examples considered thus far, there is a pattern which suggests an algorithm for multiplying rational numbers.

$$3 \times \frac{1}{2} = \frac{3}{1} \times \frac{1}{2} = \frac{3}{6}$$

$$4 \times \frac{1}{3} = \frac{4}{1} \times \frac{1}{3} = \frac{4}{3}$$

$$\frac{1}{2} \times \frac{3}{5} = \frac{3}{10}$$

$$\frac{1}{4} \times \frac{2}{3} = \frac{2}{12}$$

$$\frac{3}{4} \times \frac{3}{2} = \frac{9}{8}$$

Notice in each example that the numerator of the product is the product of the numerators of the factors, and the denominator of the product is the product of the denominators of the factors. For example,

$$\frac{1}{2} \times \frac{3}{5} = \frac{1 \times 3}{2 \times 5} = \frac{3}{10},$$

and

$$\frac{3}{4} \times \frac{3}{2} = \frac{3 \times 3}{4 \times 2} = \frac{9}{8}.$$

In general, for rational numbers $\frac{a}{b}$ and $\frac{c}{d}$, we define multiplication as

$$\frac{a}{b} \times \frac{c}{d} = \frac{a \times c}{b \times d} = \frac{ac}{bd}.$$

Note that the product of the numerators may be represented as an integer ac, where a and c are integers, and the product of the denominators may be represented as an integer bd, where b and d are integers different from zero. Thus the product of any two rational numbers is a rational number. This definition of multiplication of rational numbers includes multiplication of a rational number and an integer, since $\frac{a}{b} \times \frac{c}{d}$ may represent $\frac{a}{b} \times \frac{c}{1}$ where $\frac{c}{1}$ is any integer.

Exercises

1. Represent each equation on a number line.

   (a) $3 \times \frac{1}{7} = \frac{3}{7}$ (b) $4 \times \frac{1}{4} = \frac{4}{4}$ (c) $6 \times \frac{1}{5} = \frac{6}{5}$

2. For each part of Exercise 1, write an addition equation that is equivalent to the given multiplication equation.

3. Write an addition and a multiplication equation for each of the following.

   (a)

   (b)

   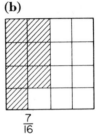 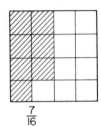

4. Write a multiplication equation for each of the following.

   (a)

   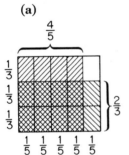

   (b)

   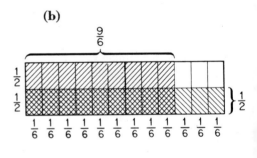

5. Draw a unit region, or unit regions, and shade properly to illustrate each of the following.

   (a) $\frac{3}{4} \times \frac{2}{7}$ (b) $\frac{1}{2} \times \frac{2}{3}$ (c) $\frac{1}{4} \times \frac{6}{5}$

6. Find each product.

   (a) $\frac{7}{9} \times \frac{5}{8}$ (b) $\frac{1}{5} \times \frac{3}{4}$ (c) $\frac{3}{5} \times \frac{5}{3}$

   (d) $^-3 \times \frac{5}{6}$ (e) $\frac{7}{4} \times \frac{6}{11}$ (f) $\frac{1}{3} \times 5$

   (g) $\frac{4}{7} \times \frac{^-3}{8}$ (h) $\frac{^-2}{3} \times \frac{^-5}{11}$ (i) $\frac{4}{12} \times \frac{24}{2}$

9-2 Properties of Multiplication of Rational Numbers

We noted earlier that increased emphasis has been placed on some of the basic patterns and properties of elementary school mathematics. We also pointed out that these underlying properties would be discussed and then accepted as assumptions.

Closure and uniqueness properties   The property of *closure* applied to the multiplication of rational numbers means that the product of any two rational numbers is also a rational number. The principle of *uniqueness* indicates that there is only one possible product when two given rational numbers are multiplied. For example, $\frac{1}{2} \times \frac{1}{3} = \frac{1}{6}$ illustrates that the factors and the product are members of the set of rational numbers. We should also note that $\frac{1}{6}$ is the only possible product when $\frac{1}{2}$ and $\frac{1}{3}$ are multiplied. Hence, the number $\frac{1}{6}$ is the *unique* product of $\frac{1}{2}$ and $\frac{1}{3}$. Notice that the number $\frac{1}{6}$ may also be expressed as $\frac{2}{12}, \frac{3}{18}, \frac{4}{24}$, and so on. In general, for the rational numbers $\frac{a}{b}$ and $\frac{c}{d}$, as explained in Section 9-1,

$$\frac{a}{b} \times \frac{c}{d} = \frac{ac}{bd},$$

where the rational number $\frac{ac}{bd}$ is the unique product of $\frac{a}{b}$ and $\frac{c}{d}$.

Commutative property   The order of the multiplication of two factors from the set of rational numbers does not change the product. In the case of $\frac{1}{2}$ and $\frac{1}{4}$, we note that

$$\frac{1}{2} \times \frac{1}{4} = \frac{1}{8} \quad \text{and} \quad \frac{1}{4} \times \frac{1}{2} = \frac{1}{8}.$$

Thus,

$$\frac{1}{2} \times \frac{1}{4} = \frac{1}{4} \times \frac{1}{2}.$$

The multiplication of rational numbers is *commutative* and, for any two rational numbers $\frac{a}{b}$ and $\frac{c}{d}$,

$$\frac{a}{b} \times \frac{c}{d} = \frac{ac}{bd} = \frac{ca}{db} = \frac{c}{d} \times \frac{a}{b}$$

since the multiplication of integers is commutative.

Associative property   If more than two rational numbers are to be multiplied, the factors are grouped for binary operations. For example, $\frac{1}{2} \times \frac{1}{3} \times \frac{1}{4}$ may be grouped for multiplication as

$$\left(\frac{1}{2} \times \frac{1}{3}\right) \times \frac{1}{4} \quad \text{or} \quad \frac{1}{2} \times \left(\frac{1}{3} \times \frac{1}{4}\right).$$

Since

$$\left(\frac{1}{2} \times \frac{1}{3}\right) \times \frac{1}{4} = \frac{1}{6} \times \frac{1}{4} = \frac{1}{24}$$

and

$$\frac{1}{2} \times \left(\frac{1}{3} \times \frac{1}{4}\right) = \frac{1}{2} \times \frac{1}{12} = \frac{1}{24},$$

we have

$$\left(\frac{1}{2} \times \frac{1}{3}\right) \times \frac{1}{4} = \frac{1}{2} \times \left(\frac{1}{3} \times \frac{1}{4}\right).$$

In general, for the rational numbers $\frac{a}{b}, \frac{c}{d}$, and $\frac{e}{f}$,

$$\left(\frac{a}{b} \times \frac{c}{d}\right) \times \frac{e}{f} = \frac{a}{b} \times \left(\frac{c}{d} \times \frac{e}{f}\right),$$

and we say that the multiplication of rational numbers is *associative*.

Identity element   The product of any rational number and the number 1 is always the original rational number. For example,

$$\frac{1}{2} \times 1 = 1 \times \frac{1}{2} = \frac{1}{2}, \quad \frac{2}{3} \times 1 = 1 \times \frac{2}{3} = \frac{2}{3}, \quad \frac{-3}{4} \times 1 = 1 \times \frac{-3}{4} = \frac{-3}{4},$$

and so on. Thus, $\frac{a}{b} \times 1 = 1 \times \frac{a}{b} = \frac{a}{b}$ since $\frac{a}{b} \times 1 = \frac{a}{b} \times \frac{1}{1} = \frac{a \times 1}{b \times 1} = \frac{a}{b}$, for any rational number $\frac{a}{b}$, and 1 is called the *identity element for multiplication* of rational numbers (also referred to as the *multiplicative identity*).

Multiplication by zero   The product of the number zero and any rational number is always zero. Thus, for any rational number $\frac{a}{b}$,

$$\frac{a}{b} \times 0 = 0 \times \frac{a}{b} = 0,$$

since

$$\frac{a}{b} \times 0 = \frac{a}{b} \times \frac{0}{1} = \frac{a \times 0}{b \times 1} = \frac{0}{b} = 0.$$

Distributive property of multiplication over addition   Multiplication is *distributive* over addition for the set of rational numbers. For example,

$$\frac{1}{3} \times \left(\frac{3}{4} + \frac{2}{4}\right) = \frac{1}{3} \times \left(\frac{5}{4}\right) = \frac{5}{12}$$

and

$$\frac{1}{3} \times \left(\frac{3}{4} + \frac{2}{4}\right) = \left(\frac{1}{3} \times \frac{3}{4}\right) + \left(\frac{1}{3} \times \frac{2}{4}\right) = \frac{3}{12} + \frac{2}{12} = \frac{5}{12}.$$

In general, we may distribute multiplication over addition from both the left and the right when we have the rational numbers $\frac{a}{b}, \frac{c}{d}$, and $\frac{e}{f}$. In other words, we have a left distributive property,

$$\frac{a}{b} \times \left(\frac{c}{d} + \frac{e}{f}\right) = \left(\frac{a}{b} \times \frac{c}{d}\right) + \left(\frac{a}{b} \times \frac{e}{f}\right),$$

and also a right distributive property,

$$\left(\frac{c}{d} + \frac{e}{f}\right) \times \frac{a}{b} = \left(\frac{c}{d} \times \frac{a}{b}\right) + \left(\frac{e}{f} \times \frac{a}{b}\right).$$

We speak of "the distributive property of multiplication over addition" when referring to either of these cases.

Distributive property of multiplication over subtraction   Multiplication is also distributive over subtraction for the set of rational numbers. For example,

$$\frac{1}{2} \times \left(\frac{4}{5} - \frac{1}{5}\right) = \frac{1}{2} \times \left(\frac{3}{5}\right) = \frac{3}{10}$$

and

$$\frac{1}{2} \times \left(\frac{4}{5} - \frac{1}{5}\right) = \left(\frac{1}{2} \times \frac{4}{5}\right) - \left(\frac{1}{2} \times \frac{1}{5}\right) = \frac{4}{10} - \frac{1}{10} = \frac{3}{10}.$$

In general, we may distribute multiplication over subtraction from both the left and the right when we have the rational numbers $\frac{a}{b}, \frac{c}{d}$, and $\frac{e}{f}$. In other words, we have a left distributive property,

$$\frac{a}{b} \times \left(\frac{c}{d} - \frac{e}{f}\right) = \left(\frac{a}{b} \times \frac{c}{d}\right) - \left(\frac{a}{b} \times \frac{e}{f}\right),$$

and a right distributive property,

$$\left(\frac{c}{d} - \frac{e}{f}\right) \times \frac{a}{b} = \left(\frac{c}{d} \times \frac{a}{b}\right) - \left(\frac{e}{f} \times \frac{a}{b}\right).$$

Exercises

1. Find each product.

(a) $\left(\frac{1}{3} \times \frac{1}{5}\right) \times \frac{3}{8}$

(b) $\frac{1}{3} \times \left(\frac{1}{5} \times \frac{3}{8}\right)$

2. What may be concluded from a comparison of the two answers in Exercise 1?

3. What property of multiplication of rational numbers is suggested in Exercises 1 and 2?

4. Give a numerical example to illustrate the assumption that multiplication of rational numbers is:
   (a) Commutative
   (b) Associative
   (c) Distributive from the left over subtraction
   (d) Distributive from the right over addition

5. (a) Find the product $\frac{1}{2} \times \frac{3}{4}$.

   (b) Find an equivalent fraction for $\frac{1}{2}$. For $\frac{3}{4}$.

   (c) Find the product using equivalent fractions for $\frac{1}{2}$ and $\frac{3}{4}$.

   (d) Is the product in part (a) equal to the product in part (c)?

   (e) Is the product of two rational numbers unique?

6. Tell whether or not multiplication is closed for each set of numbers.
   (a) Nonnegative rational numbers
   (b) Positive rational numbers
   (c) Negative rational numbers

7. Find each product.

   (a) $\frac{2}{3} \times 1$ (b) $\frac{4}{7} \times \frac{2}{2}$

   (c) $\frac{3}{3} \times \frac{1}{8}$ (d) $6 \times \frac{5}{5}$

8. What property of multiplication is suggested in Exercise 7?

9. Is it true that the multiplicative identity for rational numbers is any number that may be expressed in the form $\frac{a}{a}$ where a is any integer?

10. Find each product.

    (a) $\frac{1}{5} \times 0$ (b) $0 \times \frac{-7}{4}$

    (c) $\frac{2}{8} \times \frac{4}{9} \times \frac{0}{3}$ (d) $\frac{0}{7} \times \frac{6}{7}$

11. What property of multiplication of rational numbers is suggested in Exercise 10?

12. If $\frac{a}{b} \times \frac{c}{d} = 0$ for rational numbers $\frac{a}{b}$ and $\frac{c}{d}$, what can you say about $\frac{a}{b}$ and $\frac{c}{d}$?

13. If $\frac{a}{b} \times \frac{c}{d} = \frac{a}{b}$ for rational numbers $\frac{a}{b}$ and $\frac{c}{d}$, what can you say about $\frac{a}{b}$ and $\frac{c}{d}$?

14. If $\frac{a}{b} \times \frac{c}{d} = \frac{c}{d}$ for rational numbers $\frac{a}{b}$ and $\frac{c}{d}$, what can you say about $\frac{a}{b}$ and $\frac{c}{d}$?

9-3 Multiplicative Inverse

In the set of rational numbers, there exist pairs of numbers such that each has a product of 1. For example,

$$2 \times \frac{1}{2} = 1, \quad \frac{1}{3} \times 3 = 1, \quad \frac{-2}{5} \times \frac{-5}{2} = 1,$$

and so on. The numbers in each pair are called **reciprocals** of each other (also referred to as **multiplicative inverses**). Since $\frac{5}{8} \times \frac{8}{5} = 1$, $\frac{5}{8}$ is the reciprocal of $\frac{8}{5}$, and $\frac{8}{5}$ is the reciprocal of $\frac{5}{8}$. Therefore, the extension of our number system to the set of rational numbers enables us to state a new property. For each rational number $\frac{a}{b}$, where $\frac{a}{b} \neq 0$, there exists another rational number $\frac{b}{a}$, called the *reciprocal* or *multiplicative inverse* of $\frac{a}{b}$, such that

$$\frac{a}{b} \times \frac{b}{a} = \frac{ab}{ba} = \frac{ab}{ab} = 1.$$

Exercises

1. If possible, write the numeral for the multiplicative inverse of each rational number.

   (a) 8   (b) $\frac{5}{4}$

   (c) $\frac{-7}{-8}$ (d) $\frac{0}{3}$

(e) $\dfrac{-9}{-9}$

(f) $\dfrac{12}{-5}$

(g) $\dfrac{6}{6}$

(h) $\dfrac{-2+2}{7}$

(i) $\left(\dfrac{1}{4}+\dfrac{-1}{4}\right)$

(j) $\dfrac{-3+{}^-2}{11}$

2. Verify the correctness of each answer in Exercise 1 by writing a multiplication equation and finding the product.

3. If possible, find a rational number as a replacement for the variable, such that the statement will be true.

(a) $\dfrac{6}{7} \times \square = 1$

(b) $\dfrac{-13}{14} \times \square = 1$

(c) $\dfrac{-2}{-9} \times \square = 1$

(d) $\square \times 0 = 1$

(e) $\square \times \dfrac{6}{6} = 1$

(f) $\dfrac{97}{141} \times \square = 1$

4. When is the reciprocal of a positive rational number:
   (a) Equal to the number?   (b) Less than the number?
   (c) Greater than the number?

5. What rational number does not have a multiplicative inverse?

9-4 Division of Rational Numbers

We have learned that $3 \div 4$ may be expressed in the form $\dfrac{3}{4}$ and, in general,

$a \div b$ may be expressed in the form $\dfrac{a}{b}$, where a and b represent integers and $b \neq 0$. We shall assume that the division of rational numbers is analogous to the division of integers, and that the form $a \div b = \dfrac{a}{b}$ may be used, where a and b represent rational numbers and $b \neq 0$. For example, if $a = \dfrac{2}{3}$ and $b = \dfrac{3}{4}$, we have

$$\frac{2}{3} \div \frac{3}{4} = \frac{\dfrac{2}{3}}{\dfrac{3}{4}}.$$

By using the multiplicative identity and applying certain mathematical properties, we can determine the desired quotient as follows.

$$\frac{\frac{2}{3}}{\frac{3}{4}} = \frac{\frac{2}{3}}{\frac{3}{4}} \times 1 \qquad \text{Multiplicative identity.}$$

$$\frac{\frac{2}{3}}{\frac{3}{4}} \times 1 = \frac{\frac{2}{3}}{\frac{3}{4}} \times \frac{\frac{4}{3}}{\frac{4}{3}} \qquad \text{Substitution of } \frac{\frac{4}{3}}{\frac{4}{3}} \text{ for 1.}$$

$$\frac{\frac{2}{3}}{\frac{3}{4}} \times \frac{\frac{4}{3}}{\frac{4}{3}} = \frac{\frac{2}{3} \times \frac{4}{3}}{\frac{3}{4} \times \frac{4}{3}} \qquad \text{Definition of multiplication of rational numbers.}$$

$$\frac{\frac{2}{3} \times \frac{4}{3}}{\frac{3}{4} \times \frac{4}{3}} = \frac{\frac{2}{3} \times \frac{4}{3}}{1} \qquad \text{Product of reciprocals is 1.}$$

$$\frac{\frac{2}{3} \times \frac{4}{3}}{1} = \frac{2}{3} \times \frac{4}{3} \qquad \text{Property of dividing by 1.}$$

$$\frac{2}{3} \times \frac{4}{3} = \frac{8}{9} \qquad \text{Definition of multiplication of rational numbers.}$$

Thus we have $\frac{2}{3} \div \frac{3}{4} = \frac{2}{3} \times \frac{4}{3} = \frac{8}{9}$.

To gain further understanding of the division of rational numbers, study the following examples.

Example 1   Find the quotient $8 \div 4$.

$$8 \div 4 = \frac{8}{1} \div \frac{4}{1}$$

Any whole number a may be expressed in the form of the rational number $\frac{a}{1}$.

$$\frac{8}{1} \div \frac{4}{1} = \frac{\frac{8}{1}}{\frac{4}{1}}$$

Division of rational numbers may be represented by a fraction.

$$\frac{\frac{8}{1}}{\frac{4}{1}} = \frac{\frac{8}{1}}{\frac{4}{1}} \times 1$$

Multiplicative identity.

$$\frac{\dfrac{8}{1}}{\dfrac{4}{1}} \times 1 = \frac{\dfrac{8}{1}}{\dfrac{4}{1}} \times \frac{\dfrac{1}{4}}{\dfrac{1}{4}} \qquad \text{Substitution of } \frac{\dfrac{1}{4}}{\dfrac{1}{4}} \text{ for 1.}$$

$$\frac{\dfrac{8}{1}}{\dfrac{4}{1}} \times \frac{\dfrac{1}{4}}{\dfrac{1}{4}} = \frac{\dfrac{8}{1} \times \dfrac{1}{4}}{\dfrac{4}{1} \times \dfrac{1}{4}} \qquad \text{Definition of multiplication of rational numbers.}$$

$$\frac{\dfrac{8}{1} \times \dfrac{1}{4}}{\dfrac{4}{1} \times \dfrac{1}{4}} = \frac{\dfrac{8}{1} \times \dfrac{1}{4}}{1} \qquad \text{Product of reciprocals is 1.}$$

$$\frac{\dfrac{8}{1} \times \dfrac{1}{4}}{1} = \frac{8}{1} \times \frac{1}{4} \qquad \text{Property of dividing by 1.}$$

$$\frac{8}{1} \times \frac{1}{4} = \frac{8}{4} \qquad \text{Definition of multiplication of rational numbers.}$$

$$\text{Thus } 8 \div 4 = \frac{8}{1} \times \frac{1}{4} = \frac{8}{4}.$$

Example 2   Find the quotient $9 \div \dfrac{5}{7}$.

$$9 \div \frac{5}{7} = \frac{9}{1} \div \frac{5}{7} \qquad \begin{array}{l}\text{Any whole number may be expressed}\\ \text{in the form of the rational number } \dfrac{a}{1}.\end{array}$$

$$\frac{9}{1} \div \frac{5}{7} = \frac{\dfrac{9}{1}}{\dfrac{5}{7}} \qquad \begin{array}{l}\text{Division of rational numbers may be}\\ \text{represented by a fraction.}\end{array}$$

$$\frac{\dfrac{9}{1}}{\dfrac{5}{7}} = \frac{\dfrac{9}{1}}{\dfrac{5}{7}} \times 1 \qquad \text{Multiplicative identity.}$$

$$\frac{\dfrac{9}{1}}{\dfrac{5}{7}} \times 1 = \frac{\dfrac{9}{1}}{\dfrac{5}{7}} \times \frac{\dfrac{7}{5}}{\dfrac{7}{5}} \qquad \text{Substitution of } \frac{\dfrac{7}{5}}{\dfrac{7}{5}} \text{ for 1.}$$

$$\frac{\dfrac{9}{1}}{\dfrac{5}{7}} \times \frac{\dfrac{7}{5}}{\dfrac{7}{5}} = \frac{\dfrac{9}{1} \times \dfrac{7}{5}}{\dfrac{5}{7} \times \dfrac{7}{5}} \qquad \begin{array}{l}\text{Definition of multiplication of rational numbers.}\end{array}$$

$$\frac{\dfrac{9}{1} \times \dfrac{7}{5}}{\dfrac{5}{7} \times \dfrac{7}{5}} = \frac{\dfrac{9}{1} \times \dfrac{7}{5}}{1}$$

Product of reciprocals is 1.

$$\frac{\dfrac{9}{1} \times \dfrac{7}{5}}{1} = \frac{9}{1} \times \frac{7}{5}$$

Property of dividing by 1.

$$\frac{9}{1} \times \frac{7}{5} = \frac{63}{5}$$

Definition of multiplication of rational numbers.

Thus $9 \div \dfrac{5}{7} = \dfrac{9}{1} \times \dfrac{7}{5} = \dfrac{63}{5}.$

Example 3   Find the quotient $\dfrac{2}{9} \div \dfrac{3}{8}.$

$$\frac{2}{9} \div \frac{3}{8} = \frac{\dfrac{2}{9}}{\dfrac{3}{8}}$$

Division of rational numbers may be represented by a fraction.

$$\frac{\dfrac{2}{9}}{\dfrac{3}{8}} = \frac{\dfrac{2}{9}}{\dfrac{3}{8}} \times 1$$

Multiplicative identity.

$$\frac{\dfrac{2}{9}}{\dfrac{3}{8}} \times 1 = \frac{\dfrac{2}{9}}{\dfrac{3}{8}} \times \frac{\dfrac{8}{3}}{\dfrac{8}{3}}$$

Substitution of $\dfrac{\dfrac{8}{3}}{\dfrac{8}{3}}$ for 1.

$$\frac{\dfrac{2}{9}}{\dfrac{3}{8}} \times \frac{\dfrac{8}{3}}{\dfrac{8}{3}} = \frac{\dfrac{2}{9} \times \dfrac{8}{3}}{\dfrac{3}{8} \times \dfrac{8}{3}}$$

Definition of multiplication of rational numbers.

$$\frac{\dfrac{2}{9} \times \dfrac{8}{3}}{\dfrac{3}{8} \times \dfrac{8}{3}} = \frac{\dfrac{2}{9} \times \dfrac{8}{3}}{1}$$

Product of reciprocals is 1.

$$\frac{\dfrac{2}{9} \times \dfrac{8}{3}}{1} = \frac{2}{9} \times \frac{8}{3}$$

Property of dividing by 1.

$$\frac{2}{9} \times \frac{8}{3} = \frac{16}{27}$$

Definition of multiplication of rational numbers.

Thus $\dfrac{2}{9} \div \dfrac{3}{8} = \dfrac{2}{9} \times \dfrac{8}{3} = \dfrac{16}{27}$.

These examples provide some foundation for the familiar rule "invert the divisor and multiply." We observe that division by a rational number has the same result as multiplication by the reciprocal of the divisor. In general, if we are given two rational numbers $\dfrac{a}{b}$ and $\dfrac{c}{d}$ where $c \neq 0$,

$$\dfrac{a}{b} \div \dfrac{c}{d} = \dfrac{\dfrac{a}{b}}{\dfrac{c}{d}}$$ Division of rational numbers may be represented by a fraction.

$$\dfrac{\dfrac{a}{b}}{\dfrac{c}{d}} = \dfrac{\dfrac{a}{b}}{\dfrac{c}{d}} \times 1$$ Multiplicative identity.

$$\dfrac{\dfrac{a}{b}}{\dfrac{c}{d}} \times 1 = \dfrac{\dfrac{a}{b}}{\dfrac{c}{d}} \times \dfrac{\dfrac{d}{c}}{\dfrac{d}{c}}$$ Substitution of $\dfrac{\dfrac{d}{c}}{\dfrac{d}{c}}$ for 1.

$$\dfrac{\dfrac{a}{b}}{\dfrac{c}{d}} \times \dfrac{\dfrac{d}{c}}{\dfrac{d}{c}} = \dfrac{\dfrac{a}{b} \times \dfrac{d}{c}}{\dfrac{c}{d} \times \dfrac{d}{c}}$$ Definition of multiplication of rational numbers.

$$\dfrac{\dfrac{a}{b} \times \dfrac{d}{c}}{\dfrac{c}{d} \times \dfrac{d}{c}} = \dfrac{\dfrac{a}{b} \times \dfrac{d}{c}}{1}$$ Product of reciprocals is 1.

$$\dfrac{\dfrac{a}{b} \times \dfrac{d}{c}}{1} = \dfrac{a}{b} \times \dfrac{d}{c}$$ Property of dividing by 1.

$$\dfrac{a}{b} \times \dfrac{d}{c} = \dfrac{ad}{bc}$$ Definition of multiplication of rational numbers.

Thus, for rational numbers $\dfrac{a}{b}$ and $\dfrac{c}{d}$, where $c \neq 0$,

$$\dfrac{a}{b} \div \dfrac{c}{d} = \dfrac{a}{b} \times \dfrac{d}{c} = \dfrac{ad}{bc}.$$

We learned that for integers a, b, and c ($b \neq 0$), $a \div b = c$ if and only if $c \times b = a$. In other words, division and multiplication are inversely related operations. We shall assume that this inverse relationship is preserved in the expansion of the number systems to the set of rational numbers. Thus, if $\frac{m}{n}$ and $\frac{p}{q}$ are rational numbers with $\frac{p}{q} \neq 0$, $\frac{m}{n} \div \frac{p}{q}$ is equal to the rational number $\frac{r}{s}$ if and only if $\frac{r}{s} \times \frac{p}{q} = \frac{m}{n}$. For example,

$$\frac{2}{3} \div \frac{1}{3} = 2 \quad \text{if and only if} \quad 2 \times \frac{1}{3} = \frac{2}{3},$$

$$\frac{4}{6} \div \frac{1}{2} = \frac{8}{6} \quad \text{if and only if} \quad \frac{8}{6} \times \frac{1}{2} = \frac{4}{6},$$

and

$$\frac{5}{12} \div 6 = \frac{5}{72} \quad \text{if and only if} \quad \frac{5}{72} \times 6 = \frac{5}{12}.$$

Exercises

1. Find each quotient.

(a) $\frac{5}{9} \div \frac{2}{9}$

(b) $\frac{1}{6} \div \frac{1}{6}$

(c) $\frac{11}{13} \div \frac{13}{11}$

(d) $\frac{e}{f} \div \frac{g}{h}$

(e) $\dfrac{\frac{4}{7}}{\frac{4}{1}}$

(f) $\frac{13}{4} \div \frac{5}{8}$

(g) $\frac{2}{3} \div \frac{-4}{7}$

(h) $-3 \div \frac{-1}{9}$

2. Find a rational number as a replacement for each variable, such that the statement will be true.

(a) $8 \div \frac{4}{3} = \square$

(b) $\bigcirc \times \frac{4}{3} = 8$

(c) $\frac{23}{24} \div \frac{5}{6} = \triangle$

(d) $\frac{5}{6} \times \bigcirc = \frac{23}{24}$

3. In Exercise 2, is it true to state that
   (a) $\square = \bigcirc$?   (b) $\triangle = \bigcirc$?

4. From Exercise 2 and other similar examples, what may we generalize about the relationship between multiplication and division of rational numbers?

5. Indicate whether each statement is true or false. If false, tell why.

(a) $\dfrac{4}{7} \div \dfrac{1}{8} = \dfrac{4}{7} \times \dfrac{8}{1}$

(b) $\dfrac{3}{10} \div \dfrac{2}{10} = 3 \div 2$

(c) $\dfrac{5}{9} \div 9 = \square$ if and only if $\square \times \dfrac{5}{9} = 9$

(d) $\dfrac{t}{s} \div \dfrac{g}{g} = \dfrac{t}{s}$ where t, s, and g are integers

9-5 Properties of Division of Rational Numbers

Closure and uniqueness properties　Except for division by zero, the quotient of any two rational numbers is a *unique* rational number. Thus, we say that the set of rational numbers different from zero is *closed* with respect to the operation of division.

Commutative property　As was the case with the division of integers, the division of rational numbers is not commutative. For example,

$$\frac{1}{2} \div \frac{1}{3} \neq \frac{1}{3} \div \frac{1}{2}$$

because

$$\frac{1}{2} \times \frac{3}{1} \neq \frac{1}{3} \times \frac{2}{1}; \qquad \text{that is,} \ \frac{3}{2} \neq \frac{2}{3}.$$

Associative property　The associative property does not hold for division in the set of rational numbers. For example,

$$\left(\frac{1}{4} \div \frac{3}{8}\right) \div \frac{1}{16} \neq \frac{1}{4} \div \left(\frac{3}{8} \div \frac{1}{16}\right),$$

because

$$\left(\frac{1}{4} \times \frac{8}{3}\right) \div \frac{1}{16} \neq \frac{1}{4} \div \left(\frac{3}{8} \times \frac{16}{1}\right);$$

that is,

$$\frac{8}{12} \div \frac{1}{16} \neq \frac{1}{4} \div \frac{48}{8} \quad \text{and} \quad \frac{32}{3} \neq \frac{1}{24}.$$

Identity element　Any rational number divided by the number 1 results in the given rational number as the quotient; that is, in general, $\dfrac{a}{b} \div 1 = \dfrac{a}{b}$.

Since in general for any rational number $\dfrac{a}{b}$, $1 \div \dfrac{a}{b} \neq \dfrac{a}{b}$, we say that 1 is a *right identity* that is not a left identity for division of rational numbers.

Distributive property of division over addition   As with the division of integers, there is a distribution of division over the addition of rational numbers from the right, but not the left. Observe these examples:

Distributive from the right:

$$\left(\frac{4}{3}+\frac{2}{3}\right)\div\frac{1}{2}=\left(\frac{4}{3}\div\frac{1}{2}\right)+\left(\frac{2}{3}\div\frac{1}{2}\right),$$

because

$$\frac{6}{3}\div\frac{1}{2}=\frac{8}{3}+\frac{4}{3},$$

and

$$\frac{12}{3}=\frac{12}{3}.$$

Distributive from the left:

$$\frac{1}{2}\div\left(\frac{4}{3}+\frac{2}{3}\right)\neq\left(\frac{1}{2}\div\frac{4}{3}\right)+\left(\frac{1}{2}\div\frac{2}{3}\right),$$

because

$$\frac{1}{2}\div\frac{6}{3}\neq\frac{3}{8}+\frac{3}{4},$$

and

$$\frac{3}{12}\neq\frac{9}{8}.$$

Distributive property of division over subtraction   Division of rational numbers continues to follow a pattern established by division of integers, in that there is a distribution of division over subtraction from the right, but not the left. Consider these examples.

Distributive from the right:

$$\left(\frac{7}{8}-\frac{2}{8}\right)\div\frac{1}{4}=\left(\frac{7}{8}\div\frac{1}{4}\right)-\left(\frac{2}{8}\div\frac{1}{4}\right),$$

because

$$\frac{5}{8}\div\frac{1}{4}=\frac{7}{2}-\frac{2}{2},$$

and

$$\frac{5}{2}=\frac{5}{2}.$$

Distributive from the left:

$$\frac{8}{16}\div\left(\frac{3}{4}-\frac{1}{4}\right)\neq\left(\frac{8}{16}\div\frac{3}{4}\right)-\left(\frac{8}{16}\div\frac{1}{4}\right),$$

because

$$\frac{8}{16}\div\frac{2}{4}\neq\frac{8}{12}-\frac{8}{4},$$

and

$$1\neq\frac{-16}{12}.$$

Property of division by zero   As stated in Section 5-13, division by zero is meaningless in our number system. This also applies to the set of rational numbers.

Exercises

1. Indicate whether each statement is true or false. If false, tell why.

(a) $\dfrac{9}{11} \div \dfrac{3}{2} \neq \dfrac{3}{2} \div \dfrac{9}{11}$

(b) $\dfrac{7}{8} \div 1 = \dfrac{7}{8}$

(c) $0 \div \dfrac{5}{9} \neq \dfrac{5}{9}$

(d) $\dfrac{2}{13} \div \dfrac{1}{2}$ is a rational number

(e) $\left(\dfrac{5}{8} + \dfrac{4}{8}\right) \div \dfrac{3}{2} = \left(\dfrac{5}{8} \div \dfrac{3}{2}\right) + \left(\dfrac{4}{8} \div \dfrac{3}{2}\right)$

(f) $\left(\dfrac{1}{6} \times \dfrac{0}{3}\right) \div \dfrac{0}{7} = 0$

(g) $\dfrac{11}{16} \div 1 = 1 \div \dfrac{11}{16}$

(h) $\dfrac{1}{4} \div \left(\dfrac{1}{5} \div \dfrac{1}{6}\right) > \left(\dfrac{1}{4} \div \dfrac{1}{5}\right) \div \dfrac{1}{6}$

2. Make one counterexample to show that division of rational numbers is not:
   (a) Commutative
   (b) Associative
   (c) Closed
   (d) Distributive over addition from the left

3. Make up an example to show that:
   (a) 1 is a right identity for division of rational numbers.
   (b) 1 is not a left identity for division of rational numbers.

4. If you multiply rational number $\dfrac{a}{b}$ by $\dfrac{9}{11}$, what must you do to the product to obtain $\dfrac{a}{b}$?

5. If you divide rational number $\dfrac{p}{q}$ by $\dfrac{-2}{3}$, what must you do to the quotient to obtain $\dfrac{p}{q}$?

6. If you divide a nonzero rational number $\dfrac{s}{t}$ by $\dfrac{s}{t}$, what is the quotient?

7. Is it always true that $\dfrac{s}{t} \div \dfrac{s}{t} = 1$, when $\dfrac{s}{t}$ is a rational number? Explain.

9-6 Mixed Numerals and Computation

In Section 8-5 we learned that numerals such as $1\dfrac{7}{8}$, $2\dfrac{3}{4}$, and $^-3\dfrac{1}{5}$ are called *mixed numerals.* We also learned that mixed numerals may be expressed in fractional form. We have developed computational procedures for multiplying and dividing rational numbers. Now we may apply these same algorithms to rational numbers represented by mixed numerals, if we change the mixed numerals to fractional form.

The definition for the multiplication of two rational numbers,

$$\frac{a}{b} \times \frac{c}{d} = \frac{a \times c}{b \times d} = \frac{ac}{bd},$$

provides us with a procedure for working with mixed numerals such as $1\dfrac{7}{8}$ and $2\dfrac{3}{4}$. Since

$$1\frac{7}{8} = 1 + \frac{7}{8} = \frac{8}{8} + \frac{7}{8} = \frac{8+7}{8} = \frac{15}{8},$$

and

$$2\frac{3}{4} = 2 + \frac{3}{4} = \frac{8}{4} + \frac{3}{4} = \frac{8+3}{4} = \frac{11}{4},$$

we have

$$1\frac{7}{8} \times 2\frac{3}{4} = \frac{15}{8} \times \frac{11}{4} = \frac{165}{32}.$$

Applying the definition for the division of rational numbers

$$\frac{a}{b} \div \frac{c}{d} = \frac{a}{b} \times \frac{d}{c} = \frac{ad}{bc},$$

we have

$$1\frac{7}{8} \div 2\frac{3}{4} = \frac{15}{8} \div \frac{11}{4} = \frac{15}{8} \times \frac{4}{11} = \frac{15}{22}.$$

A second procedure for multiplication of rational numbers represented by mixed numerals is to regroup using the distributive property of multiplication over addition, in order to work with the integers and fractional numbers separately. This is illustrated in Example 1(b). Consider these examples illustrating several procedures for computation with rational numbers represented by mixed numerals. Study the explanations given for each step.

Example 1 Find the product $2\frac{1}{5} \times 3\frac{1}{4}$.

(a) $2\frac{1}{5} \times 3\frac{1}{4}$

$$= \left(2 + \frac{1}{5}\right) \times \left(3 + \frac{1}{4}\right) \qquad \text{Definition of mixed numeral.}$$

$$= \left(\frac{2}{1} + \frac{1}{5}\right) \times \left(\frac{3}{1} + \frac{1}{4}\right) \qquad \text{Rename integer } a \text{ as } \frac{a}{1}.$$

$$= \frac{10 + 1}{5} \times \frac{12 + 1}{4} \qquad \text{Definition of addition of rational numbers.}$$

$$= \frac{11}{5} \times \frac{13}{4} \qquad \text{Addition of integers in each numerator.}$$

$$= \frac{143}{20} \qquad \text{Definition of multiplication of rational numbers.}$$

$$= \frac{140 + 3}{20} \qquad \text{Rename integer 143 as } 140 + 3.$$

$$= \frac{140}{20} + \frac{3}{20} \qquad \text{Definition of addition of rational numbers with common denominators.}$$

$$= 7 + \frac{3}{20} \qquad \text{Rename } \frac{140}{20} \text{ as 7.}$$

$$= 7\frac{3}{20} \qquad \text{Definition of mixed numeral.}$$

(b) $2\frac{1}{5} \times 3\frac{1}{4}$

$$= 2\frac{1}{5} \times \left(3 + \frac{1}{4}\right) \qquad \text{Definition of mixed numeral.}$$

$$= \left(2\frac{1}{5} \times 3\right) + \left(2\frac{1}{5} \times \frac{1}{4}\right) \qquad \text{Distributive property of multiplication over addition from the left.}$$

$$= \left(2 + \frac{1}{5}\right) \times 3 + \left(2 + \frac{1}{5}\right) \times \frac{1}{4} \qquad \text{Definition of mixed numeral.}$$

$$= (2 \times 3) + \left(\frac{1}{5} \times 3\right) \qquad \text{Distributive property of multiplication over addition from the right.}$$

$$+ \left(2 \times \frac{1}{4}\right) + \left(\frac{1}{5} \times \frac{1}{4}\right)$$

$$= 6 + \frac{3}{5} + \frac{2}{4} + \frac{1}{20} \qquad \text{Definition of multiplication of rational numbers.}$$

$$= 6 + \frac{3}{5} + \left(\frac{2}{4} + \frac{1}{20}\right)$$ Associative property of addition.

$$= 6 + \frac{3}{5} + \frac{11}{20}$$ Definition of addition of rational numbers.

$$= 6 + \left(\frac{3}{5} + \frac{11}{20}\right)$$ Associative property of addition.

$$= 6 + \frac{23}{20}$$ Definition of addition of rational numbers.

$$= 6 + \frac{20 + 3}{20}$$ Rename integer 23 as 20 + 3.

$$= 6 + \frac{20}{20} + \frac{3}{20}$$ Definition of addition of rational numbers with common denominators.

$$= 6 + 1 + \frac{3}{20}$$ Rename $\frac{20}{20}$ as 1.

$$= (6 + 1) + \frac{3}{20}$$ Associative property of addition.

$$= 7 + \frac{3}{20}$$ Addition of integers.

$$= 7\frac{3}{20}$$ Definition of mixed numeral.

Example 2   Find the quotient $3\frac{3}{5} \div 2\frac{1}{3}$.

$$3\frac{3}{5} \div 2\frac{1}{3} = \left(3 + \frac{3}{5}\right) \div \left(2 + \frac{1}{3}\right)$$ Definition of mixed numeral.

$$= \left(\frac{3}{1} + \frac{3}{5}\right) \div \left(\frac{2}{1} + \frac{1}{3}\right)$$ Rename integer a as $\frac{a}{1}$.

$$= \left(\frac{15 + 3}{5}\right) \div \left(\frac{6 + 1}{3}\right)$$ Definition of addition of rational numbers.

$$= \frac{18}{5} \div \frac{7}{3}$$ Addition of integers in each numerator.

$$= \frac{18}{5} \times \frac{3}{7}$$ Definition of division of rational numbers.

$$= \frac{54}{35}$$ Definition of multiplication of rational numbers.

$$= \frac{35 + 19}{35}$$ Rename integer 54 as 35 + 19.

$$= \frac{35}{35} + \frac{19}{35}$$ Definition of addition of rational numbers with common denominators.

$$= 1 + \frac{19}{35}$$ Rename $\frac{35}{35}$ as 1.

$$= 1\frac{19}{35}$$ Definition of mixed numeral.

Exercises

1. Express each number in the form $\frac{a}{b}$ where a and b are integers, $b \neq 0$.

(a) $4\frac{2}{3}$ (b) $6\frac{3}{8}$ (c) $7\frac{5}{6}$

(d) $5\frac{1}{4}$ (e) $1\frac{1}{4}$ (f) $5\frac{3}{5}$

(g) $9\frac{11}{19}$ (h) $4\frac{3}{8}$ (i) $^-8\frac{2}{3}$

2. Find each product.

(a) $4\frac{2}{3} \times 6\frac{3}{8}$ (b) $7\frac{5}{6} \times 5\frac{1}{4}$ (c) $1\frac{1}{4} \times 5\frac{3}{5}$

(d) $9\frac{11}{19} \times 4\frac{3}{8}$ (e) $^-1\frac{6}{7} \times 6\frac{1}{3}$ (f) $^-8\frac{1}{2} \times {}^-5\frac{3}{5}$

3. Find each quotient.

(a) $4\frac{2}{3} \div 6\frac{3}{8}$ (b) $7\frac{5}{6} \div 5\frac{1}{4}$ (c) $1\frac{1}{4} \div 5\frac{3}{5}$

(d) $9\frac{11}{19} \div 4\frac{3}{8}$ (e) $^-1\frac{6}{7} \div 6\frac{1}{3}$ (f) $^-8\frac{1}{2} \div {}^-5\frac{3}{5}$

9-7 Density of the Rational Numbers

We say that the set of rational numbers is **dense**; that is, there is always at least one rational number between any two distinct rational numbers.

The density of the rational numbers may be illustrated with any two rational numbers; for example, $\frac{1}{3}$ and $\frac{2}{3}$. How many rational numbers can we find between these two numbers? Since $\frac{1}{3} = \frac{2}{6}$ and $\frac{2}{3} = \frac{4}{6}$, we can find $\frac{3}{6}$ $\left(\text{that is, } \frac{1}{2}\right)$ between $\frac{1}{3}$ and $\frac{2}{3}$. Notice that we can find the number halfway between $\frac{1}{3}$ and $\frac{2}{3}$ by adding the two numbers and dividing by 2:

$$\frac{1}{3} + \frac{2}{3} = \frac{3}{3} \quad \text{and} \quad \frac{3}{3} \div 2 = \frac{3}{3} \times \frac{1}{2} = \frac{3}{6}.$$

In a similar manner we can find a number between $\frac{1}{3}$ and $\frac{3}{6}$:

$$\frac{1}{3} + \frac{3}{6} = \frac{5}{6} \quad \text{and} \quad \frac{5}{6} \div 2 = \frac{5}{6} \times \frac{1}{2} = \frac{5}{12}.$$

Following the same procedure, we can find a number halfway between $\frac{1}{3}$ and $\frac{5}{12}$. This process can be continued indefinitely. Thus, between any two distinct rational numbers there are infinitely many rational numbers. This means that the number of points representing rational numbers on any segment of the number line is unlimited. We should note that the principle of density does not apply to the set of whole numbers. For example, what are the whole numbers between 1 and 2?

Exercises

1. Can you find an integer between $^-2$ and $^-3$?

2. Is the set of integers dense?

3. If we are given the rational number $\frac{1}{2}$, what is the next largest rational number?

4. How many rational numbers are there between 0 and 1?

5. Use the procedure described in Section 9-7 to find five rational numbers between:

   (a) $\frac{2}{5}$ and $\frac{3}{5}$ (b) $\frac{7}{10}$ and $\frac{8}{10}$

   (c) $\frac{-3}{4}$ and $^-1$

6. What is the smallest positive rational number?

7. What is the smallest nonnegative rational number?

8. What is the largest positive rational number?

9. If $\dfrac{a}{b}$ and $\dfrac{c}{d}$ are distinct rational numbers, is there a rational number half-way between them?

10. Use the procedure described in Section 9-7 to find the rational number halfway between $\dfrac{a}{b}$ and $\dfrac{c}{d}$.

9-8 Chapter Test

Indicate whether each statement is true or false. If false, tell why.

1. The identity element for the multiplication of rational numbers may be expressed in the form $\dfrac{a}{a}$, where a is any positive or negative integer.

2. Every rational number has a multiplicative inverse.

3. The set of fractional numbers is dense.

4. The reciprocal of a negative rational number may never equal the number.

5. The set of nonnegative rational numbers is closed under the operation of multiplication.

6. A meaningful statement of $3 \div \dfrac{1}{2}$ is "How many $\dfrac{1}{2}$'s are in 3?"

7. A repeated addition interpretation could be used to explain $\dfrac{1}{2} \times \dfrac{5}{6}$.

8. The definition for multiplication of two rational numbers $\dfrac{p}{q}$ and $\dfrac{r}{s}$ is $\dfrac{p \times s}{q \times r}$.

9. $\left(\dfrac{1}{3} \times \dfrac{2}{7}\right) \times \dfrac{5}{9} = \left(\dfrac{2}{7} \times \dfrac{1}{3}\right) \times \dfrac{5}{9}$ is an illustration of the associative property of multiplication for rational numbers.

10. Since $\left(4 \div \dfrac{1}{2}\right) \div \dfrac{3}{3} = 4 \div \left(\dfrac{1}{2} \div \dfrac{3}{3}\right)$, we may conclude that division of rational numbers is associative.

Select the best possible answer.

11. In using a unit region to show $\dfrac{2}{3} \times \dfrac{5}{6}$, the region would be divided into:
   - (a) Halves
   - (b) Thirds
   - (c) Sixths
   - (d) Fifteenths
   - (e) Eighteenths

12. The result for $4 \div \dfrac{3}{5}$ can best be illustrated by repeatedly removing:

    (a) $\dfrac{1}{5}$ cup from a measuring pitcher containing 12 cups of sand.

    (b) $\dfrac{1}{3}$ cup from a measuring pitcher containing 20 cups of sand.

    (c) $\dfrac{1}{5}$ cup from a measuring pitcher containing $\dfrac{4}{3}$ cups of sand.

    (d) $\dfrac{5}{3}$ cups from a measuring pitcher containing 4 cups of sand.

    (e) $\dfrac{3}{5}$ cup from a measuring pitcher containing 4 cups of sand.

13. Which of the following represents the identity element for the multiplication of rational numbers?

    (a) 0   (b) $\dfrac{2}{1}$

    (c) $\dfrac{1}{0}$ (d) $\dfrac{6}{6}$

    (e) $^-1$

14. The reciprocal of $\dfrac{0}{1}$ is:

    (a) 1   (b) 0

    (c) $\dfrac{1}{0}$ (d) $\dfrac{0}{1}$

    (e) Not defined

15. Which of these statements is *not* true about the product of two nonzero fractional numbers?
    (a) If each factor is greater than 1, the product is greater than either factor.
    (b) If one of two factors is less than 1, the product is less than the other factor.
    (c) If one factor is unchanged and the other increased, the product increases.
    (d) If one factor is unchanged and the other decreased, the product decreases.
    (e) If each factor is less than 1, the product is greater than either factor.

16. Which of the following is false?

(a) $2 \times 1\frac{1}{2} = 1 + 2$

(b) $\frac{2}{3} \times \frac{4}{4} = \frac{2}{3}$

(c) $\frac{5}{9} \times \frac{0}{5} = \frac{0}{14}$

(d) $\frac{3}{4} + \frac{0}{8} = \frac{28}{32}$

(e) $\frac{1}{6} \div \frac{1}{2} = \frac{1}{3}$

17. Which of the following is (are) equivalent to $\frac{8}{4}$?

I. $\frac{16}{8}$ II. 2 III. $10 \div 5$

(a) I only

(b) II only

(c) I and II only

(d) I and III only

(e) I, II, and III

18. The shading of the unit regions below illustrates:

(a) $\frac{2}{3}$

(b) $\frac{6}{12}$

(c) $\frac{6}{4}$

(d) $\frac{2}{4}$

(e) $\frac{1}{2}$

19. Which of these sets of numbers is dense?

(a) Counting numbers

(b) Whole numbers

(c) Natural numbers

(d) Integers

(e) None of these

20. Which set of numbers is closed with respect to division, except division by 0?

(a) Natural numbers

(b) Whole numbers

(c) Integers

(d) Rational numbers

(e) None of these

chapter 10

Ratio, Percent, and Decimals

In dealing with ratio, percent, and decimals we must be careful not to confuse the concepts of number and numeral. It is important to realize that $\frac{1}{4}$, 25%, and 0.25 are different names for the same number. Convenience and social applications make it necessary for us to develop facility in translating from one form of naming a number to another. The ability to change from the percent form to the fractional or decimal form makes it possible for us to use the basic computational algorithms for fractions and decimals. Thus, we do not need to develop special algorithms for use with percent.

10-1 Ratio and Proportion

A comparison of two numbers by division is a **ratio.** The rational number $\frac{3}{4}$ may be considered a ratio, the comparison of 3 to 4. There are several alternate ways of representing the same ratio, such as

$$3:4, 3 \div 4, \text{ and } \frac{3}{4}.$$

Each of these may be read "the ratio of three to four" and also "three is to four."

In some instances ratios are equal to each other. For example, the ratios $\frac{3}{4}$ and $\frac{18}{24}$ are equal; that is, $\frac{3}{4} = \frac{18}{24}$. A statement that two ratios are equal is a **proportion**. Thus the statement $\frac{3}{4} = \frac{18}{24}$ is a proportion and may be read "three is to four as eighteen is to twenty-four." A proportion merely indicates two different names for the same ratio.

The equality of two ratios may be determined by applying some of the ideas that we studied previously. We learned that two fractional numbers are equal if the numerator of the first times the denominator of the second is equal to the denominator of the first times the numerator of the second. For example,

$$\frac{3}{4} = \frac{18}{24}, \quad \text{since} \quad 3 \times 24 = 4 \times 18.$$

The same relationship holds for any two equal ratios. In general, for the ratios $\frac{a}{b}$ and $\frac{c}{d}$,

$$\frac{a}{b} = \frac{c}{d} \quad \text{if } a \times d = b \times c,$$

and

$$a \times d = b \times c \quad \text{if } \frac{a}{b} = \frac{c}{d}.$$

Sets are sometimes used to illustrate the ideas of ratio and proportion. Many people are currently using the term **rate pair** to indicate the relationship between two unlike sets. For example, if lemons are sold at 6 for 25¢ a relationship between sets of lemons and sets of cents may be shown as in the following figure.

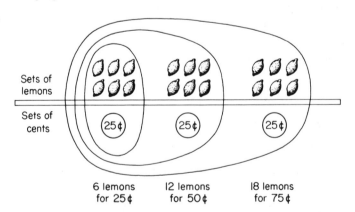

Sets of lemons		
Sets of cents		
6 lemons for 25¢	12 lemons for 50¢	18 lemons for 75¢

The following equivalence may be stated for the rate pairs determined above:

$$\frac{6}{25} = \frac{12}{50} = \frac{18}{75}.$$

Ratio and proportion situations occur primarily in problem solving which involves finding an equivalent ratio or proportion. These situations are generally derived from a one-to-many correspondence, a many-to-one correspondence, or a many-to-many correspondence.

A number line may also be used as an aid in understanding ratio and proportion. For example, if John travels 15 miles each hour, how far will he travel in 2 hours? in 3 hours?

Miles

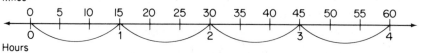

Hours

Exercises

1. Express the ratio of the first number to the second number as a fraction.
   (a) 4, 9 (b) 30, 55
   (c) 32, 56 (d) 53, 17

2. Represent each of the ratios in Exercise 1 as a rational number in simplest form.

3. In comparing two measurements, we change both measures to the same unit of measure, to avoid confusion. Expressed in simplest form, what is the ratio of:
   (a) 1 foot to 1 yard? (b) 10 minutes to 1 hour?
   (c) 3 feet to 24 inches? (d) 10 cents to 5 dollars?

4. Find three equivalent ratios for each of the following.

   (a) $\frac{1}{2}$ (b) $\frac{2}{3}$

   (c) $\frac{7}{8}$ (d) $\frac{18}{45}$

5. In each statement find a replacement for the variable such that the statement will be true.

   (a) $\frac{5}{6} = \frac{\square}{54}$ (b) $\frac{11}{13} = \frac{33}{\square}$

   (c) $\frac{\square}{1} = \frac{5}{8}$ (d) $\frac{1}{3} = \frac{\square}{7}$

6. In each proportion determine the number that the letter t represents.

(a) $\dfrac{5}{15} = \dfrac{t}{36}$

(b) $\dfrac{25}{t} = \dfrac{10}{5}$

(c) $\dfrac{70}{17} = \dfrac{30}{t}$

(d) $\dfrac{3}{8} = \dfrac{t}{15}$

7. Indicate whether each statement is true or false.

(a) $\dfrac{3}{4} = \dfrac{2}{3}$

(b) $\dfrac{6}{8} = \dfrac{9}{12}$

(c) $\dfrac{11}{14} = \dfrac{13}{17}$

(d) $\dfrac{24}{6} = \dfrac{28}{7}$

8. Make a diagram of the relationship between the elements of the first set and the elements of the second set.
(a) 5 pencils for 10¢   (b) 2 erasers for 15¢

9. Use the diagram in Exercise 8(a) to show the cost of:
(a) 1 pencil   (b) 2 pencils
(c) 9 pencils

10. Use the diagram in Exercise 8(b) to show the cost of:
(a) 1 eraser   (b) 4 erasers
(c) 5 erasers

11. Jim travels 48 miles in 1 hour. Use a number line drawing to show how far he travels in:

(a) $\dfrac{1}{3}$ hour   (b) $\dfrac{1}{2}$ hour

(c) $\dfrac{3}{4}$ hour   (d) 2 hours

12. Write a proportion and find the missing number for each of the following.
(a) If 5 pounds of hamburger will serve 9 persons, how many pounds will be needed to serve 17 persons?
(b) If 3 loaves of bread cost 78¢, what is the cost of 8 loaves?

10-2 Percent

Ratios are frequently expressed with denominators of one hundred. For example,

$$\frac{7}{10} = \frac{70}{100}.$$

The expression $\dfrac{70}{100}$ may be read as "the ratio of 70 to 100," or as "70 percent."

The term percent is derived from the Latin phrase "per centum," which means "by the hundred." Hence, 70 percent means $\frac{70}{100}$.

The symbol % may be substituted for the word "percent." It is important to note that the % symbol, division by 100, and multiplication by $\frac{1}{100}$ are different notations for the *same mathematical idea*. Any rational number may be represented by a fraction with a denominator of 100 and, if $\frac{1}{100}$ is replaced with a percent symbol, may be written as a percent. For example,

$$\frac{3}{5} = \frac{3}{5} \times \frac{20}{20} = \frac{60}{100} = 60 \times \frac{1}{100} = 60\%,$$

$$\frac{9}{25} = \frac{9}{25} \times \frac{4}{4} = \frac{36}{100} = 36 \times \frac{1}{100} = 36\%,$$

$$\frac{2}{7} = \frac{2}{7} \times \frac{100}{100} = \frac{200}{700} = \frac{200}{7} \times \frac{1}{100} = 28\frac{4}{7} \times \frac{1}{100} = 28\frac{4}{7}\%.$$

We may use the idea of proportion to express any rational number $\frac{a}{b}$ in percent notation by finding the number n such that $\frac{a}{b} = \frac{n}{100}$. Thus, $\frac{a}{b} = \frac{n}{100} = n$ percent. For example, if $\frac{2}{3}$ is to be expressed as a percent, we set up the proportion $\frac{2}{3} = \frac{n}{100}$. Then $2 \times 100 = 3 \times n$ and $\frac{200}{3} = n$, that is, $66\frac{2}{3}$. If in the original proportion we replace n by its equivalent, we have

$$\frac{2}{3} = \frac{66\frac{2}{3}}{100} = 66\frac{2}{3} \times \frac{1}{100} = 66\frac{2}{3}\%.$$

For emphasis, we recall that fractions and percents are different numeral forms that may be used to name the same number. In computation, we find a use for both forms and need a method for changing from one to the other.

Exercises

1. Use the form $\frac{a}{b} = \frac{n}{100}$ and express each rational number as a percent.

   (a) $\frac{3}{4}$ (b) $\frac{7}{7}$

   (c) $\frac{8}{5}$ (d) $3\frac{1}{2}$

(e) $\dfrac{2}{5}$ (f) 2

(g) $\dfrac{1}{2}$ (h) $\dfrac{7}{8}$

2. Express each percent as a fraction in simplest form.

   (a) 90%   (b) 42%

   (c) 63%   (d) $16\dfrac{2}{3}\%$

   (e) 130%   (f) $\dfrac{1}{2}\%$

   (g) 1%   (h) 100%

3. Find 100% of 65.   5. Find $\dfrac{1}{2}$ of 10.

4. Find 1% of 65.   6. Find $\dfrac{1}{2}\%$ of 10.

7. What percent of:
   (a) $2000 is $350?   (b) 328 is 41?

10-3 Decimal Notation

Frequently in computational work decimal notation offers an advantage over fractional notation. It is important to note from previous discussions that decimal notation is merely an extension of our decimal system of numeration and not a new system. The following place-value chart shows decimal representations both to the left and to the right of the decimal point.

In base ten notation the value of each place is ten times the value of the place to its right and one-tenth of the value of the place to its left. For example,

$$100 = 10 \times 10 \quad \text{and} \quad 100 = \frac{1}{10} \times 1{,}000.$$

Place-Value Chart

Thousands	*Hundreds*	*Tens*	*Ones*
$10 \times 10 \times 10$	10×10	10	1
10^3	10^2	10^1	10^0
1,000	100	10	1

We learned earlier that two characteristics of the decimal system of numeration are place value and a base of ten. Applying these characteristics, we represented whole numbers in expanded form. For example,

$$234 = (2 \times 100) + (3 \times 10) + (4 \times 1)$$
$$= (2 \times 10^2) + (3 \times 10^1) + (4 \times 10^0).$$

We may also use expanded notation to represent rational numbers that are not whole numbers. For example,

$$\frac{32{,}786}{1{,}000} = 32.786$$

$$32.786 = (3 \times 10) + (2 \times 1) + \left(7 \times \frac{1}{10}\right) + \left(8 \times \frac{1}{100}\right) + \left(6 \times \frac{1}{1{,}000}\right)$$

$$= (3 \times 10^1) + (2 \times 10^0) + (7 \times 10^{-1}) + (8 \times 10^{-2}) + (6 \times 10^{-3}),$$

When the number 32.786 is written in decimal notation, the **decimal point** locates the ones place, and the number is read "thirty-two *and* seven hundred eighty-six thousandths." Observe that the word "and" may designate the decimal point. Caution should be exercised when reading numerals to avoid careless use of the word "and."

We learned previously that, in general,

$$b^m \cdot b^n = b^{m+n},$$

$$b^m \div b^n = b^{m-n},$$

$$(b^m)^n = b^{m \cdot n}.$$

The idea of a negative exponent was introduced, and we agreed that, in general,

$$a^{-n} = \frac{1}{a^n}.$$

We should note that the procedures for working with multiplication, division, and powers of powers apply to both positive and negative exponents.

for Base Ten

Tenths	Hundredths	Thousandths	Ten-Thousandths
$\dfrac{1}{10}$	$\dfrac{1}{10 \times 10}$	$\dfrac{1}{10 \times 10 \times 10}$	$\dfrac{1}{10 \times 10 \times 10 \times 10}$
10^{-1}	10^{-2}	10^{-3}	10^{-4}
0.1	0.01	0.001	0.0001

For example,

$$\frac{1}{10} \times \frac{1}{10} = \frac{1}{100}$$

may also be expressed as

$$10^{-1} \times 10^{-1} = 10^{-2} = \frac{1}{10^2} = \frac{1}{100};$$

$$\frac{1}{100} \div \frac{1}{10} = \frac{1}{10}$$

may also be expressed as

$$10^{-2} \div 10^{-1} = 10^{-1} = \frac{1}{10};$$

$$\left(\frac{1}{10}\right)^2 = \frac{1}{10} \times \frac{1}{10} = \frac{1}{100}$$

may also be expressed as

$$(10^{-1})^2 = 10^{-2} = \frac{1}{10^2} = \frac{1}{100}.$$

Exercises

1. Use decimal notation and write the numeral for:

  (a) $(5 \times 10{,}000) + (7 \times 1{,}000) + (2 \times 100) + (9 \times 10) + (3 \times 1)$

  $$+ \left(6 \times \frac{1}{10}\right) + \left(4 \times \frac{1}{100}\right) + \left(1 \times \frac{1}{1{,}000}\right)$$

  (b) $(3 \times 10^2) + \left(7 \times \frac{1}{10^2}\right) + \left(4 \times \frac{1}{10^4}\right)$

  (c) $6 \times \frac{1}{10^5}$

  (d) $(2 \times 10^{-1}) + (9 \times 10^{-2}) + (5 \times 10^{-3}) + (6 \times 10^{-4})$

2. Write in expanded form.
  (a) 0.25   (b) 694.851
  (c) 300.009   (d) 5,320.7604

3. Read each numeral in words.
  (a) 246   (b) 2.46
  (c) 178.296   (d) 0.9

4. How many times greater is the value of the hundreds place than the value of the:
  (a) Ones place?   (b) Tenths place?
  (c) Thousandths place?

5. What fraction of the value of the thousands place is the value of the:
    (a) Tens place?   **(b)** Tenths place?
    (c) Thousandths place?

10-4 Terminating and Repeating Decimals

Frequently it is more convenient to work with the decimal form of a rational number than it is to work with the fractional form. Since a rational number expressed in fractional form is the quotient of two integers, we can translate from the fractional form to the decimal form by performing the indicated division and expressing the quotient as a decimal numeral. For convenience in performing an indicated division, we observe that integers may be expressed as equivalent decimals. For example,

$$1 = 1.0 = 1.00 = 1.000 = 1.0000\ldots,$$

$$2 = 2.0 = 2.00 = 2.000 = 2.0000\ldots,$$

$$3 = 3.0 = 3.00 = 3.000 = 3.0000\ldots,$$

$$^-5 = {}^-5.0 = {}^-5.00 = {}^-5.000 = {}^-5.0000\ldots.$$

Consider the following examples of the conversion from a fractional to a decimal numeral by the division process.

Example 1 Change $\dfrac{1}{2}$ to a decimal numeral.

$$
\begin{array}{r}
0.5 \\
2)\overline{\,1.0\,} \\
-1\,0 \\
\hline
0
\end{array}
$$

Thus, $\dfrac{1}{2} = 0.5$. (Note: Divisor $2 = 2^1 \times 5^0$.)

Example 2 Change $\dfrac{3}{5}$ to a decimal numeral.

$$
\begin{array}{r}
0.6 \\
5)\overline{\,3.0\,} \\
-3\,0 \\
\hline
0
\end{array}
$$

Thus, $\dfrac{3}{5} = 0.6$. (Note: Divisor $5 = 2^0 \times 5^1$.)

Example 3   Change $\frac{1}{8}$ to a decimal numeral.

$$
\begin{array}{r}
0.125 \\
8\overline{)1.000} \\
-8 \\
\hline
20 \\
-16 \\
\hline
40 \\
-40 \\
\hline
0
\end{array}
$$

Thus, $\frac{1}{8} = 0.125$. (Note: Divisor $8 = 2^3 \times 5^0$.)

Notice that in Examples 1, 2, and 3 the division process **terminates**; that is, the division process is continued until we have a remainder of 0. Notice also that in each case _the divisor is a number that is expressible as a product of a power of 2 and a power of 5_; that is, each divisor is a factor of a power of 10. Any whole number that is a factor of a power of 10 is expressible in the form $2^p \cdot 5^q$, where p and q are nonnegative integers. _Division by a factor of a power of 10 always terminates._

With many rational numbers, however, the division process does _not_ terminate. For example, examine the division process used to change $\frac{1}{3}$ to a decimal numeral:

$$
\begin{array}{r}
0.333\ldots \\
3\overline{)1.0000\ldots} \\
-9 \\
\hline
10 \\
-9 \\
\hline
10 \\
-9 \\
\hline
1
\end{array}
$$

Thus, $\frac{1}{3} = 0.333\ldots\,.$

Note that the divisor 3 may not be expressed in the form $2^p \cdot 5^q$, where p and q are nonnegative integers. Although the process of dividing does not terminate, we observe a repeating pattern in the quotient and in the differences (remainders). The numeral 3 is repeating in the quotient and the numeral 1 is repeating in the differences (remainders). The dots in $\frac{1}{3} = 0.333\ldots$ indicate that the threes continue without end in the quotient. A numeral of this type is referred to as a **repeating decimal** (also called a **periodic decimal**). The set of digits that repeats is called the **cycle**, and a bar is usually placed over the cycle to make it obvious. Thus, it is customary to write $0.333\ldots$ as $0.\overline{3}$.

To change the fraction $\frac{13}{55}$ to a decimal numeral, we may apply the division process:

$$
\begin{array}{r}
0.2363636\ldots \\
55)\overline{13.0000000\ldots} \\
-11\,0 \\
\hline
2\,00 \\
-1\,65 \\
\hline
350 \\
-330 \\
\hline
200 \\
-165 \\
\hline
350 \\
-330 \\
\hline
200 \\
-165 \\
\hline
350 \\
-330 \\
\hline
20
\end{array}
$$

Thus, $\frac{13}{55} = 0.2363636\ldots.$

Note the repeating pattern in the quotient and in the differences (remainders). In the quotient the two-digit numeral 36 is repeating, as indicated by the three-dot notation. The cycle 36 may be more clearly indicated by using the bar instead of the three dots, and we have $\frac{13}{55} = 0.2\overline{36}$. We also observe a repeating pattern in the differences (remainders) obtained at each step of the division process:

$$13, 20, 35, 20, 35, 20, 35, 20, \ldots.$$

It is important to understand that if we even once obtain a difference (remainder) in the division process such that the difference is equal to a previous difference, a cycle is repeating. Note also that only nonnegative remainders *less than the divisor* will be obtained; hence, a cycle must repeat in at most $n - 1$ steps when the divisor is represented by n. Thus, for any rational number $\frac{a}{b}$, the division process terminates if b is a factor of a power of 10 and, in general, the division process leads to a repetition of at most $b - 1$ digits.

Consider the examples below and observe the repeating pattern in each quotient.

Example 4 Change $\frac{1}{2}$ to a repeating decimal.

$$\begin{array}{r} 0.500.\,.\,. \\ 2)\overline{\;1.000.\,.\,.} \\ -1\,0 \\ \hline 00 \\ -\;0 \\ \hline 00 \\ -\;0 \\ \hline 0 \end{array}$$

Thus, $\frac{1}{2} = 0.5\bar{0}$. Note that the division process terminates, since the divisor, 2, is a factor of a power of 10. Note also that the division process leads to a cycle of one digit in the quotient, since the divisor (n) is 2 and $n - 1$ is 1.

Example 5 Change $\frac{3}{5}$ to a repeating decimal.

$$\begin{array}{r} 0.600.\,.\,. \\ 5)\overline{\;3.000.\,.\,.} \\ -3\,0 \\ \hline 00 \\ -\;0 \\ \hline 00 \\ -\;0 \\ \hline 0 \end{array}$$

Thus, $\frac{3}{5} = 0.6\bar{0}$. Note that the division process terminates, since the divisor, 5, is a factor of a power of 10. Note also that the division process leads to a cycle of at most four digits in the quotient, since the divisor (n) is 5 and $n - 1$ is 4. In this case the cycle is only one digit.

Example 6 Change $\frac{1}{8}$ to a repeating decimal.

$$\begin{array}{r} 0.12500.\,.\,. \\ 8)\overline{\;1.00000.\,.\,.} \\ -\;8 \\ \hline 20 \\ -16 \\ \hline 40 \\ -40 \\ \hline 00 \\ -\;0 \\ \hline 00 \\ -\;0 \\ \hline 0 \end{array}$$

Thus, $\frac{1}{8} = 0.125\bar{0}$. Note that the division process terminates, since the divisor, 8, is a factor of a power of 10. Note also that the division process leads to a cycle of at most seven digits in the quotient, since the divisor (n) is 8 and $n - 1$ is 7. In this case the cycle is only one digit.

In Examples 4, 5, and 6 we see that terminating decimals 0.5, 0.6, and 0.125 may also be represented as repeating decimals. Since a terminating decimal is a special case of a repeating decimal, we generalize that any rational number may be expressed as a repeating decimal.

We have discussed the procedure for changing from a fractional numeral to a decimal numeral. Now suppose we have the opposite situation; that is, we wish to change a decimal numeral to a fractional numeral. If the decimal numeral is a terminating one, the problem is simple. For example, 0.125 may be expressed as $\dfrac{125}{1,000}$ and then changed to simplest form. Then we have

$$0.125 = \frac{125}{1,000} = \frac{1}{8}.$$

If we have a repeating decimal, the problem of representing it as a fractional numeral is more difficult. For example, what is the rational number represented by $0.\overline{6}$? We shall write $0.\overline{6}$ as 0.66. . . and name it n; that is, $n = 0.66$. . . . We then multiply each member of the equality by 10, which gives us

$$10 \times n = 10 \times 0.66. \ldots$$

Completing the multiplication, we have

$$10n = 6.6. \ldots$$

If we subtract n from each side of the above equation (recall that $n = 0.66$. . .), we have $10n - n = 6.6. \ldots - 0.66. \ldots$, which is $9n = 6$, assuming that the repeated digits may be subtracted in the usual manner. Then $n = \dfrac{6}{9}$ and, in simplest form, $n = \dfrac{2}{3}$. Thus, we have determined that $0.\overline{6} = \dfrac{2}{3}$.

In general, to represent any repeating decimal as a fractional numeral, we multiply by the power of 10 whose exponent equals the number of digits in the cycle and then subtract the original number.

Example 7   Find a fractional numeral for $0.\overline{27}$.

We let $n = 0.\overline{27}$ and multiply each member of the equality by 10^2; that is, 100:

$$\begin{array}{r} 100n = 27.\overline{27} \\ - n = 0.\overline{27} \\ \hline 99n = 27 \end{array} \quad \text{and} \quad n = \frac{27}{99} = \frac{3}{11}.$$

Thus, $n = 0.\overline{27} = \dfrac{3}{11}$.

Exercises

1. Find the decimal numeral for each fraction.

(a) $\dfrac{7}{10}$ (b) $\dfrac{39}{50}$

(c) $\dfrac{13}{20}$ (d) $\dfrac{3}{8}$

2. Express each divisor in Exercise 1 as a product of a power of 2 and a power of 5; that is, in the form $2^p \cdot 5^q$, where p and q are nonnegative integers.

3. Use the division process to change each fraction to decimal form, and indicate the cycle by using a bar over one set of the repeated digits.

(a) $\dfrac{4}{7}$ (b) $\dfrac{10}{15}$

(c) $\dfrac{8}{11}$ (d) $\dfrac{29}{37}$

4. If possible, express each divisor in Exercise 3 as a product of a power of 2 and a power of 5.

5. (a) Find the decimal numeral for $\dfrac{5}{13}$.

(b) Is the decimal numeral in Exercise 5(a) a repeating or terminating decimal?

(c) What is the cycle for the decimal numeral in Exercise 5(a)?

6. Write a decimal numeral for each fraction.

(a) $\dfrac{2}{9}$ (b) $\dfrac{25}{14}$

(c) $\dfrac{47}{8}$ (d) $11\dfrac{5}{6}$

7. Find the rational number represented by each decimal numeral.
(a) 0.23   (b) 0.165
(c) 3.0738   (d) 0.009

8. Find the rational number represented by each decimal numeral.
(a) $0.\overline{8}$ (b) $0.\overline{47}$
(c) $0.\overline{235}$ (d) $0.6\overline{2}$

9. Without using the division process to convert each fraction to a decimal numeral, tell which of these fractions will have a decimal representation that terminates.

(a) $\dfrac{1}{7}$ (b) $\dfrac{5}{12}$

(c) $\dfrac{7}{15}$ (d) $\dfrac{13}{40}$

(e) $\dfrac{7}{8}$ (f) $\dfrac{25}{75}$

(g) $\dfrac{9}{25}$ (h) $\dfrac{20}{30}$

10-5 Operations with Numbers Expressed
as Terminating Decimals

Any number represented by a terminating decimal may also be represented by an equivalent fraction of the form $\frac{a}{b}$, where b is some power of 10. Consequently, we may perform addition, subtraction, multiplication, and division on such numbers by changing the decimals to fractions and applying the computational principles and procedures that we studied with the set of rational numbers. For the sake of convenience, however, we have developed algorithms for performing the fundamental operations on numbers expressed as terminating decimals.

Addition and subtraction   To add or subtract numbers expressed as terminating decimals, we arrange the numerals in vertical columns according to the concept of place value. We then add or subtract as if we were working with whole numbers. For example, the sum of 5.28 and 2.61 is illustrated below.

$$\begin{array}{r} 5.28 \\ +2.61 \\ \hline 7.89 \end{array}$$

Since $5.28 = \frac{528}{100}$ and $2.61 = \frac{261}{100}$, we may also find the sum by using fractions.

$$\frac{528}{100} + \frac{261}{100} = \frac{528 + 261}{100} = \frac{789}{100} = 7.89$$

The reason for "lining up" the decimal points in a column for addition of numbers expressed as decimals should become apparent when we study the following example.

$$\begin{array}{r} 2.78 \\ +34.2 \\ \hline \end{array}$$

Since $2.78 = \frac{278}{100}$ and $34.2 = \frac{342}{10}$, we need to obtain common denominators in order to complete the addition. Thus we change 34.2 to $34.20 = \frac{3420}{100}$ and complete the addition.

$$\begin{array}{r} 2.78 \\ +34.20 \\ \hline 36.98 \end{array} \qquad \frac{278}{100} + \frac{3420}{100} = \frac{3698}{100} = 36.98$$

Additional understanding of the algorithm may be gained from studying the role of place value in this example.

$$48.123 = (4 \times 10^1) + (8 \times 10^0) \ + (1 \times 10^{-1}) + (2 \times 10^{-2}) + (3 \times 10^{-3})$$
$$.058 = \hspace{5.5cm} (5 \times 10^{-2}) + (8 \times 10^{-3})$$
$$\underline{+\ 7.61 \ = \hspace{2.3cm} (7 \times 10^0) \ + (6 \times 10^{-1}) + (1 \times 10^{-2})}$$
$$55.791 = (4 \times 10^1) + (15 \times 10^0) + (7 \times 10^{-1}) + (8 \times 10^{-2}) + (11 \times 10^{-3})$$
$$= (5 \times 10^1) + (5 \times 10^0) \ + (7 \times 10^{-1}) + (9 \times 10^{-2}) + (1 \times 10^{-3})$$
$$= 55.791$$

The difference between 6.98 and 4.65 may be determined by any of the procedures noted above. In the decimal form we have

$$6.98 = (6 \times 10^0) + (9 \times 10^{-1}) + (8 \times 10^{-2})$$
$$\underline{-4.65 = (4 \times 10^0) + (6 \times 10^{-1}) + (5 \times 10^{-2})}$$
$$2.33 = (2 \times 10^0) + (3 \times 10^{-1}) + (3 \times 10^{-2})$$

and, in the fractional form,

$$\frac{698}{100} - \frac{465}{100} = \frac{698 - 465}{100} = \frac{233}{100} = 2.33.$$

Multiplication   To multiply numbers expressed as terminating decimals, we temporarily disregard the decimal points and multiply as if we were working with whole numbers. Then we locate the decimal point in the product. The number of decimal places in the product is determined by adding the number of decimal places in each factor. The examples

$$\frac{3}{10} \times \frac{7}{10} = \frac{21}{100} \quad \text{and} \quad 0.3 \times 0.7 = 0.21$$

illustrate that products represented by decimal numerals are equivalent to the same products represented by fractional numerals. The parallel between multiplication of numbers expressed as fractions and multiplication of the same numbers expressed as decimals is excellent for developing an understanding of the multiplication procedures used with decimals. Study the following examples.

Example 1

$$\begin{array}{r} 0.25 \\ \times\ 0.4 \\ \hline 0.100 \end{array} \qquad \frac{4}{10} \times \frac{25}{100} = \frac{4}{10^1} \times \frac{25}{10^2} = \frac{4 \times 25}{10^1 \times 10^2} = \frac{100}{10^3}$$

Example 2

$$\begin{array}{r} 2.341 \\ \times\ 0.07 \\ \hline 0.16387 \end{array}$$

$$\frac{7}{100} \times 2\frac{341}{1{,}000} = \frac{7}{100} \times \frac{2{,}341}{1{,}000} = \frac{7}{10^2} \times \frac{2{,}341}{10^3} = \frac{7 \times 2{,}341}{10^2 \times 10^3} = \frac{16{,}387}{10^5}$$

Example 3

$$\begin{array}{r} 0.0123 \\ \times\quad 0.5 \\ \hline 0.00615 \end{array}$$
$$\frac{5}{10} \times \frac{123}{10,000} = \frac{5}{10^1} \times \frac{123}{10^4} = \frac{5 \times 123}{10^1 \times 10^4} = \frac{615}{10^5}$$

A careful observation of the above examples should reveal that the number of decimal places in each product is directly related to the exponents in the denominators.

Division To divide numbers expressed as terminating decimals, we temporarily disregard the decimal points and divide as if we were working with whole numbers. Then we locate the decimal point in the quotient. The examples

$$\frac{6}{10} \div \frac{3}{10} = 2 \quad \text{and} \quad 0.6 \div 0.3 = 2$$

illustrate that quotients represented by fractional numerals are equivalent to the same quotients represented by decimal numerals. The parallel between division of numbers expressed as decimals and division of the same numbers expressed as fractions is excellent for developing an understanding of the division procedures used with decimals. Study the examples below.

Example 1 $0.6 \div 0.03 = 20$ **Example 2** $0.06 \div 0.3 = 0.2$

$$\frac{6}{10} \div \frac{3}{100} = 20 \qquad\qquad \frac{6}{100} \div \frac{3}{10} = \frac{2}{10}$$

Example 3 $10 \div 0.25 = 40$ **Example 4** $107.78 \div 1.7 = 63.4$

$$\frac{10}{1} \div \frac{25}{100} = 40 \qquad\qquad \frac{10,778}{100} \div \frac{17}{10} = \frac{634}{10}$$

We may also use a standard division algorithm to find the quotient of two numbers expressed as decimals. It is imperative that we understand why the division algorithm works before we depend on it as an answer-producing procedure.

$$\begin{array}{r} 20. \\ 0.03)\overline{0.60.} \\ -6 \\ \hline 00 \\ -00 \\ \hline 0 \end{array}$$

The arrows indicate that both 0.03 and 0.6 have been multiplied by 100; that is, $\dfrac{0.6 \times 100}{0.03 \times 100} = \dfrac{60}{3}$.

$$\begin{array}{r} 0.2 \\ 0.3)\overline{0.0.6} \\ -6 \\ \hline 0 \end{array}$$

The arrows indicate that both 0.3 and 0.06 have been multiplied by 10; that is, $\dfrac{0.06 \times 10}{0.3 \times 10} = \dfrac{0.6}{3}$.

$$\begin{array}{r} 40. \\ 0.25\overline{)\,10.00.} \\ -10\,0 \\ \hline 00 \\ -00 \\ \hline 0 \end{array}$$

The arrows indicate that both 0.25 and 10 have been multiplied by 100; that is, $\dfrac{10 \times 100}{0.25 \times 100} = \dfrac{1{,}000}{25}$.

$$\begin{array}{r} 63.4 \\ 1.7\overline{)\,107.7.8} \\ -102 \\ \hline 5\,7 \\ -5\,1 \\ \hline 6\,8 \\ -6\,8 \\ \hline 0 \end{array}$$

The arrows indicate that both 1.7 and 107.78 have been multiplied by 10; that is, $\dfrac{107.78 \times 10}{1.7 \times 10} = \dfrac{1{,}077.8}{17}$.

Since any division problem may be expressed as a multiplication problem whose product and one factor are known, we may use the "missing-factor" approach to find the quotient of two numbers expressed as terminating decimals. For example, $10 \div 0.25 = \square$ may be considered as $\square \times 0.25 = 10$, where the variable $\square$ represents the missing factor. Thus we need to determine what number used as a factor with 0.25 yields a product of 10. Recall that we may multiply numbers expressed as decimals as if they were whole numbers, and then locate the decimal point in the numeral for the product. Our knowledge of multiplication facts might lead us to guess that the missing factor is 4; however, the product $4 \times 25 = 100$, and we have a total of two decimal places in the factors, which means that $4 \times 0.25 = 1.00$. This *trial and error* procedure should lead us to realize that the missing factor is 40, since $40 \times 25 = 1{,}000$, and when the decimal places are considered we have $40 \times 0.25 = 10.00$.

Thus, when we use the missing-factor approach to find the quotient of numbers expressed as decimals, we must consider the number of decimal places as we seek the factor that produces the indicated product.

Exercises

1. Work each of these in two ways, first by using decimals and then by using fraction equivalents of the decimals.
   (a) $7.8 + 2.75 + 6$ (b) $0.123 + 1.1$
   (c) $16.88 - 9.56$ (d) $0.7 - 0.289$

2. Find each product, first by using decimal numerals and then by using fractional numerals.
   (a) 4.07×0.3 (b) 26.8×12
   (c) 0.005×0.00009 (d) 189.642×7

3. Express each quotient as a fraction with a natural number as the denominator.
   (a) $0.65 \div 1.5$ (b) $789 \div 0.0036$
   (c) $4.7896 \div 22.31$ (d) $59.2 \div 0.001$

4. Find each quotient.
   (a) $0.65 \div 1.5$ (b) $2.31 \div 1.4$
   (c) $43.8 \div 0.08$ (d) $0.0001568 \div 7$

5. Express each decimal as a percent.
   (a) 0.09 (b) 0.005
   (c) 2.78 (d) 0.75

6. Find each of the following.
   (a) 3 percent of 16   (b) 100 percent of 97

   (c) 1 percent of 28   (d) $\frac{1}{2}$ percent of 28

10-6 Scientific Notation

In the light of recent advances in science and technology it is essential that we develop a facility for working with very large and very small numbers. The decimal notation for numbers such as

$$2{,}850{,}000{,}000{,}000 \quad \text{and} \quad 0.000000000076$$

is somewhat cumbersome. These numbers may be represented by a simpler notation as

$$2.85 \times 10^{12} \quad \text{and} \quad 7.6 \times 10^{-11}.$$

This form, representing numbers as the product of an appropriate power of 10 and a number that is greater than or equal to 1 but less than 10, is called **scientific notation.** Consider these illustrations of numbers represented in scientific notation:

$$36 = 3.6 \times 10^{1},$$
$$793{,}000{,}000 = 7.93 \times 10^{8},$$
$$10{,}000{,}000{,}000{,}000{,}000 = 1.0 \times 10^{16},$$
$$0.36 = 3.6 \times 10^{-1},$$
$$0.0091 = 9.1 \times 10^{-3},$$
$$0.00000067438 = 6.7438 \times 10^{-7}.$$

Scientific notation also makes certain calculations easier. For example, the product of

$$793{,}000{,}000 \quad \text{and} \quad 10{,}000{,}000{,}000{,}000{,}000$$

may be calculated as follows:

$$793{,}000{,}000 \times 10{,}000{,}000{,}000{,}000{,}000 = (7.93 \times 10^8) \times (1.0 \times 10^{16})$$
$$= 7.93 \times 10^8 \times 1.0 \times 10^{16}$$
$$= (7.93 \times 1.0) \times (10^8 \times 10^{16})$$
$$= 7.93 \times 10^{24}.$$

Exercises

1. Represent each number in scientific notation.
   (a) 0.0000978   (b) 5,640,000
   (c) 1.265   (d) 0.000000000000431
   (e) 96,783,000   (f) 0.010005

2. Express the factors in scientific notation, then determine each product and represent it in scientific notation.
   (a) $0.00001 \times 689{,}000{,}000$ (b) $23{,}500{,}000 \times 800$
   (c) $186{,}000 \times 86{,}400$ (d) $0.000000234 \times 0.00056$

3. Divide 693,100,000 by 0.000000025 and express the quotient in scientific notation.

4. Represent each number in ordinary decimal notation.
   (a) 7.23×10^5 (b) 1.9×10^{-6}
   (c) 2.478×10^{14} (d) 5.0×10^{-14}
   (e) 6.75×10^{-8} (f) 6.75×10^8

5. Find two standard measures that can be expressed with scientific notation.

10-7 Chapter Test

Indicate whether each statement is true or false. If false, tell why.

1. A proportion is a comparison of two ratios by division.

2. $\dfrac{5}{11} = 0.45.$

3. $1.65 \times 10^{-4} = 0.000165.$

4. Any nonnegative rational number may be expressed as a percent.

5. Terminating-type decimals are a subset of repeating-type decimals.

6. If a number expressed as a decimal is multiplied by 10,000, this is equivalent to moving the decimal point in the numeral four places to the left.

7. $486\% = 0.486.$

8. $0.5\% = \dfrac{1}{200}.$

9. The ratio of 3 feet to 2 yards is $\frac{1}{2}$.

10. $2.5 \times 0.008 = 0.02$.

Select the best possible answer.

11. Which of the following translations of the "%" in 8% are correct?

    I. $\times 0.01$ II. $\times \frac{1}{100}$ III. $\times 100$

    (a) I only   (b) II only
    (c) III only   (d) I and II only
    (e) I and III only

12. Eight hundred and eighty-eight thousandths is written:
    (a) 888,000   (b) 800.088
    (c) 0.888   (d) 880.008
    (e) 800.880

13. $\frac{5}{6}$ equals $0.83\frac{1}{3}$. What does the "$\frac{1}{3}$" represent?

    (a) $\frac{1}{3}$ of 3   (b) $\frac{1}{3}$ of 83

    (c) $\frac{1}{3}$ of 0.83   (d) $\frac{1}{3}$ of 0.01

    (e) $\frac{1}{3}$ of 0.03

14. Which of these is the set of possible remainders when a counting number
    is divided by 7?
    (a) $\{1, 2, 3, 4, \ldots\}$ (b) $\{0, 1, 2, 3, 4, 5, 6, 7, 8, 9\}$
    (c) $\{1, 2, 3, 4, 5, 6\}$ (d) $\{0, 1, 2, 3, 4, 5, 6\}$
    (e) $\{0, 1, 2, 3, 4, 5, 6, 7\}$

15. The scientific notation for 2,670,000,000 is:
    (a) 2.67×10^9 (b) 267×10^7
    (c) 2.67×10^{-1} (d) 2.67×10^{-9}
    (e) 0.267×10^{-10}

16. Which of these may be expressed as a terminating decimal?

    (a) $\frac{1}{3}$ (b) $\frac{3}{11}$

    (c) $\frac{61}{80}$ (d) $\frac{97}{101}$

    (e) None of these

17. Which of the following is not correct?

(a) $150\% = 1.5$ (b) $300\% = \dfrac{300}{100}$

(c) $0.7\% = \dfrac{7}{1,000}$ (d) $\dfrac{5}{16} = 3.125\%$

(e) $\dfrac{6}{40} = 0.15$

18. The quotient 1 million $\div$ 0.01 is:
   (a) 10 thousand   (b) 100 thousand
   (c) 10 million   (d) 100 million
   (e) None of these

19. The property applied in order to justify $0.7 + 0.2$ as $0.1 \times (7 + 2)$ is:
   (a) Associative   (b) Closure
   (c) Commutative   (d) Distributive
   (e) Identity

20. Jim Yarling saves 20% of what he earns selling newspapers. Last week he saved $4.75. How much did he earn?
   (a) $38.00   (b) $28.50
   (c) $23.75   (d) $9.50
   (e) $0.95

chapter 11

Real Number System

In this chapter we extend our concept of number so that each point on a number line will have a number as its coordinate. These new numbers have most of the properties of rational numbers. From previous discussions we recall that the set of rational numbers is closed under the operations of addition, subtraction, multiplication, and division (except division by zero). Also, the operations of addition and multiplication are commutative and associative, and the operation of multiplication is distributive over the operations of addition and subtraction. In the set of rational numbers there is an additive identity and a multiplicative identity; each rational number has a rational number as its additive inverse, and each nonzero rational number has a rational number as its multiplicative inverse.

11-1 Irrational Numbers

In Chapters 8 and 9 we discussed the representation of rational numbers on a number line. We noted that the set of rational numbers is dense; that is, there is a rational number between any two given rational numbers. Consider the number line between the graphs of 0 and 1 as shown below. There are many rational numbers between 0 and 1; for example, there are the points with coordinates $\frac{1}{2}, \frac{1}{3}, \frac{2}{3}, \frac{1}{4}, \frac{3}{4}, \frac{1}{6}$, and $\frac{5}{6}$.

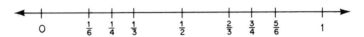

The graphs of rational numbers are spread throughout the number line. Any line segment, regardless of its length, contains an infinite number of graphs of rational numbers. In fact, one might think that every point on the number line is the graph of a rational number. *This is not so.* There are many points on the number line that are *not* graphs of rational numbers. For example, consider a square with sides of length 1 that has the line segment from 0 to 1 as a base.

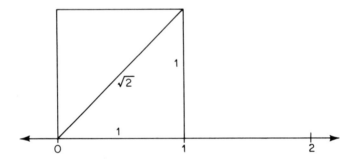

The length of the diagonal of the square may be represented by a number and, when the number is used as a factor two times, the product must be 2. We call the number $\sqrt{2}$ (read "square root of two"), and we can locate a point with coordinate $\sqrt{2}$ on the number line as shown below.

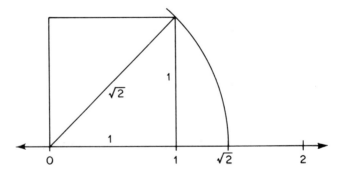

Notice that we have rotated the diagonal clockwise about the point with coordinate 0 until the other end of the diagonal marks the point on the number line with coordinate $\sqrt{2}$. Thus, if we can show that $\sqrt{2}$ is not a rational number, we have located a point on the number line that is not the graph of a rational number.

We shall assume that $\sqrt{2}$ is a rational number and use indirect reasoning to demonstrate that our original assumption is false. If $\sqrt{2}$ is a rational number, then we may express it in the form $\frac{a}{b}$, where a and b are integers, $b \neq 0$, and a and b are not both even. If $\frac{a}{b} = \sqrt{2}$, then

$$\frac{a}{b} \times \frac{a}{b} = 2 \quad \text{and} \quad \frac{a^2}{b^2} = 2.$$

Since a and b are integers, a^2 and b^2 are also integers, and $a^2 = 2b^2$. The integer $2b^2$ must be an even integer, because any integer is even if it is equal to the product of 2 and another integer. Remember that any integer is either even or odd. The product of two odd integers is an odd integer and the product $a \times a$ is even; therefore a cannot be odd. That is, a is an even integer. Thus, $a = 2k$, where k is an integer. Then the statement $a^2 = 2b^2$ may be written as

$$(2k)^2 = 2b^2 \quad \text{and} \quad (2k) \times (2k) = 2b^2$$

and as

$$k \times (2k) = b^2 \quad \text{and} \quad 2k^2 = b^2.$$

Thus b is an integer whose square is even, and b must be even. This is contrary to the assumption that a and b are not both even. Our assumption that there exists a rational number whose square is 2 has led to a contradiction, and so $\sqrt{2}$ is not a rational number.

The numbers that are coordinates of points on the number line and are not rational numbers are called **irrational numbers.** Some other examples of irrational numbers are

$$3\sqrt{2}, 5\sqrt{2}, \sqrt{3}, \sqrt[3]{7}, \sqrt[5]{19}, \pi.$$

Exercises

1. List ten positive numbers less than 10 that do not appear to be rational numbers.

2. Use a triangle with sides 1, $\sqrt{2}$, and $\sqrt{3}$ to graph $\sqrt{3}$ on a number line.

3. Assume that $\sqrt{3}$ is a rational number and use indirect reasoning to demonstrate that the original assumption is false and that therefore $\sqrt{3}$ is an irrational number. (*Note:* You may assume that, if d is an integer and d^2 is divisible by 3, then d is divisible by 3.)

4. Graph $\sqrt{5}$ on a number line.

11-2 Approximating $\sqrt{2}$ as a Decimal

We have learned that $\sqrt{2}$ is the name of a number whose square is 2. We have also demonstrated that there is a point on the number line with co-ordinate $\sqrt{2}$ and that there exist line segments of length $\sqrt{2}$. Now let us see whether we can represent $\sqrt{2}$ in decimal notation. We know from the definition of the number $\sqrt{2}$ that $(\sqrt{2})^2 = 2$. We then calculate the square of 1 and the square of 2 and quickly conclude that $\sqrt{2}$ is greater than 1 and less than 2, since $1^2 = 1$ and $2^2 = 4$. In other words, since

$$1^2 < (\sqrt{2})^2 < 2^2; \quad \text{that is, } 1 < 2 < 4,$$

we assume that

$$1 < \sqrt{2} < 2.$$

This statement, $1 < \sqrt{2} < 2$, is read "one is less than the square root of two and the square root of two is less than two." Now, to obtain a closer approximation, we find the squares of 1.1, 1.2, 1.3, 1.4, and so on, until we determine that $(1.5)^2$ is the first square that is greater than 2:

$$(1.1)^2 = 1.21,$$
$$(1.2)^2 = 1.44,$$
$$(1.3)^2 = 1.69,$$
$$(1.4)^2 = 1.96,$$
$$(1.5)^2 = 2.25.$$

An examination of these calculations should reveal that $\sqrt{2}$ is greater than 1.4 and less than 1.5, since $(1.4)^2 = 1.96$ and $(1.5)^2 = 2.25$:

$$(1.4)^2 < (\sqrt{2})^2 < (1.5)^2; \quad \text{that is, } 1.96 < 2 < 2.25,$$

we assume that

$$1.4 < \sqrt{2} < 1.5.$$

We may extend this process and find an even closer decimal approximation of $\sqrt{2}$ by squaring 1.41, 1.42, 1.43, 1.44, and so on, until we determine that $(1.42)^2 > 2$. We then conclude that $\sqrt{2}$ is greater than 1.41 and less than 1.42, since $(1.41)^2 = 1.9881$ and $(1.42)^2 = 2.0164$:

$$(1.41)^2 < (\sqrt{2})^2 < (1.42)^2; \quad \text{that is, } 1.9881 < 2 < 2.0164,$$

we assume that

$$1.41 < \sqrt{2} < 1.42.$$

The next extension of this process would reveal that $\sqrt{2}$ is greater than 1.414 and less than 1.415, since $(1.414)^2 = 1.999396$ and $(1.415)^2 = 2.002225$:

$$(1.414)^2 < (\sqrt{2})^2 < (1.415)^2; \quad \text{that is, } 1.999396 < 2 < 2.002225,$$

we assume that
$$1.414 < \sqrt{2} < 1.415.$$

We can continue this process indefinitely by checking again and again to find two values that $\sqrt{2}$ lies between. In this manner we can determine an approximation for $\sqrt{2}$ to as many decimal places as we desire. If we had continued our procedure, we would have determined that the approximation of $\sqrt{2}$ to seven decimal places is 1.4142135. For most purposes we do not need so precise a value of $\sqrt{2}$. In fact, 1.414 is commonly used as an approximation of $\sqrt{2}$.

Special note should be made of the fact that we have been calculating *decimal approximations* for $\sqrt{2}$. The process used to determine the decimal approximation for $\sqrt{2}$ suggests that we write $\sqrt{2} = 1.4142135. . .$, where the three dots are used to indicate that the digits continue without terminating.

We recall that many rational numbers are represented as decimals with digits that continue without terminating. For example,

$$\frac{1}{3} = 0.333 \ldots \quad \text{and} \quad \frac{13}{55} = 0.23636 \ldots.$$

Then how can we tell the difference between a decimal representation of a rational number and a decimal representation of an irrational number? Fortunately, there is one special characteristic of the decimal representation of a rational number that aids in making a distinction. Each rational number may be expressed as a *repeating* (periodic) decimal, and if we continue to determine new digits in the decimal representation

$$\sqrt{2} = 1.4142135 \ldots,$$

we shall find that no digit or set of digits appears to repeat indefinitely. In general, each irrational number may be represented by a **nonrepeating decimal** that continues indefinitely.

Exercises

1. Find an expression for the number n such that:
   (a) $n^2 = 5$ (b) $n^2 = 7$
   (c) $n^2 = 13$ (d) $n^2 = 16$

2. We learned that $\sqrt{2}$ is between the integers 1 and 2; that is, $1 < \sqrt{2} < 2$. Between what two consecutive integers is:
   (a) $\sqrt{3}$?   (b) $\sqrt{6}$?
   (c) $\sqrt{12}$?   (d) $\sqrt{93}$?

3. Find to the nearest tenth the decimal approximation for each of the following.
   (a) $\sqrt{5}$ (b) $\sqrt{8}$
   (c) $\sqrt{10}$ (d) $\sqrt{29}$

4. By definition, $(\sqrt{n})^2 = n$. Find a four-place decimal approximation for each.
   (a) $\sqrt{3}$ (b) $\sqrt{6}$

5. Show that there is no rational number n such that:

   (a) $n^2 = 5$ (b) $n^2 = \dfrac{3}{4}$

   (c) $n^3 = 2$ (d) $n^3 = 4$

6. Use an example to show that the sum of two irrational numbers is not necessarily an irrational number.

7. Use an example to show that the product of two irrational numbers is not necessarily an irrational number.

8. A certain number is approximated to six decimal places as 3.761256. Can you tell whether the original number was rational or irrational? Explain.

11–3 Real Numbers

Each number that we have studied thus far has been either rational or irrational. The union of the set of all rational numbers and the set of all irrational numbers is called the set of **real numbers.** We have learned that the rational and the irrational numbers are coordinates of points on the number line. In fact, every point on the number line is the graph of a real number; hence the number line is called the **real number line.** There is a one-to-one correspondence between the points on the real number line and the elements of the set of real numbers.

In Section 11-2 we stated that each rational number may be expressed as a repeating decimal and that each irrational number may be expressed as a nonrepeating decimal. Since the set consisting of all rational and irrational numbers is the set of real numbers, it follows that *any real number may be expressed either as a repeating decimal or as a nonrepeating decimal. Conversely, any repeating decimal or nonrepeating decimal represents a real number.*

A look at the following diagram may clarify the extensions used to develop the real number system. Recall that fractional numbers are also called nonnegative rational numbers.

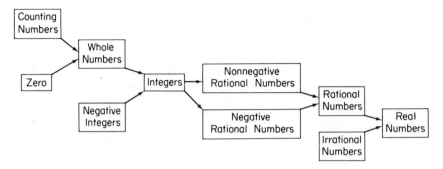

There are several ways of classifying real numbers. Each real number is

(1) positive, negative, or zero,
(2) a rational number or an irrational number,
(3) expressible as a repeating decimal or as a nonrepeating decimal.

Exercises

1. Classify each number as rational or irrational.
   (a) $\sqrt{3}$
   (b) $\sqrt{9}$
   (c) $\dfrac{17}{26}$
   (d) 0.172
   (e) π
   (f) $\dfrac{22}{7}$
   (g) $\sqrt{11}$
   (h) $\sqrt{16}$

2. Which of the numbers in Exercise 1 are real numbers?

3. Represent each of the rational numbers in Exercise 1 as a repeating decimal.

4. Represent each of the irrational numbers in Exercise 1 as a nonrepeating decimal.

5. Use a decimal equivalent for $\sqrt{3}$ and show that $2 + \sqrt{3}$ is irrational.

6. Show that $^{-}3 + \sqrt{3}$ is irrational.

7. Can you find two numbers, one rational and the other irrational, such that their sum is rational?

8. Show that $3\sqrt{2}$ is irrational.

9. Show that $^{-}2\sqrt{3}$ is irrational.

10. Can you find two numbers, one rational and the other irrational, such that their product is rational?

11-4 Properties of Real Numbers

The real numbers form a mathematical system with operations, relations, and properties. We can add, subtract, multiply, and divide real numbers. We can determine two real numbers as equal or unequal, and we have certain mathematical properties that apply to the real number system. These properties are as follows.

Closure and uniqueness properties   The real number system is closed under the operations of addition, subtraction, multiplication, and division (except division by zero).

Addition: If a and b are real numbers, then $a + b$ is a unique real number.

Subtraction: If a and b are real numbers, then $a - b$ is a unique real number.

Multiplication: If a and b are real numbers, then $a \times b$ is a unique real number.

Division: If a and b are real numbers and $b \neq 0$, then $a \div b$ is a unique real number.

Commutative property   In general for the real number system, addition and multiplication are commutative operations, whereas subtraction and division are not.

Addition: If a and b are real numbers, then $a + b = b + a$.

Subtraction: If a and b are real numbers, then $a - b \neq b - a$ when $a \neq b$.

Multiplication: If a and b are real numbers, then $a \times b = b \times a$.

Division: If a and b are real numbers different from zero, then $a \div b \neq b \div a$ when $a \neq b$.

Associative property   In general for the real number system, addition and multiplication are associative operations, whereas subtraction and division are not.

Addition: If a, b, and c are real numbers, then $(a + b) + c = a + (b + c)$.

Subtraction: If a, b, and c are real numbers, then in general $(a - b) - c \neq a - (b - c)$.

Multiplication: If a, b, and c are real numbers, then $(a \times b) \times c = a \times (b \times c)$.

Division: If a, b, and c are real numbers different from zero, then in general $(a \div b) \div c \neq a \div (b \div c)$.

Identity property   Zero is the identity element for the operation of addition and 1 is the identity element for the operation of multiplication. Zero is a right identity for subtraction and 1 is a right identity for division.

Addition: If a is a real number, then $a + 0 = 0 + a = a$.

Subtraction: If a is a real number, then $a - 0 = a$, whereas in general, $0 - a \neq a$.
Multiplication: If a is a real number, then $a \times 1 = 1 \times a = a$.
Division: If a is a real number, then $a \div 1 = a$, whereas in general, $1 \div a \neq a$.

Distributive property For the real number system, there are several instances where one operation may be distributed over another, and three cases may be considered: distributive from the left, distributive from the right, and distributive from both the left and the right. Some of the possibilities for consideration are as follows.

Multiplication over addition from the left: If a, b, and c are real numbers, then $a \times (b + c) = (a \times b) + (a \times c)$.
Multiplication over addition from the right: If a, b, and c are real numbers, then $(b + c) \times a = (b \times a) + (c \times a)$.
Multiplication over subtraction from the left: If a, b, and c are real numbers, then $a \times (b - c) = (a \times b) - (a \times c)$.
Multiplication over subtraction from the right: If a, b, and c are real numbers, then in general $(b - c) \times a = (b \times a) - (c \times a)$.
Division over addition from the left: If a, b, and c are real numbers, then in general $a \div (b + c) \neq (a \div b) + (a \div c)$.
Division over addition from the right: If a, b, and c are real numbers and $c \neq 0$, then $(a + b) \div c = (a \div c) + (b \div c)$.
Division over subtraction from the left: If a, b, and c are real numbers, then in general $a \div (b - c) \neq (a \div b) - (a \div c)$.
Division over subtraction from the right: If a, b, and c are real numbers and $c \neq 0$, then $(a - b) \div c = (a \div c) - (b \div c)$.

Inverse property If a is a real number, then ^-a is a real number, called the additive inverse (negative) of a, such that $a + {}^-a = {}^-a + a = 0$. If a is a real number and $a \neq 0$, then $\frac{1}{a}$ is a real number, called the multiplicative inverse (reciprocal) of a, such that $a \times \frac{1}{a} = \frac{1}{a} \times a = 1$.

Order property The real number system is linearly ordered; that is, if a and b are *different* real numbers, then either $a > b$ or $a < b$.

Density property The real number system is dense; that is, between any two distinct real numbers there is always another real number. Consequently, we can find an infinite number of real numbers between any two distinct real numbers.

Completeness property The real number system is complete; that is, there is a real number for every point of the number line and, conversely, there is a point of the number line for every real number.

Exercises

1. Tell whether each of the following sets of numbers is dense or not.
   (a) Whole numbers   (b) Integers
   (c) Nonnegative rational numbers   (d) Rational numbers
   (e) Irrational numbers   (f) Real numbers

2. Tell which sets of numbers in Exercise 1 are ordered.

3. Tell which sets of numbers in Exercise 1 are complete.

4. Tell which sets of numbers in Exercise 1 are closed under:
   (a) Addition   (b) Subtraction
   (c) Multiplication   (d) Division

5. Tell which sets of numbers in Exercise 1 have:
   (a) An additive identity   (b) A multiplicative identity

6. What is the additive inverse of:
   (a) $\sqrt{3}$?   (b) $^-\sqrt{6}$?

   (c) $\dfrac{1}{\sqrt{7}}$?   (d) $\sqrt{3} + \sqrt{2}$?

7. What is the multiplicative inverse of:
   (a) $\sqrt{5}$?   (b) $^-\sqrt{2}$?

   (c) $\dfrac{1}{\sqrt{11}}$?   (d) $\sqrt{6} + \sqrt{7}$?

8. If two positive real numbers are represented as infinite decimals, how can you tell which is the greater?

11-5 Chapter Test

Indicate whether each statement is true or false. If false, tell why.

1. $\sqrt{4}$ is an irrational number.

2. Every point on the number line is the graph of a real number.

3. If n may be any real number and $n^2 = 7$, then $n = \sqrt{7}$.

4. The additive inverse of $\dfrac{1}{\sqrt{5}}$ is $\sqrt{5}$.

5. If b is any integer and b^2 is divisible by 3, then b is divisible by 3.

6. The irrational numbers are closed under the operation of multiplication.

7. π is a rational number.

8. The irrational number system is complete.

9. A decimal approximation for $\sqrt{29}$ is 5.4.

10. In general, every irrational number may be represented by a repeating decimal.

Select the best possible answer.

11. Which of these numerals does *not* represent a rational number?
   (a) 2.131331333 . . .   (b) $0.98\bar{0}$
   (c) $0.2345\overline{623456}$ (d) $4.\overline{10}$
   (e) 0

12. The rational number named by $0.0089\overline{89}$ is

   (a) $\dfrac{89}{990}$ (b) $\dfrac{89}{9,900}$

   (c) $\dfrac{89}{99}$ (d) $\dfrac{8,989}{1,000,000}$

   (e) $\dfrac{9}{10}$

13. What is the additive inverse of $\sqrt{5} - \sqrt{3}$?
   (a) $\sqrt{3} - \sqrt{5}$ (b) $\sqrt{5} + \sqrt{3}$

   (c) $\dfrac{1}{\sqrt{5}} - \sqrt{3}$ (d) $\sqrt{5} - \dfrac{1}{\sqrt{3}}$

   (e) $\dfrac{1}{\sqrt{5} - \sqrt{3}}$

14. Which of these sets of numbers is (are) dense?
   I. Whole numbers.   II. Rational numbers.   III. Real numbers.
   (a) II only   (b) III only
   (c) I and II only   (d) II and III only
   (e) I, II, and III

15. Which of these statements illustrate(s) an inverse property for any real number a?
   I. $a + {}^-a = 0.$ II. $a \times \dfrac{1}{a} = 1.$ III. $a \times 1 = a.$

   (a) I only   (b) II only
   (c) I and II only   (d) I and III only
   (e) II and III only

chapter 12

Nondecimal
Systems of
Numeration

In Chapter 3 we learned that a system of numeration is a systematic method of naming numbers. In the decimal system of notation we can represent any number with only ten digits (0, 1, 2, 3, 4, 5, 6, 7, 8, 9) and the idea of place value. In this chapter we shall discuss systems of numeration with bases other than ten. The study of different number bases provides an opportunity for us to gain an appreciation of the features of our decimal system of numeration and our computational methods. We should develop a better understanding of the decimal system as a result of this study.

12-1 Collecting in Sets Other than Ten

Although the base is not specified, the idea of collecting in sets other than ten is frequently used in various kinds of measurements. For example, the idea of collecting in sets of two is used in the following dry measures (the symbol $\stackrel{m}{=}$ is read "is equal in measure to"):

$$2 \text{ cups} \stackrel{m}{=} 1 \text{ pint},$$

$$2 \text{ pints} \stackrel{m}{=} 1 \text{ quart},$$

$$2 \text{ quarts} \stackrel{m}{=} 1 \text{ half-gallon},$$

$$2 \text{ half-gallons} \stackrel{m}{=} 1 \text{ gallon}.$$

Base two, called the binary base, is used also in electronic computers.

Collecting objects by the dozen, gross, and great gross are illustrations of the use of twelve as a base. Other examples of this method of grouping are the number of months in a year and the number of inches in a foot.

The idea of base sixty is used in measuring time and angles. In the case of time,

$$60 \text{ seconds} \stackrel{m}{=} 1 \text{ minute},$$

$$60 \text{ minutes} \stackrel{m}{=} 1 \text{ hour}.$$

In the case of angles,

$$60 \text{ seconds} \stackrel{m}{=} 1 \text{ minute},$$

$$60 \text{ minutes} \stackrel{m}{=} 1 \text{ degree}.$$

12-2 Base Five Numeration

The base of any system of numeration generally establishes the method of grouping and the number of digits needed. In the base ten system we collect in sets of ten. In the base five system, we collect in sets of five, and we can represent any number with only five digits (0, 1, 2, 3, 4) and a positional notation. For example, consider the following illustrations of grouping in base five.

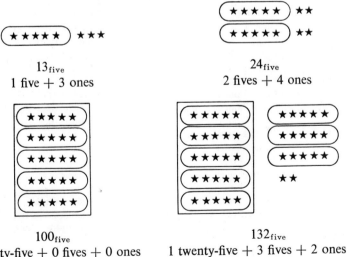

13_{five}
1 five + 3 ones

24_{five}
2 fives + 4 ones

100_{five}
1 twenty-five + 0 fives + 0 ones

132_{five}
1 twenty-five + 3 fives + 2 ones

It is important to note the subscript used in writing the numerals 13_{five}, 24_{five}, 100_{five}, and 132_{five}. The subscript "five" denotes the base, and the numeral 13_{five} is read "one three, base five." Any numeral written without a subscript will be assumed to be in base ten notation. Thus we have two names for the same number in each of the following:

$$13_{\text{five}} = 8,$$

$$24_{\text{five}} = 14,$$

$$100_{\text{five}} = 25,$$

$$132_{\text{five}} = 42.$$

Since counting is fundamental to the understanding of a system of numeration, we should study the table below, which gives the base ten and base five numerals for the numbers 1 through 26.

Base Ten		Base Five	
Numeral	*Meaning*	*Numeral*	*Meaning*
1	1 one	1_{five}	1 one
2	2 ones	2_{five}	2 ones
3	3 ones	3_{five}	3 ones
4	4 ones	4_{five}	4 ones
5	5 ones	10_{five}	1 five, 0 ones
6	6 ones	11_{five}	1 five, 1 one
7	7 ones	12_{five}	1 five, 2 ones
8	8 ones	13_{five}	1 five, 3 ones
9	9 ones	14_{five}	1 five, 4 ones
10	1 ten, 0 ones	20_{five}	2 fives, 0 ones
11	1 ten, 1 one	21_{five}	2 fives, 1 one
12	1 ten, 2 ones	22_{five}	2 fives, 2 ones
13	1 ten, 3 ones	23_{five}	2 fives, 3 ones
14	1 ten, 4 ones	24_{five}	2 fives, 4 ones
15	1 ten, 5 ones	30_{five}	3 fives, 0 ones
16	1 ten, 6 ones	31_{five}	3 fives, 1 one
17	1 ten, 7 ones	32_{five}	3 fives, 2 ones
18	1 ten, 8 ones	33_{five}	3 fives, 3 ones
19	1 ten, 9 ones	34_{five}	3 fives, 4 ones
20	2 tens, 0 ones	40_{five}	4 fives, 0 ones
21	2 tens, 1 one	41_{five}	4 fives, 1 one
22	2 tens, 2 ones	42_{five}	4 fives, 2 ones
23	2 tens, 3 ones	43_{five}	4 fives, 3 ones
24	2 tens, 4 ones	44_{five}	4 fives, 4 ones
25	2 tens, 5 ones	100_{five}	1 twenty-five, 0 fives, 0 ones
26	2 tens, 6 ones	101_{five}	1 twenty-five, 0 fives, 1 one

The value represented by each digit in a given base five numeral is determined by the position it occupies. For example, consider the numeral 32_{five}. The position of the 3 in this numeral indicates that it represents three sets of five whereas the 2 represents two sets of one.

The place values in the base five system are based on powers of five. Starting at the ones place and moving to the left, each place has a value five times as large as the value of the place to its right.

Place Values for Base Five

Six Hundred Twenty-fives	One Hundred Twenty-fives	Twenty-fives	Fives	Ones
$5 \times (5 \times 5 \times 5)$	$5 \times (5 \times 5)$	5×5	5×1	1
5^4	5^3	5^2	5^1	5^0

To change the notation for a number from base five to base ten, we express the number in terms of powers of five and simplify. Consider the examples below.

Example 1   Represent 432_{five} in base ten notation.

$$432_{\text{five}} = (4 \times 5^2) + (3 \times 5^1) + (2 \times 5^0)$$
$$= (4 \times 25) + (3 \times 5) + (2 \times 1)$$
$$= 100 + 15 + 2$$
$$= 117$$

Example 2   Represent $3{,}201_{\text{five}}$ in base ten notation.

$$3{,}201_{\text{five}} = (3 \times 5^3) + (2 \times 5^2) + (0 \times 5^1) + (1 \times 5^0)$$
$$= (3 \times 125) + (2 \times 25) + (0 \times 5) + (1 \times 1)$$
$$= 375 + 50 + 0 + 1$$
$$= 426$$

Example 3   Represent 123.42_{five} in base ten notation.

$$123.42_{\text{five}} = (1 \times 5^2) + (2 \times 5^1) + (3 \times 5^0) + (4 \times 5^{-1}) + (2 \times 5^{-2})$$
$$= (1 \times 25) + (2 \times 5) + (3 \times 1) + \left(4 \times \frac{1}{5}\right) + \left(2 \times \frac{1}{25}\right)$$
$$= (1 \times 25) + (2 \times 5) + (3 \times 1) + \left(4 \times \frac{2}{10}\right) + \left(2 \times \frac{4}{100}\right)$$
$$= 25 + 10 + 3 + \frac{8}{10} + \frac{8}{100}$$
$$= 38.88$$

Exercises

1. Write the base five numeral for each of these groupings.

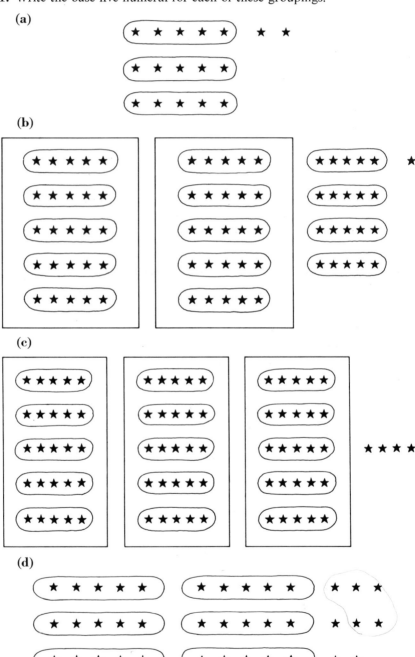

(a)

(b)

(c)

(d)

2. Illustrate with a diagram the grouping indicated by each of the following.
   (a) 12_{five} (b) 44_{five}
   (c) 113_{five} (d) 204_{five}

3. Draw a number line and graph points for the numbers one through ten. Label these points with the appropriate base five numerals.

4. List the base ten and base five numerals for the numbers 27 through 60.

5. Represent each number in terms of powers of five.
   (a) 413_{five} (b) $2{,}243_{\text{five}}$
   (c) 42.3_{five} (d) 1.32_{five}

6. Represent each number in base ten notation.
   (a) 12_{five} (b) 44_{five}
   (c) 113_{five} (d) 204_{five}
   (e) 342_{five} (f) $4{,}124_{\text{five}}$
   (g) $243{,}214_{\text{five}}$ (h) 31.4_{five}

12-3 Computation in Base Five Notation

Since the place values in base five follow the pattern of our denominations of money to a certain point, consider these illustrations of addition in base five in which pennies, nickels, and quarters are used.

Example 1 The sum of four pennies and three pennies is equal in value to one nickel and two pennies. This sum may be represented in base five notation as

$$4_{\text{five}} + 3_{\text{five}} = 12_{\text{five}},$$

where 12_{five} means 1 set of five (1 nickel) and 2 sets of one (2 pennies).

Example 2 If we add three nickels and four pennies to four nickels and three pennies, the sum is equal in value to one quarter, three nickels, and two pennies. This sum may be represented as

$$34_{\text{five}} + 43_{\text{five}} = 132_{\text{five}},$$

where 132_{five} means 1 set of twenty-five (1 quarter), 3 sets of five (3 nickels), and 2 sets of one (2 pennies).

Base Five Addition Table

+	0	1	2	3	4
0	0	1	2	3	4
1	1	2	3	4	10
2	2	3	4	10	11
3	3	4	10	11	12
4	4	10	11	12	13

$3_{\text{five}} + 4_{\text{five}} = 12_{\text{five}}$

Our understanding of addition in base five notation will be improved by an examination of the table for basic addition facts in base five notation.

Refer to the addition table to verify the sums obtained by computation in base five notation in Examples 3 and 4.

Example 3   Add 13_{five} and 22_{five}.

First method:

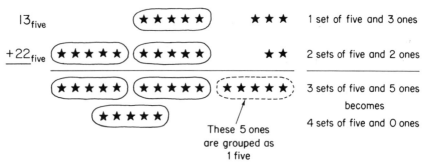

13_{five} ... 1 set of five and 3 ones

$+22_{five}$ ... 2 sets of five and 2 ones

3 sets of five and 5 ones
becomes
4 sets of five and 0 ones

These 5 ones are grouped as 1 five

Second method:

$$\begin{array}{r} \overset{1}{1}3_{five} \\ +22_{five} \\ \hline 40_{five} \end{array}$$

Procedure: From the table we determine that

$$3_{five} + 2_{five} = 10_{five}$$

and then write 0 in the ones column and rename ("carry") 1 set of five to the fives column. Next we determine from the table that

$$1_{five} + 2_{five} = 3_{five} \quad \text{and} \quad 3_{five} + 1_{five} = 4_{five}.$$

Thus, we have 4 sets of five, and we represent the sum of 13_{five} and 22_{five} as 40_{five}. We can check our result by converting the base five numerals to base ten numerals and adding.

$$\begin{array}{rcl} 13_{five} = (1 \times 5) + (3 \times 1) = & 8 \\ +22_{five} = (2 \times 5) + (2 \times 1) = & 12 \\ \hline 40_{five} = (4 \times 5) + (0 \times 1) = & \overline{20} \end{array}$$

Example 4   Add 424_{five} and 323_{five}.

$$\begin{array}{r} \overset{111}{4}24_{five} \\ +323_{five} \\ \hline 1{,}302_{five} \end{array}$$

Procedure: From the table we determine that

$$4_{five} + 3_{five} = 12_{five}$$

and then write 2 in the ones column and rename ("carry") 1 set of five to the fives column. Next we determine from the table that

$$2_{five} + 2_{five} = 4_{five} \quad \text{and} \quad 4_{five} + 1_{five} = 10_{five};$$

then write 0 in the fives column and rename ("carry") 1 set of twenty-five to the twenty-fives column. Next we determine from the table that

$$4_{five} + 3_{five} = 12_{five} \quad \text{and} \quad 12_{five} + 1_{five} = 13_{five};$$

then we write 3 in the twenty-fives column and rename ("carry") 1 set of one hundred twenty-five to the one hundred twenty-fives column. So we represent the sum of 424_{five} and 323_{five} as $1{,}302_{five}$. We can check our result by converting the base five numerals to base ten numerals and adding.

$$
\begin{aligned}
424_{five} &= & (4 \times 5^2) + (2 \times 5) + (4 \times 1) &= 114 \\
+323_{five} &= & (3 \times 5^2) + (2 \times 5) + (3 \times 1) &= \underline{88} \\
\overline{1{,}302_{five}} &= (1 \times 5^3) + (3 \times 5^2) + (0 \times 5) + (2 \times 1) &= 202
\end{aligned}
$$

The table of addition facts also may be used to solve subtraction problems, since subtraction is the inverse operation of addition.

Example 5   Subtract 32_{five} from 42_{five}.

$$
\begin{array}{r}
42_{five} \\
-32_{five} \\
\hline
10_{five}
\end{array}
\qquad \triangle + 32_{five} = 42_{five}
$$

Procedure: Recall that a subtraction problem may be considered as an addition problem in which the sum and one addend are known. In this example we are trying to find the missing addend that produces a sum of 42_{five} when added to 32_{five}. From the addition table we determine that 0 is the number that must be added to 2 to produce the sum 2; thus, we write 0 in the ones column. Next we determine what number must be added to 3 to produce the sum 4 in the fives column:

$$1_{five} + 3_{five} = 4_{five}.$$

Thus we determine that 10_{five} is the missing addend and that

$$42_{five} - 32_{five} = 10_{five}.$$

We can check our result by converting the base five numerals to base ten numerals and subtracting.

$$
\begin{aligned}
42_{five} &= (4 \times 5) + (2 \times 1) &= 22 \\
-32_{five} &= (3 \times 5) + (2 \times 1) &= \underline{17} \\
\overline{10_{five}} &= (1 \times 5) + (0 \times 1) &= 5
\end{aligned}
$$

Example 6   Subtract 14_{five} from 31_{five}.

$$
\begin{array}{r}
\overset{2\ 1}{\cancel{3}1}_{five} \\
-14_{five} \\
\hline
12_{five}
\end{array}
\qquad \triangle + 14_{five} = 31_{five}
$$

Procedure: From the addition table we determine that there is no number that may be added to 4 to produce 1; however, if we rename ("borrow") 1 of the 3 sets of five, we determine that

$$4_{\text{five}} + 2_{\text{five}} = 11_{\text{five}}.$$

So we write a 2 in the ones column and understand that we have only 2 of the 3 given sets of five left. Next we determine what number must be added to 1 to produce 2 in the fives column:

$$1_{\text{five}} + 1_{\text{five}} = 2_{\text{five}}.$$

So we represent

$$31_{\text{five}} - 14_{\text{five}} \quad \text{as} \quad 12_{\text{five}}.$$

We can check our result by converting the base five numerals to base ten numerals and subtracting.

$$
\begin{aligned}
31_{\text{five}} &= (3 \times 5) + (1 \times 1) = 16 \\
-14_{\text{five}} &= (1 \times 5) + (4 \times 1) = 9 \\
\hline
12_{\text{five}} &= (1 \times 5) + (2 \times 1) = 7
\end{aligned}
$$

To develop our understanding of multiplication in base five, a table of basic multiplication facts represented in base five notation is provided. The repeated-addition approach also may be used to perform multiplication in base five; see the table below.

Base Five Multiplication Table

×	0	1	2	3	4
0	0	0	0	0	0
1	0	1	2	3	4
2	0	2	4	11	13
3	0	3	11	14	22
4	0	4	13	22	31

$3_{\text{five}} \times 4_{\text{five}} = 22_{\text{five}}$

Refer to the multiplication table to verify the products obtained by computation in base five notation in Examples 7 and 8.

Example 7 Multiply 34_{five} by 3_{five}.

First method:

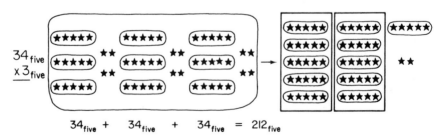

$$34_{\text{five}} + \quad 34_{\text{five}} + \quad 34_{\text{five}} = 212_{\text{five}}$$

Second method:

$$\begin{array}{r} {}^{22}\\ 34_{\text{five}} \\ \times\ \ 3_{\text{five}} \\ \hline 212_{\text{five}} \end{array}$$

Procedure: From the table we determine that

$$3_{\text{five}} \times 4_{\text{five}} = 22_{\text{five}}$$

and then write 2 in the ones column and rename ("carry") 2 sets of five to the fives column. Next we determine from the table that

$$3_{\text{five}} \times 3_{\text{five}} = 14_{\text{five}}$$

and add the 2 that we renamed:

$$14_{\text{five}} + 2_{\text{five}} = 21_{\text{five}}.$$

We write 1 in the fives column and 2 in the twenty-fives column. So we represent the product of 3_{five} and 34_{five} as 212_{five}. We can check our result by converting the base five numerals to base ten numerals and multiplying.

$$\begin{array}{rl} 34_{\text{five}} = & (3 \times 5) + (4 \times 1) = 19 \\ \times\ \ 3_{\text{five}} = & (3 \times 1) = \ \ 3 \\ \hline 212_{\text{five}} = & (2 \times 5^2) + (1 \times 5) + (2 \times 1) = 57 \end{array}$$

Example 8   Multiply 324_{five} by 12_{five}.

$$\begin{array}{r} {}^{1\,1\,1}\\ 324_{\text{five}} \\ \times\ \ 12_{\text{five}} \\ \hline 1\ 203 \\ 3\ 24\ \ \ \ \\ \hline 4{,}443_{\text{five}} \end{array}$$

Procedure: From the table we determine that

$$2_{\text{five}} \times 4_{\text{five}} = 13_{\text{five}}$$

and then write 3 in the ones column and rename ("carry") 1 set of five to the fives column. Next we determine that

$$2_{\text{five}} \times 2_{\text{five}} = 4_{\text{five}}$$

and add the 1 that we renamed:

$$4_{\text{five}} + 1_{\text{five}} = 10_{\text{five}}.$$

We write 0 in the fives column and rename ("carry") 1 set of twenty-five to the twenty-fives column. Then we determine that

$$2_{\text{five}} \times 3_{\text{five}} = 11_{\text{five}}$$

and add the 1 that we renamed:

$$11_{\text{five}} + 1_{\text{five}} = 12_{\text{five}}.$$

We write 2 in the twenty-fives column and 1 in the one hundred twenty-fives column. The partial product is then

$$2_{\text{five}} \times 324_{\text{five}} = 1{,}203_{\text{five}}.$$

We observe in the table that 1 is the multiplicative identity and that

$$1_{\text{five}} \times 324_{\text{five}} = 324_{\text{five}}.$$

The partial product 324_{five} is set one place to the left, as in decimal arithmetic, because we actually have the product

$$10_{\text{five}} \times 324_{\text{five}} = 3{,}240_{\text{five}}.$$

We add the partial products and obtain $4{,}443_{\text{five}}$. We can check this result by converting the base five numerals to base ten numerals and multiplying.

$$
\begin{aligned}
324_{\text{five}} = \quad & (3 \times 5^2) + (2 \times 5) + (4 \times 1) = 89 \\
\times\ 12_{\text{five}} = \quad & (1 \times 5) + (2 \times 1) = 7 \\
\hline
4{,}443_{\text{five}} = \ & (4 \times 5^3) + (4 \times 5^2) + (4 \times 5) + (3 \times 1) = 623
\end{aligned}
$$

The table of multiplication facts also may be used to solve division problems, since division is the inverse operation of multiplication.

Example 9   Divide 124_{five} by 3_{five}.

$$
\begin{array}{r}
23_{\text{five}} \\
3_{\text{five}}\overline{)\ 124_{\text{five}}} \\
-11_{\text{five}} \\
\hline
14_{\text{five}} \\
-14_{\text{five}} \\
\end{array}
\qquad \triangle \times 3_{\text{five}} = 124_{\text{five}}
$$

Procedure: Recall that a division problem may be considered as a multiplication problem in which the product and one factor are known. In this example we are trying to find the missing factor that produces a product of 124_{five} when used as a factor with 3_{five}. From the multiplication table we determine that 2 is the number that we use as a factor with 3 to obtain a product that is as close to 12_{five} as we can get, such that the product is less than 12_{five}; that is,

$$2_{\text{five}} \times 3_{\text{five}} = 11_{\text{five}}.$$

So we write 2 in the fives place in the quotient, subtract 11_{five} from 12_{five}, bring down the next digit. Now we must find what number used as a factor with 3_{five} will give 14_{five} and from the multiplication table we conclude that

$$3_{\text{five}} \times 3_{\text{five}} = 14_{\text{five}}.$$

Thus we determine that 23_{five} is the missing factor and that

$$124_{\text{five}} \div 3_{\text{five}} = 23_{\text{five}}.$$

We can check the result by converting the base five numerals to base ten numerals and dividing.

$$124_{\text{five}} \div 3_{\text{five}} = [(1 \times 5^2) + (2 \times 5) + (4 \times 1)] \div 3$$
$$= 39 \div 3$$
$$= 13$$

and

$$23_{\text{five}} = (2 \times 5) + (3 \times 1) = 13$$

We can also use a repeated-subtraction process to determine the quotient $124_{\text{five}} \div 3_{\text{five}}$. Consider the following illustration of the repeated-subtraction process, and note that each subtraction is performed in base five notation.

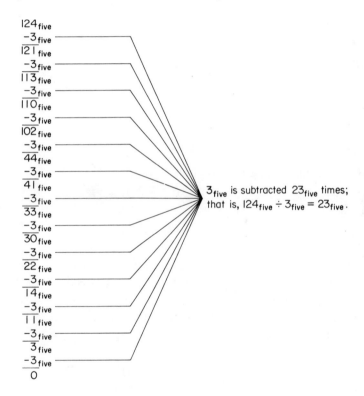

3_{five} is subtracted 23_{five} times; that is, $124_{\text{five}} \div 3_{\text{five}} = 23_{\text{five}}$.

Example 10   Divide $134,201_{\text{five}}$ by 23_{five}.

$$
\begin{array}{r}
3{,}202_{\text{five}} \\
23_{\text{five}} \overline{)\ 134{,}201_{\text{five}}} \\
-124_{\text{five}} \\
\hline
10\ 2_{\text{five}} \\
-10\ 1_{\text{five}} \\
\hline
101_{\text{five}} \\
-101_{\text{five}} \\
\hline
\end{array}
$$

Procedure: As observed in Example 9, we may divide numbers represented in base five notation by using the same computational patterns that we use to divide numbers represented in base ten; however, the operations of multiplication and subtraction are performed in base five notation. We can check the result by converting the base five numerals to base ten numerals and dividing.

$$134{,}201_{\text{five}} = (1 \times 5^5) + (3 \times 5^4) + (4 \times 5^3) + (2 \times 5^2)$$
$$+ (0 \times 5) + (1 \times 1) = 5{,}551$$

$$23_{\text{five}} = (2 \times 5) + (3 \times 1) = 13$$

$$3{,}202_{\text{five}} = (3 \times 5^3) + (2 \times 5^2) + (0 \times 5) + (2 \times 1) = 427$$

$$
\begin{array}{r}
427 \\
13)\overline{5{,}551} \\
-5\,2 \\ \hline
35 \\
-26 \\ \hline
91 \\
-91 \\ \hline
\end{array}
$$

Exercises

1. Find each sum in base five notation.

   (a) 34_{five} (b) 234_{five}
          $+\ 3_{\text{five}}$ $+102_{\text{five}}$

   (c) $1{,}234_{\text{five}}$ (d) 101_{five}
         $+3{,}211_{\text{five}}$ 124_{five}
                                    $+432_{\text{five}}$

2. Convert the base five numerals in each part of Exercise 1 to base ten numerals and check the sum.

3. Find each difference in base five notation.

   (a) 14_{five} (b) 32_{five}
          $-\ 3_{\text{five}}$ -23_{five}

   (c) 421_{five} (d) $2{,}413_{\text{five}}$
         -243_{five} $-1{,}014_{\text{five}}$

4. Convert the base five numerals in each part of Exercise 3 to base ten numerals and check the difference.

5. Find each product in base five notation.

   (a) 40_{five} (b) 121_{five}
          $\times\ 4_{\text{five}}$ $\times\ 3_{\text{five}}$

   (c) 413_{five} (d) $2{,}132_{\text{five}}$
         $\times\ 42_{\text{five}}$ $\times\ \ 21_{\text{five}}$

6. Convert the base five numerals in each part of Exercise 5 to base ten numerals and check the product.

7. Find each quotient in base five notation.

   (a) $3_{five})\overline{102_{five}}$

   (b) $4_{five})\overline{3,333_{five}}$

   (c) $21_{five})\overline{441_{five}}$

   (d) $43_{five})\overline{324,301_{five}}$

8. Convert the base five numerals in each part of Exercise 7 to base ten numerals and check the quotient.

9. We have learned that the symbols used in a system of numeration are man-made. In our discussion of base five numeration we have been using the conventional symbols 0, 1, 2, 3, and 4 as our five digits. Suppose we had chosen the letters E, F, G, H, and J for our five basic digits, as E = 0, F = 1, G = 2, H = 3, and J = 4. Complete the following tables of addition and multiplication facts, using the nonconventional symbols E, F, G, H, and J.

+	E	F	G	H	J
E					
F					
G					
H					
J					

×	E	F	G	H	J
E					
F					
G					
H					
J					

10. Compute, using the tables in Exercise 9.

   (a) $\begin{array}{r} JG \\ +HH \\ \hline \end{array}$

   (b) $\begin{array}{r} GHJ \\ -\ JG \\ \hline \end{array}$

   (c) $\begin{array}{r} HGJ \\ \times\ \ H \\ \hline \end{array}$

   (d) $J)\overline{HG,HEJ}$

Consider the set T of numbers that can be represented using the notation introduced in Exercise 9.

11. Is the set T closed under the operation of:
   (a) Addition?   (b) Subtraction?
   (c) Multiplication?   (d) Division?

12. For the numbers in set T, does the commutative property hold for:
   (a) Addition?   (b) Subtraction?
   (c) Multiplication?   (d) Division?

13. For the numbers in set T, does the associative property hold for:
(a) Addition?   (b) Subtraction?
(c) Multiplication?   (d) Division?

14. For the numbers in set T, what is the identity element for:
(a) Addition?   (b) Subtraction?
(c) Multiplication?   (d) Division?

15. Use the numbers in set T and give an example of the distributive property of multiplication over:
(a) Addition   (b) Subtraction

12-4 Change of Base

Translating from one base to another involves a regrouping process. As noted in Section 12-2, we change the notation for a number from base five to base ten by expressing the number in terms of powers of five and simplifying. For example,

$$3{,}214_{\text{five}} = (3 \times 5^3) + (2 \times 5^2) + (1 \times 5^1) + (4 \times 5^0)$$
$$= (3 \times 125) + (2 \times 25) + (1 \times 5) + (4 \times 1)$$
$$= 375 + 50 + 5 + 4$$
$$= 434.$$

Suppose we desire to represent the number 434 in base five notation, that is, in terms of powers of five. First, it is necessary to determine the highest power of five that we can subtract from 434. Since $5^4 = 625$ and $5^3 = 125$, we conclude that 5^3 is the highest power of five that is not greater than 434. Thus we can subtract 125 from 434 three times, leaving a difference of 59.

$$
\begin{array}{rl}
434 & \\
-125 & \quad (1 \times 5^3) \\
\hline
309 & \\
-125 & \quad (1 \times 5^3) \\
\hline
184 & \\
-125 & \quad (1 \times 5^3) \\
\hline
59 & \quad (3 \times 5^3)
\end{array}
$$

Our base five numeral begins to take form as

$$(3 \times 5^3) + (? \times 5^2) + (? \times 5^1) + (? \times 5^0).$$

Next we observe that the next lower power of five is 5^2, and $5^2 = 25$. Thus we can subtract 25 from 59 two times leaving a difference of 9.

$$\begin{array}{r} 59 \\ -25 \\ \hline 34 \\ -25 \\ \hline 9 \end{array}$$ (1×5^2)
(1×5^2)
(2×5^2)

We now have

$$(3 \times 5^3) + (2 \times 5^2) + (? \times 5^1) + (? \times 5^0).$$

The next lower power of five is 5^1, and we can subtract 5 from 9 one time, leaving a difference of 4.

$$\begin{array}{r} 9 \\ -5 \\ \hline 4 \end{array}$$ (1×5^1)

We now have

$$(3 \times 5^3) + (2 \times 5^2) + (1 \times 5^1) + (? \times 5^0)$$

and, since $4 = 4 \times 5^0$, we conclude that

$$\begin{aligned} 434 &= (3 \times 125) + (2 \times 25) + (1 \times 5) + 4 \\ &= (3 \times 5^3) + (2 \times 5^2) + (1 \times 5^1) + (4 \times 5^0) \\ &= 3{,}214_{\text{five}}. \end{aligned}$$

To improve our understanding of the procedure for changing from base ten to base five we may study the following examples.

Example 1   Represent 8 in base five notation.

Since $5^2 = 25$ and $5^1 = 5$, the highest power of five that we can subtract from 8 is 5^1 and the difference is 3.

$$\begin{array}{r} 8 \\ -5 \\ \hline 3 \end{array}$$

Thus we have

$$\begin{aligned} 8 &= (1 \times 5) + 3 \\ &= (1 \times 5^1) + (3 \times 5^0) \\ &= 13_{\text{five}}. \end{aligned}$$

Example 2   Represent 173 in base five notation.

Since $5^4 = 625$ and $5^3 = 125$, the highest power of five that we can subtract from 173 is 5^3.

$$\begin{array}{r} 173 \\ -125 \\ \hline 48 \end{array}$$

Next we observe that $5^2 = 25$ and 5^2 may be subtracted from 48 one time.

$$\begin{array}{r} 48 \\ -25 \\ \hline 23 \end{array}$$

Finally, we subtract 5 from 23 four times and have a remainder of 3.

$$\begin{array}{r} 23 \\ -\ 5 \\ \hline 18 \\ -\ 5 \\ \hline 13 \\ -\ 5 \\ \hline 8 \\ -\ 5 \\ \hline 3 \end{array}$$

Thus we have

$$173 = (1 \times 125) + (1 \times 25) + (4 \times 5) + 3$$
$$= (1 \times 5^3) + (1 \times 5^2) + (4 \times 5^1) + (3 \times 5^0)$$
$$= 1{,}143_{\text{five}}.$$

Another procedure for changing from base ten to base five is shown in the next two examples.

Example 3   Represent 89 in base five notation.

$$\begin{array}{r} 5\underline{|\ 89} \\ 5\overline{|\ 17} \quad \text{sets of 5 with remainder of 4 ones} \\ 5\overline{|\ \ 3} \quad \text{sets of } 5^2 \text{ with remainder of 2 fives} \\ \overline{|\ \ 0} \quad \text{sets of } 5^3 \text{ with remainder of 3 twenty-fives} \end{array}$$

Thus, if we take the remainders in reverse order from the way they were computed, we have

$$89 = (3 \times 5^2) + (2 \times 5^1) + (4 \times 5^0)$$
$$= 324_{\text{five}}.$$

Example 4   Represent 237 in base five notation.

$$\begin{array}{r} 5\underline{|\ 237} \\ 5\overline{|\ 47} \quad \text{sets of 5 with remainder of 2 ones} \\ 5\overline{|\ \ 9} \quad \text{sets of } 5^2 \text{ with remainder of 2 fives} \\ 5\overline{|\ \ 1} \quad \text{set of } 5^3 \text{ with remainder of 4 twenty-fives} \\ \overline{|\ \ 0} \quad \text{sets of } 5^4 \text{ with remainder of 1 one hundred twenty-five} \end{array}$$

Thus, if we take the remainders in reverse order from the way they were computed, we have

$$237 = (1 \times 5^3) + (4 \times 5^2) + (2 \times 5^1) + (2 \times 5^0)$$
$$= 1{,}422_{\text{five}}.$$

Exercises

1. Represent each number in base ten notation.
   (a) 123_{five} (b) $1{,}000_{\text{five}}$
   (c) $434{,}001_{\text{five}}$ (d) $1{,}000{,}000_{\text{five}}$

2. Represent each number in base five notation.
   (a) 29   (b) 93
   (c) 785   (d) 456
   (e) 18.2   *33 .1*   (f) 5.6   *10,3*
   (g) 42.08

12-5 Base Nine Numeration

In the base nine system of numeration we group in sets of nine, and we can represent any number with nine digits (0, 1, 2, 3, 4, 5, 6, 7, 8) and a positional notation. We shall rely on our understanding of base ten and base five to develop the base nine system.

Exercises

1. Illustrate the grouping of each of these sets of objects in sets of nine by a diagram, and fill in the blanks to make each statement true.

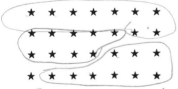

(a) 1 set of nine and *3* ones, written 13_{nine}.

(b) *0* sets of nine and *8* ones, written *8*$_{\text{nine}}$.

(c) *3* sets of nine and *1* ones, written *31*$_{\text{nine}}$.

(d) *4* sets of nine and *4* ones, written *44*$_{\text{nine}}$.

2. List the base ten and base nine numerals for the numbers 1 through 50.

3. Complete the place-value chart for base nine.

		Eighty-ones	Nines	Ones
	$9 \times (9 \times 9)$		9×1	1
9^4		9^2		9^0

4. Represent each number in base ten notation.
  (a) 38_{nine} (b) 527_{nine}
  (c) 354_{nine} (d) $1,831_{nine}$

5. Complete the following tables of basic addition and multiplication facts for base nine.

+	0	1	2	3	4	5	6	7	8
0									
1									
2									
3									
4									
5									
6									
7									
8									

×	0	1	2	3	4	5	6	7	8
0									
1									
2									
3									
4									
5									
6									
7									
8									

6. Find each sum in base nine notation.

(a) 76_{nine}
      $+12_{nine}$

(b) 38_{nine}
      $+46_{nine}$

(c) 657_{nine}
      $+302_{nine}$

(d) $2,485_{nine}$
      $+5,406_{nine}$

7. Find each difference in base nine notation.

(a) 45_{nine}
      -23_{nine}

(b) 281_{nine}
      -144_{nine}

(c) 736_{nine}
      -546_{nine}

(d) $1,000_{nine}$
      $-\ 888_{nine}$

8. Find each product in base nine notation.

(a) 82_{nine}
      $\times\ 5_{nine}$

(b) 674_{nine}
      $\times\ 8_{nine}$

(c) 78_{nine}
      $\times 43_{nine}$

(d) $8,765_{nine}$
      $\times\ 167_{nine}$

9. Find each quotient in base nine notation.

(a) $6_{nine}\overline{)323_{nine}}$

(b) $4_{nine}\overline{)2,213_{nine}}$

(c) $57_{nine}\overline{)7,117_{nine}}$

(d) $82_{nine}\overline{)145,646_{nine}}$

10. Represent each number in base nine notation.

(a) 13

(b) 97

(c) 860

(d) 12,789

12-6 Base Two Numeration

Many electronic computing machines use a numeration system with a base of two. This system, often called the **binary system,** involves only two digits, 0 and 1. Consider the following table, which compares the base ten and base two numerals for the numbers 1 through 20.

Base Ten		Base Two	
Numeral	*Meaning*	*Numeral*	*Meaning*
1	1 one	1_{two}	1 one
2	2 ones	10_{two}	1 two, 0 ones
3	3 ones	11_{two}	1 two, 1 one
4	4 ones	100_{two}	1 four, 0 twos, 0 ones
5	5 ones	101_{two}	1 four, 0 twos, 1 one
6	6 ones	110_{two}	1 four, 1 two, 0 ones
7	7 ones	111_{two}	1 four, 1 two, 1 one
8	8 ones	$1,000_{two}$	1 eight, 0 fours, 0 twos, 0 ones
9	9 ones	$1,001_{two}$	1 eight, 0 fours, 0 twos, 1 one
10	1 ten, 0 ones	$1,010_{two}$	1 eight, 0 fours, 1 two, 0 ones
11	1 ten, 1 one	$1,011_{two}$	1 eight, 0 fours, 1 two, 1 one
12	1 ten, 2 ones	$1,100_{two}$	1 eight, 1 four, 0 twos, 0 ones
13	1 ten, 3 ones	$1,101_{two}$	1 eight, 1 four, 0 twos, 1 one
14	1 ten, 4 ones	$1,110_{two}$	1 eight, 1 four, 1 two, 0 ones
15	1 ten, 5 ones	$1,111_{two}$	1 eight, 1 four, 1 two, 1 one
16	1 ten, 6 ones	$10,000_{two}$	1 sixteen, 0 eights, 0 fours, 0 twos, 0 ones
17	1 ten, 7 ones	$10,001_{two}$	1 sixteen, 0 eights, 0 fours, 0 twos, 1 one
18	1 ten, 8 ones	$10,010_{two}$	1 sixteen, 0 eights, 0 fours, 1 two, 0 ones
19	1 ten, 9 ones	$10,011_{two}$	1 sixteen, 0 eights, 0 fours, 1 two, 1 one
20	2 tens, 0 ones	$10,100_{two}$	1 sixteen, 0 eights, 1 four, 0 twos, 0 ones

Since the binary system has only two digits, the tables of basic addition and multiplication facts are relatively simple, as shown.

Base Two Addition Table   Base Two Multiplication Table

+	0	1
0	0	1
1	0	10

×	0	1
0	0	0
1	0	1

Exercises

1. Represent each number in base ten notation.
   - (a) $10,101_{two}$
   - (b) $11,111_{two}$
   - (c) $100,110_{two}$
   - (d) $100,000,010,000_{two}$

2. Represent each number in base two notation.

(a) 23   (b) 32

(c) 65   (d) 130

3. Perform the indicated operation in base two notation.

(a) $100_{two} + 101_{two}$ (b) $101_{two} \times 10{,}101_{two}$

(c) $1{,}110_{two} - 101_{two}$ (d) $11{,}001_{two} \div 101_{two}$

(e) $10{,}110_{two}$ (f) $10{,}000_{two}$ (g) $1{,}011_{two})\overline{11{,}010_{two}}$

            $1{,}001_{two}$ $\underline{-11_{two}}$

            $10{,}110_{two}$

        $\underline{+111{,}111_{two}}$

12-7 Base Twelve Numeration

The base twelve system of numeration is often called the **duodecimal system.** Our present civilization contains several evidences that our ancestors
collected things in sets of twelve. For example, eggs are sold by the dozen,
twelve inches are equal in measure to one foot, and our calendar is based
on a twelve-month year.

The duodecimal system has several advantages over the decimal system.
One of these is that the number twelve has more counting numbers as factors
than the number ten: twelve has 1, 2, 3, 4, 6, and 12 as factors, while ten
has only 1, 2, 5, and 10 as factors. Another advantage is that twelve is more
closely related to many of our common units of measure, such as those mentioned in Section 12-1.

An interesting problem arises when we begin our consideration of a
numeration system in base twelve. We have already noted that a numeration
system generally has the same number of digits as its base. Thus, the duodecimal system must have twelve digits. If we use the standard decimal digits,
0, 1, 2, 3, 4, 5, 6, 7, 8, and 9, we have only ten digits, and it becomes necessary
to invent two new symbols to make the required total of twelve.

We shall use the letter T as a digit to represent the number ten and the
letter E as a digit to represent the number eleven; that is,

$$T_{twelve} = 10 \quad \text{and} \quad E_{twelve} = 11.$$

We should also note that 10_{twelve} is another name for the number twelve.
In the case of T_{twelve} and E_{twelve} we are using one-digit numerals to represent
numbers that we are accustomed to representing with two digits.

The place values in the base twelve system are based on powers of twelve.
Each place to the left of a given place has a value twelve times as large as the

value of the place on its right; each place to the right of a given place has a value one-twelfth as large as the value of the place on its left.

Place Values for Base Twelve

One Thousand Seven Hundred Twenty-eights	One Hundred Forty-fours	Twelves	Ones	One-twelfths
$12 \times (12 \times 12)$	12×12	12×1	1	$\dfrac{1}{12}$
12^3	12^2	12^1	12^0	12^{-1}

Exercises

1. Represent each number in base ten notation.
   (a) 175_{twelve} (b) $1E_{\text{twelve}}$
   (c) $8,T90_{\text{twelve}}$ (d) T,ETE_{twelve}

2. Complete the following tables of addition and multiplication facts for base twelve.

+	0	1	2	3	4	5	6	7	8	9	T	E
0												
1												
2												
3												
4												
5												
6												
7												
8												
9												
T												
E												

×	0	1	2	3	4	5	6	7	8	9	T	E
0												
1												
2												
3												
4												
5												
6												
7												
8												
9												
T												
E												

3. Find each sum in base twelve notation.

(a) 17_{twelve}
   $+\ 4_{\text{twelve}}$

(b) 36_{twelve}
   $+2E_{\text{twelve}}$

(c) $9E0_{\text{twelve}}$
   $+T59_{\text{twelve}}$

(d) 238_{twelve}
   $T10_{\text{twelve}}$
   67_{twelve}
   $+9E5_{\text{twelve}}$

4. Find each difference in base twelve notation.

(a) $6T7_{\text{twelve}}$
   $-3T8_{\text{twelve}}$

(b) 100_{twelve}
   $-\ EE_{\text{twelve}}$

(c) 520_{twelve}
   $-3E9_{\text{twelve}}$

(d) $72,431_{\text{twelve}}$
   $-\ E,T48_{\text{twelve}}$

5. Find each product in base twelve notation.

(a) 23_{twelve}
   $\times\ 5_{\text{twelve}}$

(b) 519_{twelve}
   $\times\ 4T_{\text{twelve}}$

(c) $4,34E_{\text{twelve}}$
   $\times\ 8T9_{\text{twelve}}$

(d) T,ETE_{twelve}
   $\times\quad ET_{\text{twelve}}$

6. Find each quotient in base twelve notation.

    (a) $9_{twelve}\overline{)76_{twelve}}$ (b) $57_{twelve}\overline{)29,0T7_{twelve}}$

    (c) $T_{twelve}\overline{)7T2_{twelve}}$ (d) $10_{twelve}\overline{)8,9T6,5E0_{twelve}}$

7. Represent each number in base twelve notation.

    (a) 23   (b) 109
    (c) 295   (d) 3,467

8. What base notation is used if:

    (a) $2 + 3 + 1 = 12$?   (b) $35 + 64 = 132$?   (c) $34 + 56 + 27 = 141$?

9. Indicate whether each statement is true or false.

    (a) $0_{twelve} = 0_{two}$ (b) $3_{seven} \times 5_{seven} = 21_{six}$

    (c) $\dfrac{5_{twelve}}{10_{twelve}} = \dfrac{1}{2}$

    (d) $6_{eight} \times 15_{eight} > 4,526_{seven} \div 25_{seven}$

10. Indicate whether each statement is true or false.

    (a) $403_{five} > 163$ (b) $3,442_{five} < 379$
    (c) $249_{twelve} < 101,110,111_{two}$ (d) $1,200_{four} = 140_{eight}$

12-8 Chapter Test

Indicate whether each statement is true or false. If false, tell why.

1. The place values in a base seven system are based on multiples of seven.

2. The symbol $\overset{m}{=}$ has the same meaning as the symbol $=$.

3. The number named by $5,430_{eight}$ is eight times larger than the number represented by 543_{eight}.

4. $3,201.42_{five} = 426.88$.

5. 234_{seven} is an odd number.

6. The numeral 34_{six} should be read "thirty-four, base six."

7. In the binary system, the next (whole) number after 101 is 110.

8. $249_{eleven} < 101,110,111_{two}$.

9. Counting is one of the experiences that is fundamental to the understanding of a system of numeration.

10. In general, the larger the base, the more digits it takes to represent a particular number.

Select the best possible answer.

11. The product of $6_{\text{seven}} \times 5_{\text{seven}}$ is:
   (a) 30   (b) 30_{seven}
   (c) 42_{seven} (d) 11_{seven}
   (e) 245_{seven}

12. In what base are the numerals written if $2 \times 2 = 10$?
   (a) Base two   (b) Base three
   (c) Base four   (d) Base five
   (e) Base ten

13. The value of the digit 4 in the numeral 426_{seven} is:
   (a) 196   (b) 400
   (c) 28   (d) 4,000
   (e) None of these

14. Which numeral represents the larger number?
   (a) 43_{five} (b) 212_{three}
   (c) $10,110_{\text{two}}$ (d) 24_{nine}
   (e) $10_{\text{twenty-five}}$

15. What is the sum of $625_{\text{seven}} + 344_{\text{seven}}$?
   (a) 492_{seven} (b) 969_{seven}
   (c) $1,002_{\text{seven}}$ (d) $1,302_{\text{seven}}$
   (e) None of these

16. Let bells be a kind of number and jingle be an operation on these numbers. To say that the set of bells is closed under jingle is to say:
   (a) Two bells under the operation of jingle will always give us a bell.
   (b) No more bells can get in under jingle.
   (c) We are assured that jingle is a commutative operation.
   (d) We are assured that jingle is an associative operation.
   (e) None of these.

chapter 13

Informal
Nonmetric
Geometry

The learning of geometric concepts and relations in the elementary school years is of increasing concern to the mathematics educator. In this chapter we will encounter some of the geometric topics that might be introduced in an informal manner to elementary school children. Geometrical ideas related to measurement are often called **metric** concepts. Concepts that are developed independently from or prior to measurement concepts are often called **nonmetric.** In this chapter, we will be concerned primarily with nonmetric ideas.

13-1 Historical Background

The study of geometrical concepts has evolved from early beginnings as a collection of observations about relative positions and sizes of physical objects to modern abstract systems with axiomatic formulations. Geometry has practical applications in the modern world much as it did thousands of years ago when ancient Egyptians, Babylonians, and Greeks built temples and studied the motions of stars and planets. But the character of geometry today is more in the realm of the organization of ideas than in the realm of the study of objects.

Ancient Greek geometers were among the first to consider geometry as an organization of ideas as opposed to a study of physical things. Rather than thinking about relations among ropes, tiles, or wheels, they thought in terms of abstractions such as lines, polygons, and circles. These abstractions were considered to have an existence apart from the corresponding physical objects. The ideas were not to be confused with the objects or even confused with drawings representing them. From the three-dimensional world were abstracted the concepts of *space* and *solid*. From the two-dimensional world was abstracted the idea of *surface*, especially that of *plane*. From the one-dimensional world was abstracted the notion of *curve*, especially that of *line*. Fundamental to all of these ideas was the notion of *point*.

Axiomatic geometry has rules like a game. One of the rules in this game is that we will think in terms of *points, lines, planes, and space*. Since points, lines, planes, and space are concepts rather than objects, we sometimes draw pictures of them or use models to represent them. As we look at the models and pictures, we should remember that they represent abstract ideas that do not exist in a physical sense.

13-2 Points

The concept of **point** originated as an abstraction for a sizeless, single location. The abstraction was suggested by such things as pinpoints of light such as stars, small dots, or the place where three edges of a box meet.

The concept of **space** is used when we wish to think of the set of all possible locations. Space has the characteristic of being a set of points.

In arithmetic we developed ideas about sets of objects, called these ideas numbers, and then used symbols to represent the ideas. In geometrical discussions we often use pictures to communicate ideas, but we agree that these pictures do not prove things about these ideas. (Geometric relationships will be investigated in this text, but their mathematical proofs will be left for your future study.)

We use dots to represent points as is done on maps to show locations. The dots *are not* points but *are* pictures of points. Since we often want to consider sets which contain more than one point, we need a way to distinguish one point from another. Capital letters are used to name points. The capital letter which names a given point may be written beside the dot which represents that point. Thus a dot represents the point, and a capital letter identifies the individual point of a set of points. For example, the following picture is the representation of a set of four points, *A*, *B*, *C*, and *D*.

Exercises

1. (a) Which of these two dots is the larger?

   (b) Which of these two points is the larger?

2. What do we mean by writing "$C = D$" for the picture below?

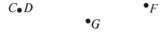

3. Indicate whether each statement is true or false.
   (a) A dot is a point.
   (b) Space is a set of dots.
   (c) A point may be represented by a dot.
   (d) Different points may be named with different capital letters.
   (e) Different capital letters may be used to name different points.

4. (a) Why might there be a strong tendency to say that points can be moved?
   (b) Can a dot be moved from one location to another? Explain.
   (c) Can a point be moved from one location to another? Explain.

13-3 Curves

If you sit and watch an ant move about for a short while, you are sure to ponder over the course of his travels from one place to another. To think about the path he has taken is roughly to think in terms of the geometric notion of **curve**. If we let A name the point at which we first saw the ant and B name the point at which we last saw him, the following picture might represent the path of his movements.

Notice that the picture has flattened out the actual surface over which the ant crawled. The actual curve is a set of points in space. Note also that the ant might have traveled many other paths over the surface to get from *A* to *B*. In these cases, we call *A* and *B* the **endpoints** of the curve. We name the curve *AB* and often say that the curve *joins A* and *B*. In the next picture, how many curves join *C* and *D*? What might we do in order to distinguish the four curves?

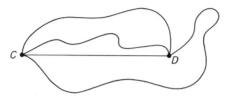

To reconsider the ant's path, note that we have no information about where he has been before nor information about where he goes after we leave him. It may be that his path extends indefinitely from both *A* and *B*. To indicate this we often place arrowheads on extensions of the picture of his path. In the next picture we are assuming that his path has no endpoints.

In going from point *E* to point *H* a path might be as in the next picture. In this case the ant has crossed his own path at *F*. Curves like *AB* above, for which there is no retracing or crossing of any part of the curve, are called **simple curves.** Curves like *EFGH*, for which there is crossing, are referred to as **not simple.**

If an ant were to start at one place and return to the same place, his path might be like one of those pictured below. Such curves are called **closed curves.**

Since no part of curve *IJK* crosses itself, it is also a simple curve and is

classified as a **simple closed curve.** Curve *LMN* is closed but not simple.
Curve *EFGH* above is neither simple nor closed.

Suppose that curve *XY* represents the path of an ant and *M* represents
the point in the curve at which the ant took a bite of food. Point *M* is
between *X* and *Y* on the given path. The point *M* **separates** the curve into

two disjoint parts. The first part corresponds to the path before the ant took
the morsel; the second part corresponds to the path after the ant took the
food. Note that the point at which he actually took the food is in neither of
these two parts. Observe that it is not possible to think of *N* in the following
curve as separating the curve in the same way as *M* separates curve *XY*
above.

Exercises

1. Why is a curve considered to be a set of points?

2. Which of these pictures represent simple curves?

(a)   (b)

(c)   (d)

3. Which of the pictures in Exercise 2 represent closed curves?

4. Which of the pictures in Exercise 2 represent simple closed curves?

5. (a) Find the Cartesian product (Section 2-8) of the sets {simple, not
simple} and {closed, not closed}.

   (b) What relationship does the answer to part (a) have to the figures in
Exercise 2?

6. (a) To get from *A* to *B* in the picture below, must you go through *C*?

   (b) In going from *A* to *B* along the curve pictured above, must you go through *C*?

7. (a) In going from *X* to *Y* in the picture below must you go through *C*? Through *D*?

   (b) In going from *X* to *Y* along the curve pictured above, must you go through *C*? Through *D*? Through at least one of *C* and *D*?

8. In which of the pictures below might the named point be said to separate the curve into two parts?

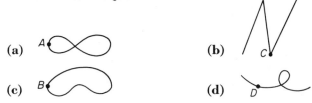

(a)   (b)

(c)   (d)

13-4 Surfaces and Regions

In the previous section we mentioned that the ant's path was only one of many possible paths on the surface on which he was traveling. If you think of spreading a sheet over some object you begin to get an idea of what is meant by a surface. If the ant can roam anywhere on top of the sheet, the set of points corresponding to all of the possible positions he might occupy corresponds to the notion of **surface.** Surfaces are special sets of points.

Just as there are different ways to classify curves, there are different kinds of surfaces. A distinction may be made between those surfaces that are flat and those that are not. To the unaided eye a window pane or a table top suggests flat surfaces (although each is actually quite bumpy). The ground, the outside of a balloon, and the sides of a tin can suggest surfaces that are not flat.

At times the surfaces considered are limited in extent such as those suggested by table tops. At other times it is appropriate to think of surfaces that are unlimited in extent. For example, think about extending the top of a table (rectangular or round) indefinitely.

Surfaces that are both flat and unlimited in extent suggest the geometric idea of a **plane.** A plane may be *represented*, as in the next drawing, by a picture of a flat surface. Note that a flat surface is *not* a plane. A plane cannot be drawn or constructed physically. The next picture actually shows only part of a plane. Planes are often named by three letters naming any three points that are points of the plane but not points of a line. The following picture represents plane *MNR*.

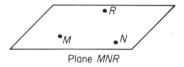

Plane *MNR*

The next picture shows a limitless simple curve *XY* and three points, *A*, *B*, *C*, in a limitless surface.

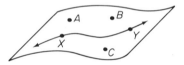

We usually say that *A* and *B* are on the same side of curve *XY*. Note that it is possible to find a curve (in the surface) which contains *A* and *B* but which does not contain any point of curve *XY*. We also say that *A* and *C* are on opposite sides of the curve *XY*. Curve *XY* separates *A* and *C*. Note that it is impossible to find a curve (in the surface) joining *A* and *C* that does not contain a point of curve *XY*. It is said that curve *XY* **separates** the surface into two parts. One part contains *A* and the other part contains *C*. The curve *XY* is in neither part. A curve separates a surface into two portions if there are at least two points of the surface for which any curve joining these two points also contains a point of the given curve.

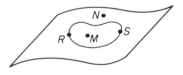

A simple closed curve, such as curve *RS* in the following figure, is also an example of a curve which separates the surface which contains it. Every curve (in the surface) joining M and N must contain a point of curve RS. So curve RS separates the surface. In cases where the surface is limitless, the portion of the separated surface containing point M is called the **interior** of the simple closed curve. The portion containing N is called the **exterior**.

A subset of a surface is sometimes called a **region** of the surface. The interior and the exterior of a simple closed curve are examples of regions. The simple closed curve is called the **boundary** of these regions. In general, the union of a simple closed curve and its interior is called a **closed region.** The shading in the following pictures illustrates the relationship between a simple closed curve, its interior, its exterior, and a closed region. Note that the boundary of the set of points is represented by a dashed line when the boundary is not contained in the set.

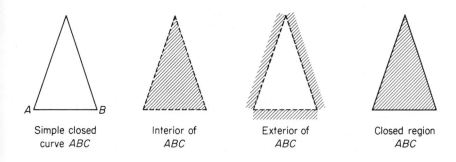

| Simple closed curve *ABC* | Interior of *ABC* | Exterior of *ABC* | Closed region *ABC* |

Exercises

1. Which curve in the pictures below separates the surface that contains it? Why don't both curves separate their respective surfaces?

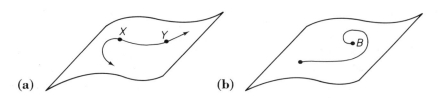

(a)   (b)

2. Draw a model of a simple closed curve and shade to represent the following.
   (a) Exterior points   (b) Interior points
   (c) Closed region

3. Consider the picture below and the indicated points.

(a) Does the picture represent a simple closed curve?

(b) Draw a picture of a curve joining A and D. How many points of the given curve are contained in the curve you drew?

(c) Draw a picture of a curve joining A and B that contains more than one point of the given curve. How many points of the given curve are contained in the curve you drew?

(d) Draw a picture of a curve joining B and C that contains more than one point of the given curve. How many points of the given curve are contained in the curve you drew?

(e) Where are A and D in relation to the given curve? A and B? B and C? Is there any relationship between the answers to these three questions and the numbers obtained in (b), (c), and (d)?

4. If a curve separates a plane into two regions, is the union of the two regions the plane?

13-5 Solids and Space Regions

To think about the set of points suggested by a rock or a block of wood is to have in mind the geometric notion of **solid.**

An inflated balloon with its entrapped air also suggests a solid. However, thinking only of the set of points corresponding to the balloon itself is to think of a surface. The solid is the set of points in space enclosed by and including the balloon's surface.

Sometimes we speak of a subset of space as a **region** in space. A solid is an example of a region in space. Regions in space are also formed when space is separated by a surface. A surface separates space if there exist at least two points in space (not in the surface) for which every curve joining the two points also contains a point of the surface. For example, consider extending a wall indefinitely. To get from one side of the wall to the other requires that you go through the wall. The separating surface (the extended wall) determines two regions in space. The surface is in neither region. It is the **boundary** of the two regions.

Regions like solids are often called **closed regions** since they are often regions formed as the union of a boundary (closed surface) and the region in space that is the interior of the closed surface. For example, a closed box and its inside correspond to a closed region in space. The closed box corresponds to the boundary and the inside corresponds to the interior.

Note that a solid is a set of points in space. The boundary of a solid is a surface. In this surface may be found many curves. For example, the path of an ant roaming on the outside of a box suggests one such curve.

13-6 Lines, Line Segments, and Rays

In Section 13-3 the ant was free to follow any path joining the two points. He did not have to take the most direct route. The most direct route between any two points would be along the *straight* path that contains the two points. A straight path corresponds to a special kind of curve.

The limitless curve that corresponds to a straight path is called a **line.** Lines are usually pictured as below to indicate their properties of straightness and unlimited extent.

It is a property of lines that given any two points, there is only one line which contains both of them. (How many "straight lines" can you draw with a ruler to connect two dots on a piece of paper? How many straight wires can you string between two given positions?) Another way to say this is that two points determine a line.

This property of lines is the basis for one system of naming lines. Any two points on a line are named by letters and a special symbol for lines is placed over the letters. The following line may be named $\overleftrightarrow{GH}$. The notation $\overleftrightarrow{GH}$ is read "the line *GH*." This line may also be named $\overleftrightarrow{HG}$. It is important to note that a line may be named by *any two* of its points. Now suppose an ant does actually take the most direct route joining two points. His actual path does not correspond to the notion of line since the real path is not limitless in extent. His path corresponds to a curve that is a subset of a line.

In the previous picture it might be the part of the line between and including
G and H. Such a curve is called a **line segment,** as in the following diagram.

G   H

A line segment is a subset of a line. It consists of two points, called
endpoints of the line segment, and all the points of the line between
these two points. Given any two points there is only one line segment joining
them. Line segments are named by placing a special symbol over the letters
corresponding to the two endpoints. The line segment above is named
either $\overline{GH}$ or $\overline{HG}$. The notation $\overline{GH}$ is read "the line segment GH."

There are some other subsets of a line which are given special names
because of their usefulness as concepts in geometry. Since a line is a simple
curve, any point in the line separates it into two disjoint curves as in the
picture below. For lines these curves are called **half lines.** Neither half line
contains the separating point. The line AB is actually the union of three sets:
the two half lines and the point of separation.

| Half-line on the | P | Half-line on the |
| left of point P | | right of point P |

There are times when we want to refer to the set of points (in a line)
that consists of the union of a half line and the point which determines the
half line. Such a set is called a **ray.** The following picture shows a ray.

The separating point is called the **endpoint of the ray.** We denote
a ray by its endpoint and some other point on the ray. In the following
figure, we designate the endpoint as P and another point on the ray as Q.
The ray is named $\overrightarrow{PQ}$. The notation $\overrightarrow{PQ}$ is read "the ray PQ." If T is a point
on $\overrightarrow{PQ}$ and is distinct from P, the $\overrightarrow{PQ}$ may also be named $\overrightarrow{PT}$. Note that $\overrightarrow{PQ}$
and $\overrightarrow{QP}$ do *not* name the same ray. They have different endpoints.

Exercises

1. How many endpoints does each of the following have?
   (a) Line   (b) Line segment
   (c) Ray   (d) Half line

2. Given two different points *M* and *N*,
   (a) How many paths are there from *M* to *N*?
   (b) How many straight paths are there from *M* to *N*?
   (c) How many line segments are determined by *M* and *N*?
   (d) How many lines are determined by *M* and *N*?
   (e) How many rays pass through *M* and have *N* as endpoint?

3. For the picture below:

   (a) Give three names for the line.   (b) Name three segments.
   (c) Name three rays.

4. Consider the set of points *A*, *B*, *C*, and *D* as pictured.

   A.   .B

   .C   .D

   (a) Draw a picture to represent all possible line segments determined by
       these points, each segment having two of the given points as end-
       points.
   (b) How many line segments are determined by these points?
   (c) Name each line segment indicated in part (a).

5. Draw a picture of:
   (a) $\overleftrightarrow{CD}$ (b) $\overrightarrow{CD}$
   (c) $\overrightarrow{DC}$ (d) $\overline{DC}$

6. Represent a point *G* on a piece of paper.
   (a) Draw a picture to represent four lines containing *G*.
   (b) How many lines exist which contain *G*?
   (c) Are there any lines containing *G* for which it is not possible to draw
       representations on the paper?

7. Represent a point F on a sheet of paper.
   (a) Draw a picture to represent four rays with endpoint F.
   (b) How many rays are possible with endpoint F?

8. (a) Draw a picture to represent $\overrightarrow{JK}$.
   (b) How many rays are determined by endpoint J such that they pass through point K?
   (c) Is $\overline{JK}$ contained in $\overrightarrow{JK}$?

9. Indicate whether each statement is true or false.
   (a) A line segment is a curve.
   (b) No point of a half line is an endpoint of the half line.
   (c) The endpoints of $\overleftrightarrow{XY}$ are X and Y.
   (d) $\overline{AB}$ is the same line segment as $\overline{BA}$.

13-7 Points, Lines, and Planes

The following picture shows some of the unlimited number of lines containing the single point K. The point K is the intersection of all the lines.

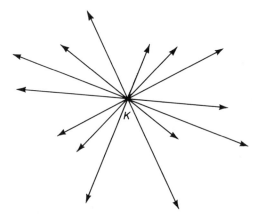

Another important property of lines is that if two lines intersect (intersection is not empty), the intersection is a single point. Observe the next picture.

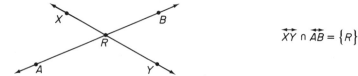

$$\overleftrightarrow{XY} \cap \overleftrightarrow{AB} = \{R\}$$

However, two lines do not always intersect. Sometimes the intersection is empty. Before considering this case we need to examine a relationship of lines and planes.

Planes are related to flatness and lines to straightness. Straight curves can be found in flat surfaces. For example, poke two small holes in a flat piece of cardboard. Put one end of a piece of string through one hole, the other end in the second hole and pull the string taut. The string between the holes comes tight against the cardboard and represents a straight curve. The cardboard represents a flat surface. Together, the string and cardboard are a model of a straight curve on a flat surface. From this situation it is not hard to see why it is said that if the endpoints of a line segment lie on a plane, then every point of the line segment lies on that plane. Similarly, if two points of a line lie on a plane, then every point of the line lies on that plane. Lines are contained on planes.

A plane contains an infinite number of points and each point of the plane is said to be *on* the plane. Since two distinct points determine a line, a plane contains an infinite number of lines. To illustrate this we may use a sheet of paper as a model for a plane and mark two dots on the paper to represent points *A* and *B* as in the picture.

Now we may fold the paper in such a way that the crease falls on the dots. The crease represents a line on the plane passing through *A* and *B*, and there are infinitely many points on the crease. If *C* is a point that is not on the crease, then each of the points of the crease may be used with *C* to determine another line of the plane as in the next picture. Thus there are infinitely many lines on the plane that contain *C*. Of course, there are also infinitely many lines on the plane that do not contain *C*!

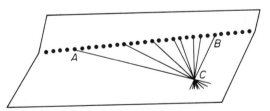

It is important to note (as we have assumed) that two distinct points on a plane may determine a line, every point of which is on the plane. In the drawing above, $\overleftrightarrow{AB}$ is said to lie in the plane and the plane is said to contain $\overleftrightarrow{AB}$.

Now if two lines lie in the same plane and the intersection of the two lines is empty, the lines are called **parallel lines.** See the next figure.

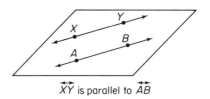

$\overleftrightarrow{XY}$ is parallel to $\overleftrightarrow{AB}$

If two lines do not lie in the same plane, then the intersection of the two lines must be empty and the two lines are called **skew lines.** The picture below shows the lines along the edges of the top and bottom of an ordinary box or along the floor and ceiling of an ordinary room. In this picture $\overleftrightarrow{AB}$ and $\overleftrightarrow{DH}$ are skew lines, whereas $\overleftrightarrow{AB}$ and $\overleftrightarrow{DC}$ are parallel lines. There are also other skew lines and parallel lines in this picture. Can you name a pair of each?

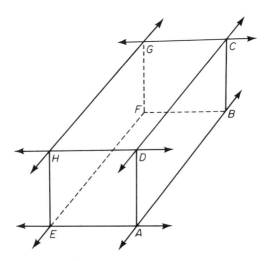

Thus, given two lines, their intersection is either a single point or it is empty. If the intersection is a single point, the lines are in a plane. If the lines are in a plane, they are parallel or their intersection is a single point.

Consider a folded sheet of paper again as a model. We observe that the two parts of the folded paper could represent two planes and the crease a line. The line is contained in both planes. Thus the intersection of the two planes

is a line. This is illustrated in the next picture where the planes *FGH* and *JKL*
intersect in $\overleftrightarrow{AB}$.

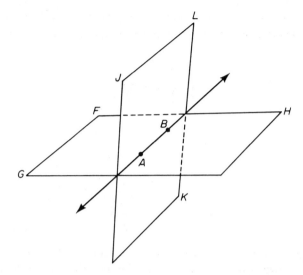

If two planes do not intersect they are called parallel planes. The floor
and ceiling of an ordinary room may be considered a model of parallel
planes.

Thus, given two planes, they either intersect in a line or they are parallel.

What is the nature of the intersection of a line and a plane? The following
pictures show the three possibilities. A line may intersect a plane in a single
point. The intersection may be the entire line. Finally, the intersection may
be empty in which case the line and plane are said to be parallel.

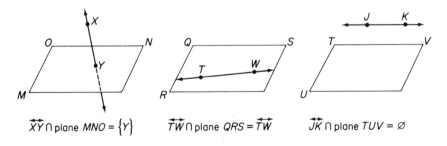

Note also that in the case above where the line lies on the plane, the line
separates the plane into two disjoint regions. These regions are called **half
planes.** For $\overleftrightarrow{TW}$ and plane *QRS* above, the two half planes are the *Q* side
of $\overleftrightarrow{TW}$ and the *R* side of $\overleftrightarrow{TW}$. The line is not contained in either half plane.

A single point may be contained in an unlimited number of lines. In how many planes may a single line be contained? Sheets of paper may again be used as models of planes. Think of many sheets of paper fastened together along a common edge. See the next picture. The common edge represents a single line. The line is contained in an uncountable number of planes.

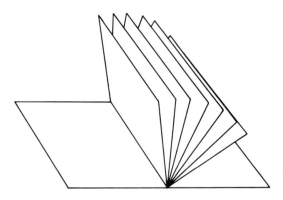

We have observed that two points determine a line. Is there such a relationship between points and planes? Let us think about three points in space and name these points D, E, and F. We know that points D and E determine $\overleftrightarrow{DE}$. Is the point F on $\overleftrightarrow{DE}$? With a little thought, we should conclude that F may or may not be on $\overleftrightarrow{DE}$. From the set of all points (space), there are many points on $\overleftrightarrow{DE}$, and there are many points not on $\overleftrightarrow{DE}$. We shall consider the case in which F is not on $\overleftrightarrow{DE}$. Then, of the uncountable number of planes that contain $\overleftrightarrow{DE}$, how many also contain F? We may think

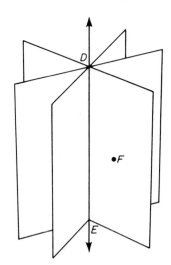

of a revolving door, its axis fastened at the top and bottom, and let the points *D* and *E* be represented by the connections of the axis at the top and bottom of the door. We may imagine the axis extended through *D* and *E*, as a repre-

sentation of $\overleftrightarrow{DE}$. The many positions of the revolving door suggest planes that contain $\overleftrightarrow{DE}$. If point *F* is not on $\overleftrightarrow{DE}$, then exactly one of the many planes represented by the revolving door contains the point *F*. Observe the immediately preceding picture.

It is important to note that three points not on the same line determine a unique plane. If three points are on the same line, then the line determined by two of the points also contains the third point, and this line is contained in an infinite number of planes.

A plane is a surface unlimited in extent. Therefore a plane separates space into two disjoint regions. These regions are called **half spaces.**

Exercises

1. Draw a picture of each of the following.
   (a) A plane and a line in the plane
   (b) A plane and two intersecting lines in the plane
   (c) Three planes containing the same line, $\overleftrightarrow{AB}$
   (d) Three planes containing the same point, *C*

2. Indicate whether each statement is true or false.
   (a) Two lines either intersect or are parallel
   (b) Two planes either intersect or are parallel
   (c) A line and a plane either intersect or are parallel
   (d) Two distinct lines may intersect in more than one point

3. Consider this picture and think of the lines and planes suggested.

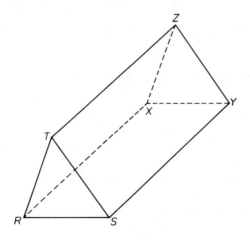

From the picture name the following:

(a) Two lines that are skew

(b) Two lines that are parallel

(c) Two lines that intersect in a point

(d) Two planes that are parallel

(e) Two planes that intersect in a line

(f) A line and a plane that intersect in a point

(g) A line and a plane that intersect in a line

(h) A line and a plane that do not intersect

(i) Three lines that intersect in a point

(j) Three planes that intersect in a point

4. Two distinct points in space are contained in how many planes?

5. Indicate whether each statement is true or false.

(a) One point in space is contained in more than one plane.

(b) Three points in space always determine a plane.

(c) A plane is a flat surface.

(d) A plane contains only a finite number of points.

6. (a) How many different lines may be determined by three distinct points not on the same line?

(b) Do all of the lines indicated in part (a) lie in the same plane?

7. Consider the following picture.

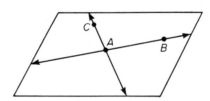

(a) Is it true that $\overleftrightarrow{AC}$ intersects $\overleftrightarrow{AB}$ in only one point; that is, that $\overleftrightarrow{AC} \cap \overleftrightarrow{AB} = A$?

(b) Is it true that $\overleftrightarrow{AC}$ intersects every plane containing $\overleftrightarrow{AB}$ in the one point, A?

8. Indicate whether each statement is true or false.

(a) If a ray has its endpoint interior to a simple closed curve and is located in the plane of the curve, it will always intersect the simple closed curve.

(b) If a plane separates space into two half spaces, every curve in space will intersect the separating plane.

(c) A half line and a ray are the same idea.

(d) If a line separates a plane into two half planes, then every line in the plane either intersects the separating line or is parallel to the separating line.

13-8 Angles, Polygons

In the previous section we examined some relations among points, lines, and planes in terms of set intersection. In this section we will consider several geometric figures which are unions of some familiar figures.

If a simple closed curve in a plane is the union of three or more line segments it is called a **polygon**. Note that a curve can be the union of three or more line segments without being a polygon, that is, without being simple or without being closed. Each of the following pictures represents a polygon. Note that the polygons have special names depending on the number of line segments in the figure.

Triangle   Quadrilateral   Pentagon   Hexagon   Decagon

None of the following pictures represents a polygon. Why?

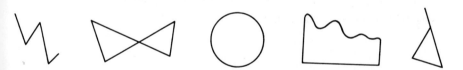

The endpoints of the segments of a polygon are called the **vertices** of the polygon, and the line segments are called the **sides** of the polygon. For example, the vertices of the octagon pictured below are A, B, C, D, E, F, G, and H, and the sides are $\overline{AB}$, $\overline{BC}$, $\overline{CD}$, $\overline{DE}$, $\overline{EF}$, $\overline{FG}$, $\overline{GH}$, and $\overline{HA}$.

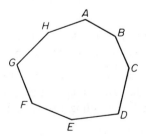

A segment which joins two vertices of a polygon but that is not a side of the polygon is called a **diagonal** of the polygon. In the next picture, $\overline{TV}$ is a diagonal of heptagon $TUVWXYZ$. Segment BD is a diagonal of polygon $ABCD$.

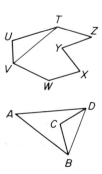

Because of the frequency with which we use triangles, there is a special notation for naming triangles. A triangle with vertices A, B, and C is named $\triangle ABC$. The notation $\triangle ABC$ is read "the triangle ABC."

In Section 13-6 we learned that a ray is the union of a half line and the point determining the half line. Consider the following picture in which the two rays BA and BC have a common endpoint B. Since A, B, and C are three points not on the same line, they determine a plane. We should note that $\overrightarrow{BA}$ and $\overrightarrow{BC}$ lie in that plane.

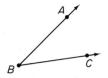

Any figure that is the union of two rays that have a common endpoint is called a **plane angle.** The common endpoint is the **vertex** of the angle, and the rays are **sides** of the angle. Thus, B is the vertex, $\overrightarrow{BA}$ and $\overrightarrow{BC}$ are the sides, and the angle may be named either $\angle ABC$ or $\angle CBA$, where we agree to write the letter corresponding to the vertex between the other two letters. Thus,

$$\angle ABC = \angle CBA,$$

$$\angle ABC = \overrightarrow{BA} \cup \overrightarrow{BC}.$$

Are plane angles formed by the sides of a triangle? Remember that the sides of a triangle are line segments. Thus, we should realize that each side of a triangle is only part of a ray. We may extend the sides of a triangle from a particular vertex to form rays and thus associate a plane angle with each vertex of the triangle. For example, we may extend $\overline{XZ}$ to obtain $\overrightarrow{XZ}$, extend $\overline{XY}$ to $\overrightarrow{XY}$, and thus associate $\angle ZXY$ with vertex X of $\triangle XYZ$ as shown in the following picture.

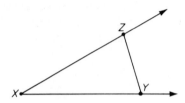

Since the points of the extended segments $\overrightarrow{XZ}$ and $\overrightarrow{XY}$ are not points of $\triangle XYZ$, we must agree that $\angle ZXY$ is not a part of $\triangle XYZ$. We do agree to call $\angle ZXY$ *an angle of* $\triangle XYZ$. However, we realize that the angle consists of points that are not points of the triangle, and thus we actually think of the angle as *determined by* the sides of the triangle.

A plane angle with distinct rays as sides separates the plane into two sets of points. If the sides of a plane angle are not on the same line, it is possible to determine a set of points that is interior to the angle and a set of points that is exterior to the angle. In the next picture, $\angle DEF$ is determined by $\overrightarrow{ED}$ and $\overrightarrow{EF}$, which are not on the same line. If we imagine $\overrightarrow{ED}$ extended to form $\overleftrightarrow{ED}$, the plane is separated into two half planes and F is in one of them. The half plane containing F is indicated by horizontal shading. Next we imagine $\overrightarrow{EF}$ extended to form $\overleftrightarrow{EF}$ and indicate the half plane containing D by vertical shading. The **interior points** of $\angle DEF$ are indicated by the double shading (shaded both horizontally and vertically). The rays are depicted by dashed marks that indicate that the points on the angle are neither interior nor exterior to the angle. All points of the plane that are neither interior points nor points of the rays are **exterior points** of $\angle DEF$.

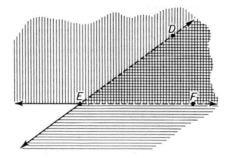

We should note that it is not always possible to determine the interior and exterior of an angle. In the case in which the two rays forming the angle also form a straight line, we have two half planes and either half plane may be thought of as the interior of the angle. Since we cannot determine which half plane to choose, we do not define interior for such angles. Such is the case in the following drawing where K represents the vertex of $\angle JKL$.

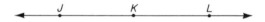

Exercises

1. Which of the figures below represent polygons?

(a)   (b)

(c)

(d)

(e)

2. Write the special name for each of the polygons in Exercise 1.

3. For the polygon represented at the right:
   (a) Name all the vertices
   (b) Name all the sides
   (c) Name all the possible diagonals
   (d) Name all the angles of the polygon

4. Indicate whether each statement is true or false.
   (a) A polygon cannot have less than three sides.
   (b) All diagonals of a polygon are contained in the interior of the polygon.
   (c) Every polygon has the same number of sides as it has vertices.
   (d) In naming a triangle the order of writing the letters makes no difference.
   (e) There are no angles in a polygon.

5. Draw pictures to show polygons with 3, 4, 5, 6, 7, 8, 9, and 10 sides. Draw and count all the possible diagonals for each polygon. Is there a relationship between the number of sides of a polygon and the number of diagonals?

6. Why is an angle a plane figure?

7. Consider the following picture and name the following.

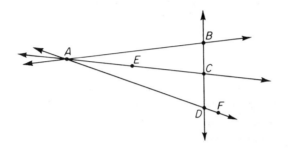

(a) $\overrightarrow{AB} \cap \overrightarrow{DC}$

(b) $\overrightarrow{AB} \cap \overrightarrow{CD}$

(c) $\overleftrightarrow{AF} \cap \overleftrightarrow{BC}$

(d) $\overleftrightarrow{AC} \cap \overleftrightarrow{DF}$

(e) $\overline{AE} \cap \overline{BD}$

(f) $\overrightarrow{BA} \cup \overrightarrow{BC}$

(g) $\overline{AE} \cup \overline{CE}$

(h) $\overrightarrow{EA} \cup \overrightarrow{EC}$

8. Refer to the preceding picture and name the following:
  (a) Three different triangles
  (b) A line segment that is not the side of a triangle
  (c) A point of the exterior of $\triangle ABD$
  (d) A point of the interior of $\angle ABD$
  (e) A point of the exterior of $\angle ABD$
  (f) A point of the exterior of $\angle AEC$
  (g) A point of the interior of $\angle AEC$

13-9 Chapter Test

Indicate whether each statement is always true, sometimes true, or never true.

1. Closed curves are simple curves.

2. A polygon has more than three sides.

3. The intersection of two planes is a single point.

4. There is only one line segment containing two given points.

5. Three points, not all on the same line, determine a plane.

6. $\overline{AB} = \overline{BA}$.

7. $\overrightarrow{AB} = \overrightarrow{BA}$.

8. The sides of an angle of a polygon are line segments.

9. A piece of rope is a model of a curve.

10. The union of two half lines is a line.

Select the best possible answer.

11. A polygon with eight sides is called a(an):
    (a) Octahedron   **(b)** Heptagon
    (c) Decagon   **(d)** Octagon
    (e) Quadrilateral

12. Skew lines always refer to:
    (a) Any curves which are not straight
    (b) Any lines which do not intersect
    (c) Any lines in two different planes
    (d) Any lines which are not parallel
    (e) Lines which do not intersect and which are not parallel

13. An angle is:
    (a) A closed curve
    (b) A figure with two segments
    (c) The union of any two rays
    (d) A figure which lies in a plane
    (e) A figure with two endpoints

14. Which of the following is *not* a simple curve?

    (a)   **(b)**

    (c)

                          (d)

    (e)

15. Which of the following can *not* be the intersection of two rays?
    (a) Point   **(b)** Line segment
    (c) Ray   **(d)** Line
    (e) Empty set

chapter 14

Informal Metric Geometry

One of the characteristics of physical objects which often interests us is the size of the object. Size determination always involves some kind of comparison and usually involves measurement. In the previous chapter we were classifying geometric figures without consideration of size or measurement. For this reason those classifications were called *nonmetric*.

In this chapter we will discuss some *metric* properties of line segments, plane regions, space regions, and angles. These properties are known respectively as length, area, volume, and angular measure.

14-1 Congruence and Similarity

Fundamental to the concept of measurement is the concept of congruence. We usually say that two geometric figures are **congruent** if they have the same size and shape. To get a better idea of the meaning of congruence, let us consider the five following curves.

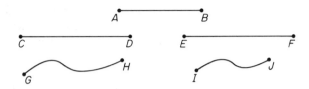

Curves *AB*, *CD*, and *EF* are line segments. They have the same shape. However, they are not the same size. Only $\overline{CD}$ and $\overline{EF}$ are the same size. Segments *CD* and *EF* are both the same size and same shape. (Make a model of $\overline{CD}$ using tracing paper and place the model on top of the picture of $\overline{EF}$. They match exactly.) We say that $\overline{CD}$ and $\overline{EF}$ are congruent. We write $\overline{CD} \cong \overline{EF}$ which is read "$\overline{CD}$ is congruent to $\overline{EF}$." (Note: $\overline{CD} \neq \overline{EF}$ since they are different segments.)

Curves *EF* and *GH* are the same size, but they have different shapes. (You can test the size relationship by laying a string along the picture of *EF* and cutting the string to match the endpoints. Then lay this piece of string along the picture of curve *GH*.) These curves are not congruent.

Curves *GH* and *IJ* are the same shape but not the same size. (One way to show that they are the same shape is to trace the picture of curve *IJ* on a clear acetate sheet. Then hold this sheet above the picture of *GH* so that the sheet is parallel to the page. By adjusting the distance between the acetate and the page, you will find a position of the acetate where if you look directly down on the page through the acetate, the copy of *IJ* on the acetate will exactly match the picture of *GH* on the page.)

Figures that have the same shape but not necessarily the same size are said to be **similar**. Segments *AB*, *CD*, and *EF* are similar figures. Curves *GH* and *IJ* are similar figures.

In the following drawing, △*ABC*, △*DEF*, and △*GHI* are similar to each other. To indicate that △*ABC* is similar to △*DEF* we write △*ABC* ∼ △*DEF*.

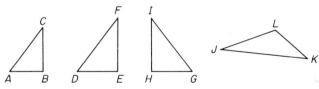

Triangle *JKL* is not similar to any of the other three triangles in the picture. Triangles *DEF* and *GHI* also are congruent to each other. (Trace a copy of one and flip this tracing over to see that they match.) Figures that are congruent are also similar.

The size of a curve is called its **length.** If the curve is a closed figure, the length of the curve is called its **perimeter.** If the curve is a line segment the length of the line segment is the **distance** between its two endpoints.

Exercises

1. Which of the curves represented below are congruent? (Use tracing paper to help you decide.)

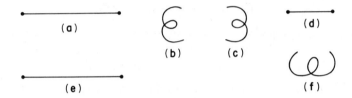

(a)

(b)   (c)

(d)

(e)

(f)

2. Which curves in Exercise 1 are similar?

3. Which curves in Exercise 1 are the same length?

4. Indicate whether each statement is true or false.
   (a) Figures that are congruent have the same shape.
   (b) A quadrilateral and a triangle that have equal perimeters are also congruent.
   (c) Triangles that have equal perimeters are congruent.

5. Refer to the curve represented at the right and draw a figure that is:
   (a) Congruent
   (b) Similar but not congruent
   (c) Not similar

6. Indicate whether each statement is true or false.
   (a) If $\overline{GH} = \overline{JK}$, then $\overline{GH} \cong \overline{JK}$.
   (b) If $\overline{LM} \cong \overline{RS}$, then $\overline{LM} = \overline{RS}$.
   (c) If $\overline{PQ} \cong \overline{EF}$ and $\overline{EF} \cong \overline{XY}$, then $\overline{PQ} \cong \overline{XY}$.
   (d) Every line segment is congruent to itself.

7. Use straws and pipe cleaners as fasteners to make two models of triangles. Each triangle is to have one side of 5 inches, one side of 6 inches, and one side of 7 inches. Do the two models represent congruent triangles?

14-2 Measurement of Segments

Consider the problem of comparing the two line segments pictured.

In our comparison we must agree that only one of the following three possibilities is true:

$\overline{AB}$ is longer than $\overline{CD}$,

$\overline{AB}$ is shorter than $\overline{CD}$,

or

$\overline{AB}$ is the same length as $\overline{CD}$.

One method of comparing these two segments is to use a compass and copy a picture of one segment onto the representation of a ray formed by extending the second line segment indefinitely in one direction. In the next picture we extend $\overline{CD}$ through D to form $\overrightarrow{CD}$, so that we may copy $\overline{AB}$ onto $\overrightarrow{CD}$.

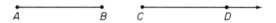

If the compass is set so that the metal tip is on A and the pencil point is on B, then the line segment between the metal tip and the pencil point represents $\overline{AB}$. To copy $\overline{AB}$ onto $\overleftarrow{CD}$, we place the metal tip of the compass on C and then mark on the picture of $\overrightarrow{CD}$ the location of the pencil point. If the mark of the pencil point is on the extension of $\overline{CD}$ (that is, beyond point D), then $\overline{AB}$ is longer than $\overline{CD}$. If the mark is between C and D, then $\overline{AB}$ is shorter than $\overline{CD}$. If the mark is on D, then $\overline{AB}$ **is congruent to** $\overline{CD}$; that is, $\overline{AB}$ and $\overline{CD}$ have the same length.

Using this procedure with a compass and picture of a ray, it is possible to order any group of line segments using the relation *is longer than*.

A more refined process of comparing lengths of line segments involves the assignment of a number to the size of a segment. This number is called the **measure** of the segment. To do this we must first agree that the size of a particular line segment (and any line segment congruent to it) is assigned the value 1. This segment is called a **unit segment.** Some of our standard units for length are the inch, foot, and meter.

In the following figure the size of $\overline{PQ}$ is 1. To determine the size of any

other segment as compared to the unit segment we may use a process related
to that described above. For example, to find the measure of $\overline{XY}$ in the
following illustration, we first make a copy of $\overline{PQ}$ and extend it through Q
to form $\overrightarrow{PQ}$. We associate the value 1 with the point Q. We then make
successive non-overlapping adjacent copies of $\overline{PQ}$ on $\overrightarrow{PQ}$. (We are using

subscripts on the letter Q to indicate the successive location of the endpoint
Q.) We associate the values 2, 3, 4, and so on with the successive Q endpoints. With the endpoint of the ray we associate the value 0. We now make
a copy of $\overline{XY}$ on $\overrightarrow{PQ}$ with endpoint X at P. Since endpoint Y corresponds to
Q_3 and the value three, we say that the measure of $\overline{XY}$ in terms of $\overline{PQ}$ is
three.

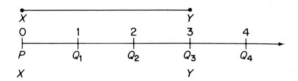

We have, in effect, made a ruler using $\overrightarrow{PQ}$ and $\overline{PQ}$ as a unit segment.

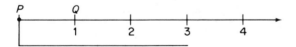

Exercises

1. Use a compass to determine which of the line segments pictured below
   are congruent, and then state your conclusion in symbol form.

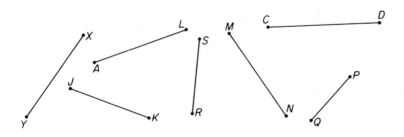

2. Extend one of the line segments given below through an endpoint to form a ray, and then copy a picture of the other line segment onto the representation of the ray. Determine whether $\overline{FG}$ is longer than, shorter than, or congruent to $\overline{BC}$.

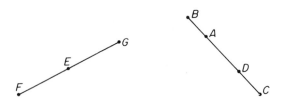

3. Use $\overline{AB}$ below as a unit segment to make a ruler and find the measure of each of the other segments.

4. Find the length of each curve below using $\overline{AB}$ in Exercise 3 as the unit. (String may help.)

14-3 Circles

The geometric figure known as a circle is an example of a set of points whose description depends on metric concepts. In particular we need the concept of distance between two points.

Think about a polygon of many sides. We shall begin by representing a point on a plane by a dot on a piece of paper. We name the point C. Next, we represent a line segment with C as one endpoint and R as the other. Then we represent a large number of different line segments of the same length as $\overline{CR}$ and with C as one endpoint as in (a) of the picture below.

If the representation of each line segment were erased except for the endpoints, we would have point C and a set of points such that each point was the same distance from C. We may connect the endpoints that are different from C by line segments and obtain a many-sided polygon as in (b) of the following picture. Each vertex of this polygon would be at the same distance from C.

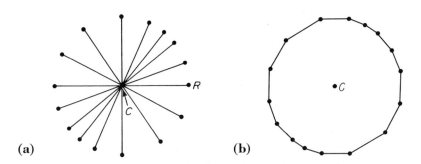

(a)   **(b)**

Imagine a process in which we return to the set of line segments, each having C as an endpoint, and locate another such segment between each pair; suppose we do this over and over again. If every time we had done so we constructed a picture of the polygon like the preceding drawings, we would find that the number of sides of the polygon was increasing and that the length of each side was decreasing. As the process continued, it would suggest the set of *all* points that are at a given distance (length of $\overline{CR}$) from a point (C) in a plane.

Thus we accept the existence of a simple closed curve consisting of a set of points in a plane such that each point is at a given distance from a given point in the plane. This simple closed curve is called a **circle**. The given point (C in the previous illustration) is called the **center** of the circle. It is important to note that the center is *not* an element of the set of points forming the circle. Any line segment with one endpoint on the circle and the other endpoint at the center of the circle is called a **radius** of the circle. The given distance (length of $\overline{CR}$ in the previous illustration) is the measure of a radius of the circle. A radius of a circle is not a point and thus is not a part of the circle. Only the point R of $\overline{CR}$ is a point of the circle. We should also note that a circle is a plane curve; that is, a set of points on a plane.

We may use a compass to draw a model of a circle with any point as center and any line segment as a radius. Remember, we cannot actually draw a circle, because a circle is an idea that we represent by a model.

A line segment that passes through the center of a circle and has both endpoints on the circle is called a **diameter** of the circle. We observe in the next figure that the length of the diameter, $\overline{AB}$, of the circle is twice the length of the radius, $\overline{AO}$, of the same circle (since the radii $\overline{AO}$ and $\overline{OB}$ have the same length). Notice that the diameter is not a part of the circle. Only the points A and B of $\overline{AB}$ are points of the circle.

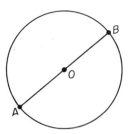

Recall that a curve is a particular set of points passed through in going from one point to another. A curve which is part of a circle is called an **arc.** In other words, an arc is the set of points on a circle that are passed through in going from one point on the circle to another point on the circle. In (a) of the following picture, the arc DXE is represented in heavier ink than the arc DYE. The notation $\overset{\frown}{DXE}$ may be used to represent arc DXE.

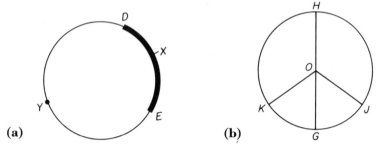

(a)   (b)

The special case of an arc of a circle with endpoints the same as the endpoints of a diameter of the circle is called a **semicircle.** The semicircle does not include any points of the diameter except the endpoints. In (b) of the above picture $\overset{\frown}{GKH}$ is a semicircle, $\overset{\frown}{GJH}$ is a semicircle, $\overline{GH}$ is a diameter of the circle, $\overline{GO}, \overline{KO}, \overline{JO},$ and $\overline{HO}$ are radii of the circle O.

Exercises

1. Consider the circle represented below with the center at point S.

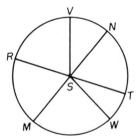

    (a) Name three radii of the circle.
    (b) Name four different arcs with endpoint *R*.
    (c) Name two diameters of the circle.
    (d) Name five points on the circle.

2. Refer to the drawing for Exercise 1.
    (a) Is point *S* on the circle?
    (b) Is the length of $\overline{RS}$ equal to the length of $\overline{VS}$?
    (c) Is $\overset{\frown}{VMW}$ a semicircle?
    (d) Is point *R* on $\overset{\frown}{VMW}$?

3. Indicate whether each statement is true or false.
    (a) All radii of the same circle have the same length.
    (b) A line segment with both endpoints on the circle is always called a diameter.
    (c) All diameters have the same length.
    (d) An arc of a circle is a line segment whose endpoints are also endpoints of a diameter of the circle.
    (e) All diameters of the same circle intersect at the center of the circle.
    (f) A semicircle is half of a circle.
    (g) A circle is a plane curve.
    (h) A circle is a polygon.

4. Use a compass for the following.
    (a) Draw a model of a circle with center at point *P* and with $\overline{PR}$ as a radius of length $\frac{1}{2}$ inch.
    (b) On the same drawing made for part (a), draw a model of a circle with *R* as center and $\overline{PR}$ as a radius.
    (c) Determine how many points are common to the two circles.
    (d) Draw a model of a circle with center at one of the points of intersection of the two circles pictured in part (b) and with a radius of the same length as $\overline{PR}$.
    (e) Determine whether points *P* and *R* are on the circle suggested in part (d).

5. Is it possible to construct a model of a circle such that it contains:
    (a) One given point?   (b) Two given points?
    (c) Three given points?

6. The perimeter of a circle is usually called the *circumference* of the circle. The following experiment should suggest a way to approximate the circumference of any circle given information about its diameter.

    Locate five different circular objects. Using a tape measure, find the circumference and the length of the diameter for each circle.

Divide the circumference by the length of the diameter in each case. What is the approximate result in each case?

Suppose that you know only the length of the diameter of a circle. What do the above results suggest as a way to obtain an approximation of the circumference of the circle?

14-4 Congruence and Measurement of Plane Angles

We can compare two plane angles in much the same way that we compared line segments. Recall that a plane angle is defined as any figure formed by two rays with a common endpoint. Consider the following two plane angles.

In making a comparison of these two angles we must agree that only one of the following three possibilities is true:

$$\angle CDE \text{ is larger than } \angle FGH,$$

$$\angle CDE \text{ is smaller than } \angle FGH,$$

or

$$\angle CDE \text{ is the same size as } \angle FGH.$$

One method of comparing these two angles is to use a compass and copy a picture of one angle onto the representation of the other angle. To copy $\angle CDE$ onto $\angle FGH$ we set the compass to any convenient radius, place the metal tip on D, and draw with the pencil point an arc that intersects $\angle CDE$ in two points, X and Y. We retain the same compass opening (length of $\overline{DY}$), place the metal tip on G, and draw with the pencil point an arc that intersects $\angle FGH$ in two points, X' and Y'. See the following figure.

Next we set the compass so that the metal tip is on X and the pencil point is on X, place the metal tip on Y', and draw with the pencil point an arc (in the same half plane as X' relative to $\overleftrightarrow{GH}$) that intersects $\overset{\frown}{X'Y'}$. See the following figure.

If the arc intersects $\overset{\frown}{X'Y'}$ in the exterior of $\angle FGH$, then $\angle CDE$ is larger than $\angle FGH$. If the arc intersects $\overset{\frown}{X'Y'}$ in the interior of $\angle FGH$, then $\angle CDE$ is smaller than $\angle FGH$. If the arc intersects $\overset{\frown}{X'Y'}$ at X', then $\angle CDE$ is congruent to $\angle FGH$, and we write $\angle CDE \cong \angle FGH$. Consider the following three drawings which picture the three possible cases.

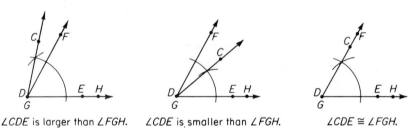

∠CDE is larger than ∠FGH.   ∠CDE is smaller than ∠FGH.   ∠CDE ≅ ∠FGH.

Consider the following picture of two congruent angles, $\angle RJK$ and $\angle SJK$.

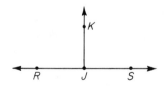

This drawing represents a special case of congruent angles. Observe that $\overrightarrow{JR}$ and $\overrightarrow{JS}$ form $\overleftrightarrow{RS}$ with R and S on opposite sides of J. Observe also that K is not on $\overleftrightarrow{RS}$ and, thus, $\overrightarrow{JK}$ intersects $\overleftrightarrow{RS}$ and forms two congruent angles. Each of the angles formed in this manner is called a **right angle.** Models of right angles exist all about us. The corner of a page in this book is a model of a right angle, the hands of a clock at exactly nine o'clock represent a right angle, and the printed letter L is often a model of a right angle.

A more refined process of comparing the sizes of angles involves the assignment of a number to the size of an angle. To do this we must first agree that the size of a particular angle (and any angle congruent to it) will be assigned the value 1. This angle is called a **unit angle**. The standard unit for angular measurement in elementary work is the degree.

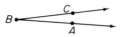

In the following figure, the size of $\angle ABC$ is 1. The size of any angle with distinct sides may be taken as 1. To determine the measure of any other angle as compared to this unit angle we may use a process related to that described above. For example, to find the measure of $\angle XYZ$ illustrated below we first make a copy of $\angle ABC$ and associate the value 0 with A and the value 1 with C. We then make successive non-overlapping, adjacent copies of $\angle ABC$ using B as the vertex for each copy. (We are using subscripts on the letter C to indicate the successive positions corresponding to that point.) We associate the values 2, 3, 4, and so on with the successive points C on the copies of the unit angle. We now make a copy of $\angle XYZ$ such that vertex Y corresponds to vertex B and $\overrightarrow{YZ}$ corresponds to $\overrightarrow{BA}$. Since $\overrightarrow{YX}$ corresponds to $\overrightarrow{BC_3}$ and the value three, we say the measure of $\angle XYZ$ in terms of the unit $\angle ABC$ is three.

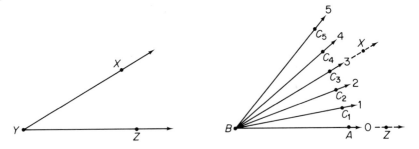

We have, in effect, made a protractor with $\angle ABC$ as the unit angle.

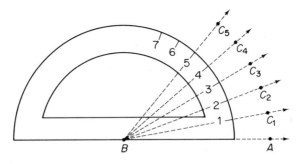

When the unit angle used is the degree, a **right angle** such as $\angle ABC$ in the next figure has a measure of 90. An angle whose measure is between 0 and 90 is called an **acute angle.** An angle whose measure is between 90 and 180 is called an **obtuse angle.**

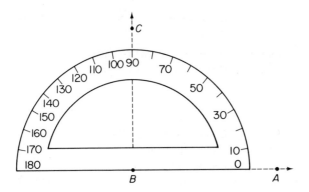

Exercises

1. Determine which of the angles pictured below are congruent, and then state your conclusion in symbol form.

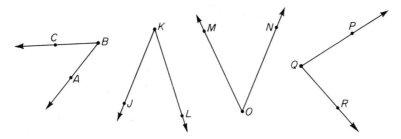

2. Represent one of the angles given below and then copy a picture of the other angle onto the representation, to determine whether $\angle DEF$ is larger than, smaller than, or congruent to, $\angle XYZ$.

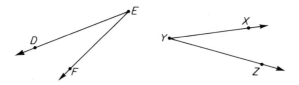

3. Use a model of a right angle to determine which of the following drawings of angles represent right angles.

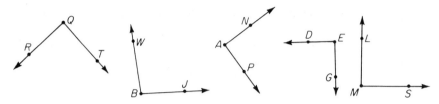

4. Indicate whether each statement is true or false.
   (a) Two congruent angles always form a right angle.
   (b) Two rays form a plane angle.
   (c) If $\angle ABC = \angle DEF$, then $\angle ABC \cong \angle DEF$.
   (d) If $\angle GHJ \cong \angle KLM$, then $\angle GHJ = \angle KLM$.
   (e) If $\angle NOP \cong \angle QRS$ and $\angle QRS \cong \angle TUX$, then $\angle NOP \cong \angle TUX$.
   (f) Every angle is congruent to itself.
   (g) Any two right angles are congruent.
   (h) If two angles are not equal, then one angle is either larger or smaller than the other.

5. Find the measure of each angle pictured below by making a protractor with $\angle DEF$ as the unit angle.

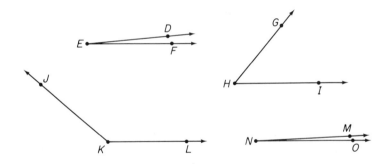

6. Draw pictures of each of the following.
   (a) Acute angle   (b) Right angle
   (c) Obtuse angle

7. Use straws and pipe cleaners to make models of two triangles, each of which has an angle of 50 degrees. The sides of the triangle corresponding to the sides of this angle are to be 4 inches and 5 inches in each triangle. Make the third side in each case so that it fits. Do the two models represent congruent triangles?

8. Use straws and pipe cleaners to make models of two triangles, each of which has angles of 50, 60, and 70 degrees. Can you make the models so that they do not represent congruent triangles? Can you make the models so that they do not represent similar triangles?

14-5 Classification of Triangles and Quadrilaterals

Knowledge of metric concepts allows us to distinguish various kinds of triangles and quadrilaterals.

In Chapter 13 we learned that a triangle is a polygon that is the union of three line segments. These line segments are called sides of the triangle, and the endpoints of the segments are called vertices of the triangle. We also noted that the sides of a triangle do not form angles; however, we agreed that we may extend the sides of a triangle from a particular vertex to form rays and thus associate a plane angle with each vertex of the triangle. We can classify triangles in two ways: by comparing the lengths of their sides and by comparing the measures of their angles. (*Note:* We are actually comparing the angles formed by extending the sides of a triangle from a particular vertex to form rays.)

First, we shall consider the classification of triangles by comparing the sides of a triangle. See the following figure. A triangle with three sides congruent is called an **equilateral triangle**. A triangle with at least two sides congruent is called an **isosceles triangle**. A triangle with no two sides congruent is called a **scalene triangle**. The symbol $\not\cong$ means "is not congruent to."

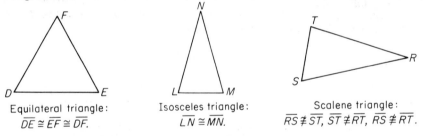

Equilateral triangle:
$\overline{DE} \cong \overline{EF} \cong \overline{DF}$.

Isosceles triangle:
$\overline{LN} \cong \overline{MN}$.

Scalene triangle:
$\overline{RS} \not\cong \overline{ST}, \overline{ST} \not\cong \overline{RT}, \overline{RS} \not\cong \overline{RT}$.

Now consider the classification of triangles by comparing the angles of a triangle. See the next figure. A triangle with a right angle is called a **right triangle**. A triangle with each angle smaller than a right angle is called an **acute triangle**. A triangle with one angle larger than a right angle is called an **obtuse triangle**.

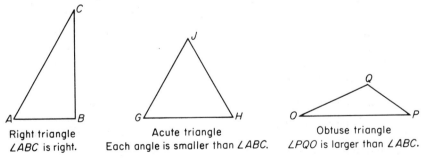

Right triangle
∠ABC is right.

Acute triangle
Each angle is smaller than ∠ABC.

Obtuse triangle
∠PQO is larger than ∠ABC.

Classification systems for quadrilaterals (polygons with four sides) use the concepts of parallelism, segment congruence, and right angles.

Quadrilaterals may have two pairs of parallel sides, one pair of parallel sides, or no pair of parallel sides as illustrated below. A quadrilateral which has no parallel sides is called a **trapezium.** A quadrilateral which has only one pair of parallel sides is called a **trapezoid.** A quadrilateral which has two pairs of parallel sides is called a **parallelogram.**

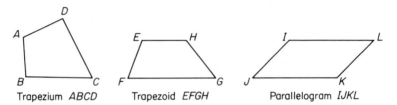

Trapezium *ABCD*   Trapezoid *EFGH*   Parallelogram *IJKL*

Quadrilaterals may have various arrangements of congruent sides, some of which are illustrated below. A quadrilateral with no congruent sides is called a **scalene quadrilateral.** (All scalene quadrilaterals happen to be trapeziums.) A quadrilateral for which all sides are congruent is called a **rhombus.** (All rhombi happen to be parallelograms.) A quadrilateral in which opposite sides are congruent also happens to be a parallelogram. There are other possibilities which we will not consider here.

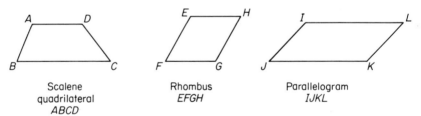

Scalene   Rhombus   Parallelogram
quadrilateral   *EFGH*   *IJKL*
ABCD

A parallelogram for which all angles are right angles is called a **rectangle.** A rhombus for which all angles are right angles is called a **square.** (Note that a square is also a rectangle. Why?)

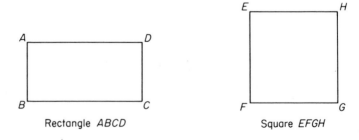

Rectangle *ABCD*   Square *EFGH*

Exercises

1. Classify each of the triangles pictured below by comparing their sides.

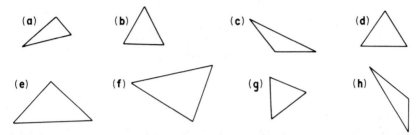

2. Classify each of the triangles pictured in Exercise 1 by comparing their angles.

3. Draw a picture of a triangle that is:
   (a) Obtuse and scalene   (b) Acute and scalene
   (c) Right and scalene   (d) Obtuse and isosceles
   (e) Acute and isosceles   (f) Right and isosceles

4. Can you draw a picture of a triangle that is:
   (a) Obtuse and equilateral?   (b) Acute and equilateral?
   (c) Right and equilateral?

5. Indicate whether each statement is always true, sometimes true, or never true.
   (a) All angles of a rectangle are congruent.
   (b) All angles of a parallelogram are congruent.
   (c) Rectangles are squares.
   (d) Squares are rhombi.
   (e) Opposite sides of parallelograms are congruent.
   (f) Opposite sides of trapezoids are congruent.

14-6 Area

At times we find it necessary to compare the sizes of two regions of a surface. For example, when we want to cover one surface with another we usually need to compare their sizes. The size of such a region is called the **area** of the region. In most elementary work we are usually interested in the area of closed regions in a plane.

One of these regions is the rectangular region such as those in the following picture. Recall that a rectangular region is the union of a rectangle and its interior.

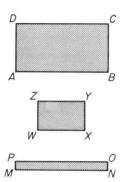

Sometimes it is possible to compare the sizes of two regions by making a copy of one and placing it over a copy of the other. For example, a copy of rectangular region *WXYZ* above fits within a picture of rectangular region *ABCD*. The area of region *WXYZ* is less than that of region *ABCD*.

Sometimes it is not so easy to make the comparison. If you make copies of regions *WXYZ* and *MNOP* above, neither one fits within the picture of the other.

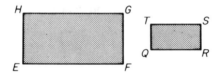

A more refined method of comparing plane regions associates numbers with the sizes of the regions. A unit region is established and its size is assigned the value 1. The areas of the regions in question are then compared with the area of the unit region. For example, to determine the area of the following rectangular region *EFGH* in terms of the unit *QRST*, we could make copies of the unit region and place them on the representation of region *EFGH*. In this case the area of region *EFGH* is 4.

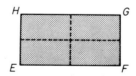

In working with area, the unit regions are generally square regions, such as the inch square region or the foot square region. The area measurement of a square region each of whose sides has measurement of 1 inch is 1 square inch. The square region is an especially convenient unit for determining area of rectangular regions. If the lengths of the sides of the rectangular region are based on the same unit of length as the length of a side of the unit square region, a simple formula for the area of a rectangular region is obtained. One simply multiplies the number of units in the length and the number of units in the width of the region to obtain the number of square regions which will cover the corresponding rectangular region. This formula is usually written

$$A = lw.$$

The area of the rectangular region in the following picture is obtained by multiplying 3 times 4 which yields 12.

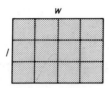

Formulas for the areas of many other closed plane regions can be obtained by studying the relationship between the region in question and a rectangular region. For example, to find the area of right triangular region ABC in the following picture, we may first find the area of rectangular region $ABCD$.

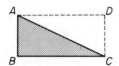

The area of the triangular region is half the area of the rectangular region, since region $ABCD$ is the union of the two congruent regions ABC and CDA. The areas of two congruent regions are always equal.

Exercises

1. Use closed region XYZ below as the unit and try to find the area of closed region $ABCD$.

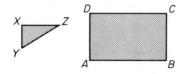

2. Use closed region JKL as the unit and try to estimate the area of closed region $MNOPQ$. What problems do you encounter that did not occur in Exercise 1?

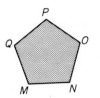

3. Use closed region *STUV* as the unit and try to estimate the area of closed region *FGH*.

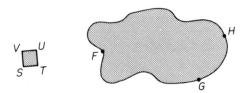

4. Compute the area for each of the rectangular regions whose dimensions (length and width) are listed below.
(a) $l = 2, w = 18$ (b) $l = 4, w = 9$
(c) $l = 6, w = 6$ (d) $l = 12, w = 3$
(e) $l = 9, w = 4$ (f) $l = 18, w = 2$

5. (a) Do all of the regions in Exercise 4 have the same area?
(b) Are all of the regions in Exercise 4 congruent?
(c) Are some of the regions in Exercise 4 congruent? Which ones?
(d) Do all of the regions in Exercise 4 have the same perimeter?
(e) Do the regions in Exercise 4 that have the same perimeter also have the same area?
(f) Do all rectangular regions that have the same perimeter also have the same area?

6. Use the following diagram to describe a way to find the area of a parallelogram region. *ABCD* is a parallelogram. *XYCD* is a rectangle. *AXD* and *BYC* are congruent triangles.

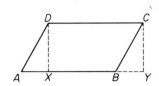

14-7 Volume

It is also possible to compare the sizes of space regions. The size of a space region is called its **volume.**

Typically we are interested in finding the volume of geometric figures called solids. We usually want to know how much of something it will take to fill a certain container, or how much space a certain object occupies. Of special interest is the solid known as a **rectangular solid.** The boundary of a rectangular solid is the union of six rectangular regions. The regions are called **faces** of the solid. If two faces intersect, the intersection is a segment and is called an **edge** of the solid. See the next figure. Such solids correspond to ordinary boxes which have been filled and closed.

Rectangular solid

The standard unit region for volume measure is the cubical solid. A **cubical solid** is a rectangular solid each of whose faces is a square region. All edges of a cubical solid are congruent. The value 1 is assigned to the volume of a cubical solid each of whose edges has a length of one unit. The volume measurement of a cubical solid each of whose edges has a measurement of one inch is one cubic inch.

Suppose we want to find the volume of the rectangular solid represented below. Values for the length, width, and height of the solid are 4, 2, and

3 respectively. We will try to find the volume in terms of a cubical solid each of whose edges has a length of one unit. We will be able to place four unit cubical solids in a row along the bottom rear of the rectangular solid. We will be able to place two such rows or eight cubical solids on the bottom. We will be able to place three layers each with eight cubical solids to fill the solid. In all it takes $4 \times 2 \times 3$, or 24 cubical solids, to fill the rectangular solid. The volume of the solid is 24. To find the volume of a rectangular solid we need simply find the product of its length, width, and height. This is usually written as the formula

$$V = lwh.$$

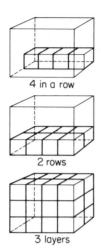

4 in a row

2 rows

3 layers

Exercises

1. Lump sugar comes in two shapes: cubical and rectangular. Find a small box (perhaps a box in which paper clips are sold) and a supply of both kinds of lump sugar.
   (a) Use the cubical lumps as the units of volume and determine the approximate volume of the box.
   (b) Use the rectangular lumps as the units of volume and determine the approximate volume of the box.
   (c) Why do the results for (a) and (b) not agree?

2. Suppose you have a cubical block. You score each edge of the front, top, and right side at its midpoint as shown in the picture. You then cut through the block using the midpoints of four parallel edges to determine the cutting plane. If you complete all three possible cuts:
   (a) How many pieces of the block will you have?
   (b) What is the shape of each piece?
   (c) How does the volume of each piece compare to the volume of the original block?

3. What is a cubic inch? A cubic foot?

4. How many cubic inches are equivalent to a cubic foot?

5. Find the volume of each of the rectangular solids whose dimensions are listed below.

   (a) $l = 6, w = 5, h = 10$ (b) $l = 10, w = 10, h = 10$

   (c) $l = 3, w = 6, h = 0.5$

14-8 Chapter Test

Indicate whether each statement is true or false. If the statement is false, tell why.

1. All radii are congruent line segments.

2. If the perimeters of two rectangular regions are equal, then the areas of the two regions are equal.

3. If two triangles are similar, they are congruent.

4. On a circle an arc with both endpoints as endpoints of a diameter is a semicircle.

5. Congruent angles have congruent sides.

6. All right triangles have 2 acute angles.

7. A square is a rhombus.

8. A rectangle is a parallelogram.

9. If two rectangular regions have equal areas, then the regions are congruent.

10. A circle is a many-sided polygon.

Select the best possible answer.

11. All parallelograms are:
    - (a) Squares   (b) Triangles
    - (c) Trapezoids   (d) Rectangles
    - (e) Quadrilaterals

12. No triangle is a:
    - (a) Curve   (b) Polygon
    - (c) Pentagon   (d) Simple closed curve
    - (e) Boundary of a region

13. An equilateral triangle may also be:
    - (a) Scalene and obtuse   (b) Acute and isosceles
    - (c) Right and isosceles   (d) Isosceles and obtuse
    - (e) Acute and scalene

14. Which of these statements is false?
   (a) All quadrilaterals are four-sided figures.
   (b) No obtuse triangle is an equilateral triangle.
   (c) Some rectangles are squares.
   (d) No polygon is a curve.
   (e) Some right triangles are isosceles.

15. What is the volume of a rectangular solid with length, width, and height of 8, 5, and $2\frac{1}{2}$ respectively?
   (a) 80 cubic units   (b) 96 cubic units
   (c) 100 cubic units   (d) 250 cubic units
   (e) None of these

Mathematical Symbols and Abbreviations

	Meaning
$+$	Plus; add; also used to indicate positive direction, as in $^+2$ (read "positive two").
$-$	Minus; subtract; also used to indicate negative direction, as in $^-2$ (read "negative two").
$\times$, $\cdot$	Times; multiply, as in $a \times b$, $a \cdot b$, and ab (read "a times b").
$\div$, $-$	Divide, as in $a \div b$, $\dfrac{a}{b}$ (read "a is divided by b").
$:$	Ratio, as in $a:b$ (read "a is to b").
$\ldots$	The three-dot notation means to complete, as in $1, 2, 3, \ldots, 12$, or to continue indefinitely in the same manner, as in $1, 2, 3, \ldots$.

337

SYMBOL	MEANING
$=$	Is equal to, as in $a = b$ (read "a is equal to b").
$\neq$	Is not equal to, as in $a \neq b$ (read "a is not equal to b").
$>$	Is greater than, as in $a > b$ (read "a is greater than b").
$\ngtr$	Is not greater than, as in $a \ngtr b$ (read "a is not greater than b").
$<$	Is less than, as in $a < b$ (read "a is less than b").
$\nless$	Is not less than, as in $a \nless b$ (read "a is not less than b").
$\geq$	Is greater than or equal to, as in $a \geq b$ (read "a is greater than or equal to b").
$\leq$	Is less than or equal to, as in $a \leq b$ (read "a is less than or equal to b").
$\cong$	Is congruent to, as in $a \cong b$ (read "a is congruent to b").
$\ncong$	Is not congruent to, as in $a \ncong b$ (read "a is not congruent to b").
$\leftrightarrow, \sim$	Is equivalent to, as in $R \leftrightarrow S$, $R \sim S$ (read "set R is equivalent to set S" or "set R has the same number of elements as set S").
$\sqrt{\ }, \sqrt[3]{\ }, \sqrt[5]{\ }$	Square root, cube root, fifth root, respectively.
$\overset{m}{=}$	Is equal in measure to.
$( ), [ ], \{ \}$	Symbols of inclusion, called parentheses, brackets, and braces, respectively.
$\{x, y\}$	The set whose elements are x and y.
$\in$	Is an element of, as in $a \in A$ (read "a is a member of set A").
$\subseteq$	Is a subset of, as in $A \subseteq B$ (read "set A is a subset of set B"). Each element of A is an element of B.
$\subset$	Is a proper subset of, as in $D \subset E$ (read "set D is a proper subset of set E"). Each element of D is an element of E and there is at least one element of E that is not an element of D.
$\nsubseteq$	Is not a subset of, as in $F \nsubseteq G$.
$\not\subset$	Is not a proper subset of, as in $J \not\subset K$.
$\cup$	Union, as in $A \cup B$ (read "set A union set

SYMBOL	MEANING
	B"). $A \cup B$ is the set of all elements that are elements of at least one of the two given sets.
$\cap$	Intersection, as in $D \cap E$ (read "set D intersection set E"). $D \cap E$ is the set of all elements in both D and E.
$\{\,\}, \varnothing$	The empty (null) set.
U	The universal set.
$\{1, 2, 3, 4, \ldots\} = C$	The set of counting (natural) numbers.
$\{0, 1, 2, 3, 4, \ldots\} = W$	The set of whole numbers.
$\{\ldots, -3, -2, -1, 0, 1, 2, 3, \ldots\} = I$	The set of integers.
$n(T)$	The number of elements in set T.
A'	The complement of set A.
$\square = 3 + 2$	An example of an open number sentence; the symbol $\square$ is a placeholder (also referred to as a variable).
a^n	The letter a represents the base of a power, the letter n an exponent, and a^n represents a power.
π	Pi, $\pi = 3.14159\ldots.$
$\%$	Percent; by the hundred.
10_{five}	The subscript "five" indicates the base of the system of numeration.
$0.\overline{6}$	A repeating decimal; the digit 6 is repeated indefinitely.
$\overline{AB}$	Line segment with endpoints A and B.
$\overrightarrow{AB}$	Ray with endpoint A, extending in the direction of B from A.
$\overleftrightarrow{AB}$	Line determined by the points A and B.
$\overarc{CD}$	Arc with endpoints C and D.
$\triangle ABC$	Triangle with vertices at A, B, and C.
$\angle ABC$	Angle with vertex at point B and sides $\overrightarrow{BA}$ and $\overrightarrow{BC}$.
L.C.M.	Least common multiple.
G.C.F.	Greatest common factor.
G.C.D.	Greatest common divisor.
$A \times B$	The Cartesian product of sets A and B.

Glossary of
Mathematical
Terms

The purpose of this glossary is to provide a convenient and useful explanation of mathematical words and phrases used in this book. This list is by no means complete and the explanations given are not intended to be precise definitions. Additional discussion of these terms as well as examples and illustrations may be found in the book by referring to the index.

Abstract—That which exists in the mind, as opposed to that which is concrete and physically in existence.

Abundant number—A number with proper divisors whose sum is greater than the number.

Acute angle—An angle whose measure is between 0 and 90 degrees.

Acute triangle—A triangle with each angle smaller than a right angle.

Addend—The name given to each of the numbers to be added.

Addition—The process of finding the cardinal number of a set formed by the union of two disjoint sets; also a binary operation defined for numbers.

Additive identity—The number zero is the additive identity because the sum of any number and zero is always the original number; that is, for every number a, $a + 0 = 0 + a = a$. The additive identity is also called the identity element for addition.

Additive inverse of a number—The unique number which produces a sum of 0 when added to a given number; that is, for every number a there is a unique number ^-a such that $a + {}^-a = {}^-a + a = 0$. The additive inverse of a number is often called the negative of the number.

Additive system of numeration—A system of numeration in which the number represented by a particular set of symbols is the sum of the numbers represented by the symbols in the set.

Algorithm—A computational procedure; in general, an algorithm is a method of arranging numerals in such a way as to reduce the number of steps necessary to determine a correct answer to a particular problem.

Amicable numbers—A relationship between two numbers such that each number equals the sum of the proper divisors of the other.

Arbitrary—Fixed at the discretion of an individual without reference to any established pattern.

Arc—The set of points on a circle that are passed through in going along the circle from one point on the circle to another point on the circle.

Area—The size of a region.

Array—An arrangement of symbols, often in columns and rows.

Associative property of addition—For every three numbers to be added together the sum is not affected by grouping the numbers differently (without changing their order); that is, for all numbers a, b, and c, $(a + b) + c = a + (b + c)$.

Associative property of multiplication—For every three numbers to be multiplied together the product is not affected by grouping the numbers differently (without changing their order); that is, for all numbers a, b, and c, $(a \cdot b) \cdot c = a \cdot (b \cdot c)$.

Assumption—Any statement that is accepted without proof in order to provide a basis for discussion.

Base of a power—A number used a given number of times as a factor; the product is called the power, and the repeated factor is the base of the power.

Base of a system of numeration—A number that establishes the method of grouping and in general the number of digits needed for a particular system of numeration.

Between—A particular order relation; specifically, for three different numbers a, b, and c, the number a is between the numbers b and c if $b > a > c$ or $c > a > b$.

Binary operation—A rule that assigns to two given elements a third element.

Borrowing—A colloquial term used to describe the renaming (regrouping) of a number that is sometimes necessary in order to subtract one number from another. Generally, it is suggested that the term *renaming* (or regrouping) be taught instead of the term *borrowing*.

Boundary of a region—A simple closed curve.

Braces, {, }—Symbols used in this book to indicate sets of elements. The members of a set are represented within the braces.

Brackets, [,]—Symbols used to indicate that the enclosed numerals or symbols should be considered together.

Cardinal number—A number used to indicate how many elements are contained in a set, irrespective of the order in which the elements are arranged.

Carrying—A colloquial term used to describe the renaming (regrouping) of a number that is sometimes necessary in order to add two or more numbers together. Generally, it is suggested that the term *renaming* (or regrouping) be taught instead of the term *carrying*.

Cartesian product of sets—The Cartesian product of sets A and B (denoted by $A \times B$) is the set of all possible pairs (a, b) formed by selecting the first element a from set A and the second element b from set B.

Center of a circle—The point in the plane of the circle that is the same distance from every point on the circle.

Check—Verify the correctness of a solution.

Circle—A simple closed curve consisting of a set of points in a plane such that each point is at a given distance from a given point in the plane.

Circular region—The union of the interior of a circle and the circle; that is, the set of all points inside and on a circle.

Circumference of a circle—The perimeter of the circle.

Closed curve—A plane curve consisting of a path that begins at a point and comes back to the same point.

Closed region—The union of a simple closed curve and its interior.

Closure property of addition—For every two elements of a given set of numbers the sum of the two elements also is an element of the given set.

Closure property of multiplication—For every two elements of a given set of numbers the product of the two elements also is an element of the given set.

Column—A vertical arrangement of symbols.

Common factors—Factors of each of two or more given numbers.

Common multiples—Multiples of each of two or more given numbers.

Commutative property of addition—For every two numbers the sum is not affected by the order of addition; that is, for every number a and every number b, $a + b = b + a$.

Commutative property of multiplication—For every two numbers the product is not affected by the order of multiplication; that is, for every number a and every number b, $a \cdot b = b \cdot a$.

Compass—An instrument used to draw pictures of circles and arcs; also used to make comparisons of geometric figures.

Complement of a set—If A is a subset of a universal set, then the complement of A with respect to the universal set is the set of all elements in the universal set that are not elements of A.

Completeness—A property of the set of real numbers. The real numbers are complete, because there is a real number for every point of the number line; conversely, there is a point of the number line for every real number.

Composite number—A whole number greater than 1 that is not prime; that is, a whole number greater than 1 that has whole numbers other than 1 and the number itself as factors.

Congruent—Two geometric figures are congruent if they have the same size and shape; that is, they can be matched exactly. In such cases one figure can be copied onto the other and the two figures are exactly the same.

Conjecture—An inference formulated from a small amount of evidence.

Consecutive—Following one after the other in a regular order.

Coordinate of a point—The number that is associated with a particular point on a number line.

Correspondence—A matching (that is, pairing) of the elements of one set with the elements of another set.

Counterexample—An example or case in which a particular mathematical idea does not hold true.

Counting—A procedure whereby the counting numbers beginning with 1 are matched in order with the elements of a set, to determine that the set contains as many elements as the last number named.

Counting number—Any element of the set $\{1, 2, 3, 4, 5, 6, \ldots\}$. The set of counting numbers is also called the set of natural numbers.

Cubical solid—A rectangular solid, each of whose faces is a square region.

Curve—The set of points of a path.

Cycle—The set of digits that repeats in a periodic decimal.

Decagon—A polygon with ten sides.

Decimal—A numeral in decimal notation.

Decimal notation—A notation which extends the base ten place-value system to the right of the ones place by making use of a decimal point to separate the ones place from the tenths place immediately to the right of the ones place.

Decimal point—A dot used in decimal notation to separate the ones place from the tenths place.

Decimal system of numeration—A system of numeration that uses the number ten as a base and requires ten digits. The decimal system employs place value and is also called the base ten system of numeration.

Deficient number—A number with proper divisors whose sum is less than the number.

Denominator—The number that is represented by b in a rational number expressed in the form $\frac{a}{b}$, where $b \neq 0$.

Dense—A set of numbers is dense if there is always at least one number between any two different numbers of the set.

Diagonal of a polygon—A line segment which joins two vertices of a polygon but that is not a side of the polygon.

Diameter of a circle—A line segment that passes through the center of the circle and has both endpoints on the circle.

Difference—The result obtained when one number is subtracted from another.

Digit—Any one of the single symbols that is used to write a numeral in a system based on place value. The ten decimal digits are 0, 1, 2, 3, 4, 5, 6, 7, 8, and 9.

Directed line segment—A line segment pictured with an arrowhead at one endpoint to indicate the direction associated with the line segment. Although the choice of directions for positive and negative is arbitrary, on the number line we generally represent a positive number by a line segment directed to the right and a negative number by a line segment directed to the left.

Disjoint sets—Any two sets with no elements in common; that is, two sets whose intersection is the empty set.

Distinct—Not the same.

Distributive property of multiplication over addition—For all numbers a, b, and c, $a \times (b + c) = (a \times b) + (a \times c)$.

Distributive property of multiplication over subtraction—For all numbers a, b, and c, $a \times (b - c) = (a \times b) - (a \times c)$.

Divisible—In general, a number n is divisible by a number b if b is a factor of n.

Division—The process of finding the quotient of two numbers. Division is the inverse of multiplication; division may be considered as repeated subtraction.

Division, measurement-type—A division problem which may be interpreted as finding how many subsets of a certain number of elements each are contained in a given set of elements.

Division, partition-type—A division problem which may be interpreted as separating a given set into a certain number of parts, each containing the same number of elements.

Division by zero—A meaningless operation that is excluded from consideration in our number system.

Divisor—Any one of the numbers that are multiplied to form a product. A divisor of a number is also called a factor of the number. In a division problem the number that we divide by is the divisor.

Dot—A small mark that is used to picture a point.

Edge of a solid—The line segment formed where two faces of a solid intersect.

Element of a set—An object contained in a set. An element of a set is also called a member of the set.

Empty set—A set containing no elements. An empty set is also called a null set.

Endpoint—Either of the two points that are connected by a path; also the initial point of a ray.

Equality—A relation between two numbers that are exactly the same. The symbol = is written between two numerals to indicate that both numerals represent the same number; it is also used in other cases to indicate that different names are being used for the same thing. For example, two sets are equal if they have exactly the same members.

Equation—A number sentence expressing the relation of equality. The statement may involve specified numbers, or numbers represented by variables, or both.

Equilateral triangle—A triangle with three sides congruent.

Equivalent fractions—Two or more fractions that name the same number.

Equivalent sets—Two sets whose elements can be placed in one-to-one correspondence.

Even number—Any element of the set $\{\ldots, {}^-8, {}^-6, {}^-4, {}^-2, 0, 2, 4, 6, 8, \ldots\}$.

Expanded form of a numeral—A form of a numeral that clearly illustrates the place value of each digit making up the numeral. For example, an expanded form of 321 is $(3 \times 10^2) + (2 \times 10^1) + (1 \times 10^0)$.

Exponent (positive integral)—A number represented by a small numeral written above and to the right of another numeral representing the base. The exponent indicates how many times the base is to be used as a factor.

Exterior of a plane curve (or angle)—One of the sets of points into which the curve (or angle) separates the plane in which it lies. Generally, the set of points is considered to be "outside" the curve (or angle).

Faces of a solid—The regions that comprise the boundary of a solid.

Factor—Any one of the two or more numbers that may be multiplied to form a product. A factor of a product is also called a divisor of the product.

Factor tree—An arrangement of numerals which may be used to display the factors of a number.

Finite set—A set with a whole number as its cardinal number.

Fraction—A numeral that names a fractional number. A fraction is also called a fractional numeral.

Fractional number—Any number that may be expressed in the form $\frac{a}{b}$ where the numerator a and the denominator b are whole numbers and $b \neq 0$.

Fractional numeral—A numeral that names a fractional number. A fractional numeral is also called a fraction.

Fundamental Theorem of Arithmetic—Every composite number can be expressed as a product of prime numbers which is unique except for the order of the factors. The Fundamental Theorem of Arithmetic is also called the Unique Factorization Theorem.

Geometry—The part of mathematics that deals with the study of points, lines, planes, and space.

General—Not specific; covering all known special cases.

Graph of a number—The point on a number line that is associated with a particular number.

Greater than—An order relation of numbers. For any two numbers a and b, a is greater than b (written $a > b$) if $a - b$ is a positive number.

Greatest common factor (greatest common divisor)—The largest natural number that is a common factor (divisor) of two or more given numbers.

Half line—One of the sets of points that is formed when a point separates a line. A point separates a line into two half lines, neither one of which contains the point.

Half plane—One of the sets of points that is formed when a line separates a plane. A line separates a plane into two half planes, neither one of which contains the line.

Half space—One of the two sets of points that is formed when a plane separates space. A plane separates space into two half spaces, neither one of which contains the plane.

Heptagon—A polygon with seven sides.

Hexagon—A polygon with six sides.

Hindu-Arabic system of numeration—A name associated with the decimal system of numeration.

Identical sets—Two sets whose elements are exactly the same. Identical sets are also called equal sets.

Identity element—A number that does not change the value of a given number when a certain operation is performed on the two numbers. Zero is the identity element for addition because, for every number a, $a + 0 = 0 + a = a$. The number 1 is the identity element for multiplication because, for every number a, $a \times 1 = 1 \times a = a$.

Inequality—A relation between two numbers that are not the same. Each of the symbols $\neq$ (is not equal to), $<$ (is less than), and $>$ (is greater than) represents an inequality relation when written between two numerals.

Inequation—A number sentence expressing an inequality. The statement may involve specified numbers, or numbers represented by variables, or both.

Infinite decimal—A decimal that does not terminate; that is, the digits continue indefinitely.

Infinite set—A set that does not have a whole number as its cardinal number.

Informal geometry—A study of geometric concepts that makes no attempt to substantiate the concepts by formal proof. This approach to the study of geometry is dependent upon intuition, experimentation, observation, reasoning by analogy, and reasoning by induction.

Integer—Any element of the set $\{\ldots, ^-3, ^-2, ^-1, 0, 1, 2, 3, \ldots\}$.

Integral—Having the property of being an integer.

Interior of a plane curve (or angle)—One of the sets of points into which the curve (or angle) separates the plane in which it lies. Generally, the set of points is considered to be "inside" the curve (or angle).

Intersect—To have a point, a nonempty set of points, or other elements in common.

Intersection of sets—The intersection of two sets M and N is the set consisting of the elements that are common to both M and N.

Inverse operations—Two operations that are such that one operation undoes what the other operation does; that is, the net effect of performing the two operations is the same as that of not having performed either operation.

Irrational number—Any real number that cannot be expressed in the form $\frac{a}{b}$, where a and b are integers and $b \neq 0$, that is, the numbers that are coordinates of points on the real number line and are not rational numbers. Each irrational number may be expressed as an infinite nonrepeating decimal.

Isosceles triangle—A triangle with at least two congruent sides.

Least common denominator—The least common multiple of the denominators under consideration.

Least common multiple—The smallest of the common multiples of two or more numbers.

Length—The measure of a curve.

Less than—An order relation of numbers. For any two numbers a and b, a is less than b (written $a < b$) if $b - a$ is a positive number.

Like fractions—Fractions whose numerals representing denominators are identical.

Line (straight line)—A set of points that can be thought of informally as a line segment that is extended in both directions without end.

Line segment—The set of all points contained in the straight path connecting two points.

Lowest terms—A fractional number is expressed in lowest terms when the greatest common factor of its numerator and denominator is 1. A fractional number in lowest terms is also said to be in simplest form.

Match—To pair an element of one set with an element of another set.

Mathematical system—A set of elements, one or more binary operations, one or more relations, and some rules which the elements, operations, and relations satisfy.

Measure of a line segment—The number assigned to the size of a line segment.

Member of a set—An object contained in a set. A member of a set is also called an element of the set.

Metric geometry—Aspects of geometry that are related to measurement.

Missing addend—A name given to the number to be determined when one number is subtracted from another.

Missing factor—A name given to the number to be determined when one number is divided by another.

Mixed numeral—A numeral which includes representations of both an integer and a fractional number.

Model (geometric)—A representation of an abstract geometrical idea.

Modern mathematics—An approach to the study of mathematics that emphasizes the importance of concepts, patterns, and mathematical structure, as well as the development of mathematical skills.

Multiple of a number—Any number that is obtained by multiplying a given number by a positive integer.

Multiplication—The process of finding the product of two numbers. Multiplication may be considered as repeated addition.

Multiplication property of zero—For every number a, $a \times 0 = 0 \times a = 0$.

Multiplicative identity—The number 1 is the multiplicative identity, because the product of any number and 1 is always the original number; that is, for every number a, $a \times 1 = 1 \times a = a$. The multiplicative identity is also called the identity element for multiplication.

Multiplicative inverse of a number other than zero—The unique number which produces a product of 1 when multiplied by a given number (other than 0); that is, if $a \times b = b \times a = 1$, then b is the multiplicative inverse of a and a is the multiplicative inverse of b. Two numbers whose product is 1 are also called reciprocals of each other.

Multiplicative system of numeration—A system of numeration in which the number represented by a particular set of symbols is derived from the multiplication of numbers represented by symbols in the set.

Natural number—Any element of the set $\{1, 2, 3, 4, 5, 6, \ldots\}$. The set of natural numbers is also called the set of counting numbers.

Negative integer—Any element of the set $\{^-1, ^-2, ^-3, ^-4, ^-5, \ldots\}$.

Negative number—Any number less than zero. On a number line the negative numbers are generally represented to the left of zero.

Nondecimal system of numeration—A system of numeration in which a number other than 10 is used as a base.

Nonmetric geometry—Aspects of geometry that are developed independently from measurement concepts.

Nonnegative integer—An element of the set $\{0, 1, 2, 3, 4, 5, \ldots\}$.

Nonnegative number—Any number that is not negative; that is, any number greater than or equal to zero.

Nonrepeating decimal—An infinite decimal with no digit or set of digits that repeats indefinitely.

Notation—Any systematic convention for expressing ideas, operations, relations, etc. by means of symbols.

Null set—A set containing no elements. A null set is also called an empty set.

Number—An abstract idea of quantity.

Number line—A representation of numbers by points on a line. The line is usually pictured horizontally and is thought of as extending, without end, both to the left and to the right.

Number sentence—A statement in symbols describing a relation between two or more numbers. The relation may be an equality or an inequality, and it may be either true or false.

Numeral—A symbol used to represent a number.

Numeration system—An organized procedure for arranging a set of symbols so that they can be used effectively to name numbers.

Numerator—The number that is represented by a in a rational number expressed in the form $\frac{a}{b}$, where $b \neq 0$.

Obtuse angle—An angle whose measure is between 90 and 180 degrees.

Obtuse triangle—A triangle with one angle larger than a right angle.

Octagon—A polygon with eight sides.

Odd number—Any element of the set $\{\ldots, ^-7, ^-5, ^-3, ^-1, 1, 3, 5, 7, \ldots\}$. The odd numbers may be obtained by adding 1 to each element of the set of even numbers.

One-to-one correspondence—An arrangement whereby the elements of two sets are matched so that each element of the first set is matched to one and only one element of the second set, and each element of the second set is matched to one and only one element of the first set.

Open number sentence—A number sentence containing one or more variables.

Order—A set of numbers is ordered if, for any two numbers in the set, one is less than, greater than, or equal to the other; that is, for every pair of numbers a and b in a given set exactly one of the relations $a > b$, $a < b$, $a = b$ must hold.

Ordered pair—A pair of elements which have been assigned a specific order; that is, one element is designated as the first element and the other is designated as the second element.

Ordinal number—A natural number used to indicate a particular position of an element in a set.

Parallel lines—Lines in the same plane that do not intersect.

Parallel planes—Planes that do not intersect.

Parallelogram—A quadrilateral with two pairs of parallel sides.

Parentheses, (,)—Symbols used to indicate that the enclosed numerals or symbols should be considered together.

Partial product—Part of a product, as illustrated below.

$$\begin{array}{r}
123 \\
\times 321 \\
\hline
123 \\
2\,46 \\
36\,9 \\
\hline
39{,}483
\end{array}$$

Each of these numbers is a partial product.

Path—A set of points in space that are passed through in going from one point to another.

Pentagon—A polygon with five sides.

Percent—A numeral form that means by the hundred; that is, 8 percent means $\dfrac{8}{100}$.

Perfect number—A number with proper divisors whose sum equals the number.

Perimeter—The length of a closed curve.

Periodic decimal—Any decimal in which a single digit or set of digits repeats indefinitely. A periodic decimal is also called a repeating decimal.

Place value—A notation wherein the value represented by each digit in a given numeral is determined by the position it occupies. Place value is also referred to as positional notation.

Placeholder—A symbol used to represent an unspecified number. A placeholder is also called a variable.

Plane—A set of points in space that can be thought of as an indefinite extension of a flat surface.

Plane angle—The union of two rays that have a common endpoint.

Plane curve—A curve consisting of a set of points that lie in the same plane.

Point—An exact location in space. A point has no size because it is only an idea.

Polygon—A simple closed curve that is the union of three or more line segments.

Positional notation—A notation wherein a value is assigned to each position in a numeral; thus, the positional value of each digit in a numeral can be determined by multiplying the value of the digit by the value assigned to the particular place it occupies. Positional notation is also called place value.

Positive integer—Any element of the set $\{1, 2, 3, 4, 5, \ldots\}$.

Positive number—Any number greater than zero. On a number line positive numbers are generally represented to the right of zero.

Power—Any product in which all the factors are the same; the common factor is the base, and the number of such factors is indicated by the exponent. For example, 2^4 is the fourth power of the base 2, and $2^4 = 2 \times 2 \times 2 \times 2$.

Prime number—A whole number greater than 1 which has only itself and 1 as whole number factors.

Prime twins—Two consecutive prime numbers whose difference is 2.

Product—The result obtained when two or more numbers are multiplied.

Proof—A pattern of reasoning used to establish the validity of a statement. The procedure is based on undefined terms, definitions, assumptions, and previously proved statements.

Proper divisors—All the divisors of a number except the number itself.

Proper subset—A proper subset of set K is any subset of K that is not the entire set K; that is, there is at least one element of set K that is not an element of a proper subset of K.

Proportion—A statement that two ratios are equal.

Quadrilateral—A polygon with four sides.

Quotient—The result obtained when a number is divided by another number different from zero.

Radius of a circle (plural, radii)—A line segment with the center of the circle as one endpoint and a point on the circle as the other endpoint.

Rate pair—A comparison of two numbers by division. The numbers generally indicate a relationship between two unlike sets.

Ratio—A comparison of two numbers by division.

Rational number—Any number that can be expressed in the form $\frac{a}{b}$, where the numerator a and the denominator b are integers and $b \neq 0$. Each rational number may be expressed as a repeating decimal.

Ray—The union of a half line and the point determining the half line.

Real number—A coordinate of a point on the number line. Any real number is either a rational number or an irrational number.

Reciprocal—One of a pair of numbers whose product is 1. The reciprocal of a number is also called the multiplicative inverse of the number.

Rectangle—A parallelogram with four right angles.

Rectangular region—The union of the interior of a rectangle and the rectangle; that is, the set of all points inside and on a rectangle.

Rectangular solid—A solid whose surface is the union of six rectangular regions.

Reduce—Change the name of a fractional number by dividing both the numerator and the denominator by the same number (usually a natural number). The value of a fractional number is not changed when its name is changed in this manner.

Region of a plane—The union of the interior of a simple closed curve and the curve.

Relation—A statement of the association between two or more numbers, ideas, operations, etc.

Relatively prime numbers—Two or more numbers that have no common factors other than 1.

Remainder—A number that is "left over" in a division example when the quotient is not an exact factor of the number being divided. The remainder is greater than or equal to 0, but less than the divisor.

Repeating decimal—Any decimal in which a single digit or set of digits repeats indefinitely. A repeating decimal is also called a periodic decimal.

Repeats—Continues on and on; that is, the same element (or set of elements) appears over and over again in some pattern.

Repetitive system of numeration—A system of numeration in which a particular symbol always represents the same value, regardless of position or frequency of use.

Rhombus—A quadrilateral with each pair of sides congruent.

Right angle—Either of the two angles formed when a ray extends from a point on a line in such a way that the two angles are congruent. An angle whose measure is 90 degrees.

Right identity that is not a left identity—An identity element that holds only when it is on the right, that is, when it follows the operational symbol. Zero is a right identity for subtraction because for every number a, $a - 0 = a$, whereas in general $0 - a \neq a$. One is a right identity for division because, for every number a, $a \div 1 = a$, whereas in general $1 \div a \neq a$.

Right triangle—A triangle with a right angle.

Row—A horizontal arrangement of symbols.

Rule—A defined method of performing some operation.

Scalene quadrilateral—A quadrilateral with no two sides congruent.

Scalene triangle—A triangle with no two sides congruent.

Scientific notation—A form for representing any given number as the product of an appropriate power of 10 and some number that is greater than or equal to 1 but less than 10.

Semicircle—An arc of a circle such that the endpoints of the arc are the endpoints of a diameter of the circle.

Separate—A geometric concept that means to divide a given set of points into two disjoint subsets such that the dividing point (or set of points) is not a member of either subset and any path with one endpoint in each of the two disjoint subsets contains the separating point (or a point of the separating set of points).

Set—Any well-defined collection, class, or aggregate of objects or ideas.

Sides of a polygon—The line segments forming the polygon.

Signed number—A directed number, that is, a number whose direction is indicated by a + or − sign prefixed to the numeral that names the number.

Similar figures—Geometric figures that have the same shape, but not necessarily the same size.

Simple closed curve—A closed curve that does not intersect itself.

Simplest form—A fractional number is expressed in simplest form when the greatest common factor of its numerator and denominator is 1. A fractional number that is expressed in simplest form is said to be in lowest terms.

Simplify—To transform a given expression into a simpler form that is equal in value to the given expression.

Skew lines—Lines that do not intersect and are not parallel. Skew lines are not in the same plane.

Solution set—The set of numbers that make an open number sentence true.

Space—The set of all points; the universal set of points.

Square—A rectangle with all four sides congruent. Also may be thought of as a rhombus with four right angles.

Square of a number—The number obtained by multiplying a number by itself; that is, the second power of the number.

Straight path—The most direct path from one point to another.

Subset—A subset of A is a set B such that each element of B is an element of A.

Subscript—A word, numeral, or letter that is written below and (in this text) to the right of a symbol to distinguish it from other symbols.

Substitution—A process of replacing one symbol by another symbol representing the same thing; the symbols may represent numbers, or variables, or both.

Subtraction—The process of finding the difference between two numbers; subtraction is the inverse of addition.

Subtractive system of numeration—A system of numeration in which the number represented by a particular set of symbols is derived from the subtraction of numbers represented by symbols in the set.

Sum—The result obtained when two or more numbers are added.

Symbol—A mark of any sort that represents a number, an operation, a relation, etc.

Terminates—Comes to an end; the division process is said to terminate when the division continues until there is a remainder of 0.

Terminating decimal—A decimal in which the digit 0 repeats indefinitely; that is, since the digit 0 repeats indefinitely in the numeral, we may terminate the decimal with the last digit prior to the repeating 0.

Trapezium—A quadrilateral which has no parallel sides.

Trapezoid—A quadrilateral which has exactly one pair of parallel sides.

Triangle—A polygon with three sides.

Union of sets—The union of two sets is the set consisting of all the elements that are elements of either or both of the two given sets.

Unique—One and only one.

Unique Factorization Theorem—Every composite number can be expressed as a product of prime numbers which is unique except for the order of the factors. The Unique Factorization Theorem is also called the Fundamental Theorem of Arithmetic.

Uniqueness property for an operation—When an operation is performed on two elements of a set, there is one and only one result produced. The result is said to be the unique result of the operation.

Unit angle—An arbitrary plane angle that has been assigned the value 1. This angle is used as a reference for the comparison of angle measures and for the construction of a protractor.

Unit segment—An arbitrary line segment that has been assigned the value 1. This segment is used as a reference for the comparison of lengths and for the construction of a number line.

Universal set—The set from which all subsets under discussion are derived. The universal set is also referred to as the universe.

Unlike fractions—Fractions whose numerals representing denominators are not the same.

Variable—A symbol used to represent an unspecified number. A variable is also called a placeholder.

Venn diagram—A picture illustrating relations and operations on sets. Usually, a rectangular region is used to represent a universal set and circular regions are used to represent subsets of the universal set.

Vertex of an angle—The common endpoint of the two rays forming the angle.

Vertices of a polygon—The endpoints of the line segments forming the polygon. Each endpoint is a vertex of the polygon.

Volume—The size of a region in space.

Whole number—Any element of the set $\{0, 1, 2, 3, 4, 5, \ldots\}$.

Zero—The number designated by the numeral 0; the cardinal number of the empty set; the identity element for addition and subtraction; the integer that is neither positive nor negative; the point on the real number line that separates the positive real numbers and the negative real numbers.

Answers for
Odd-Numbered
Exercises

CHAPTER 1—INTRODUCTION TO MATHEMATICAL IDEAS

1-2 Concept of Sets

1. Among the many correct answers are:
   Show the members in a loop: (△ □)

   List the names of the members: triangle, square.
   Show the members by spacing: 人 人   △ □

   Show the members in braces: {△, □}

3. (a) The members of the set are a basketball and a football.
   (b) The members of the set are a comb, a brush, and a mirror.
   (c) G is the set whose members are the numbers 1, 3, 5, 7, and 9.
   (d) H is the set whose members are the words candy and cake.

5. (a) {30, 40, 50, 60, 70, 80}.   **(b)** {George Washington, John Adams}.
(c) {11, 13, 15, 17}.   **(d)** {0, 19, 38, 57, 76, 95}.

1-3 Empty Set

1. (a) { }.   **(b)** { }.
3. Among the many correct answers are: the set of natural numbers less than 0; the set of pictures of pink elephants on this page.
5. No; B is not the empty set because it contains one member, namely the element 0.

1-4 One-to-One Correspondence

1. (a) Yes.   **(b)** No.   **(c)** Yes.   **(d)** Yes.
3. (a) One.   **(b)** Among the many correct diagrams are: **(c)** No.

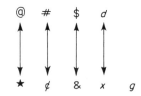

1-7 Whole Numbers

1. 0.   **3.** There is no largest whole number.
5. Yes. 1 2 3 4 . . .
             ↕ ↕ ↕ ↕
             0 1 2 3 . . .
7. (a) True.   **(b)** False.   **(c)** True.
(d) False.   **(e)** False.   **(f)** False.

1-8 Cardinal Numbers

1. (a) 3.   **(b)** 6.   **(c)** 9.   **(d)** 14.

1-9 Ordinal Numbers

1. (a) Cardinal.   **(b)** Ordinal.   **(c)** Cardinal, ordinal.   **(d)** Cardinal, cardinal.
3. (a) Ordinal.   **(b)** Cardinal.   **(c)** Cardinal.   **(d)** Ordinal.

1-10 Distinction between Number and Numeral

1. The number 7.
3. (a), **(d)**, and **(f)** are all names for the number two.

1-11 Finite and Infinite Sets

1. Infinite. **3.** Finite. **5.** Finite (also empty). **7.** Infinite. **9.** Finite.
11. (a) $\{1, 2\}$. **(b)** $\{16, 17, 18, 19\}$. **(c)** $\{0\}$. **(d)** $\{6, 12, 18, 24, \ldots, 108\}$.

1-12 Equal and Equivalent Sets

1. (a) Equal and equivalent. **(b)** Neither. **(c)** Equivalent.
(d) Neither. **(e)** Equivalent.
3. Among the many possible correct answers for each part are:
(a) $\{1, 2, 3\}$. **(b)** $\{a, b, c, d, e, f, g, h, i\}$. **(c)** $\{2, 4, 6, 8, \ldots, 100\}$.
5. (a) Yes. **(b)** No. **(c)** No. **(d)** Yes.
7. $\{$John, Ruth, Bob$\}$.

1-13 Relations of Equality and Inequality

1. (a) $17 > 11$. **(b)** $25 < 50$. **(c)** $9 = 6 + 3$. **(d)** $G = \{\triangle, \square, \diamondsuit\}$.
(e) $2 \neq 3$. **(f)** $8 \geq 7$. **(g)** $12 \leq 9 + 3$.
3. $6 + n < 10 + n$. **5.** $a + n < b + n$. **7.** $41 \times n > 40 \times n$.
9. $5 < 7$. **11.** $a < c$. **13.** $47 \times 52 > 23 \times 47$.

1-14 Number Sentences

1. (a) $\neq$. **(b)** $\neq$. **(c)** $=$.
(d) $\neq$. **(e)** $\neq$. **(f)** $\neq$.
3. (a) $\{8\}$. **(b)** $\{4\}$. **(c)** $\{0\}$.
(d) $\{19, 20, 21, \ldots\}$. **(e)** $\square = \{6\}, \triangle = \{4\}$. **(f)** $\{0, 1, 2, 3, \ldots, 11\}$.

1-15 Number Line

1. (a) In general, the graph of the smaller number is located at the left of the graph
of the larger number.
(b) In general, the graph of the larger number is located at the right of the graph
of the smaller number.
(c) In general, the graph of the two numbers is located at the same point.
3. (a) N. **(b)** P. **(c)** T. **(d)** R. **(e)** M. **(f)** Q.

CHAPTER 2—SETS, RELATIONS, AND OPERATIONS

2-1 Sets and Subsets

1. (a) B and C. **(b)** B and C.
(c) D is not a subset of A because D contains members that are not members of A.

3. (a) $\{1\}$, $\{2\}$, $\{3\}$, $\{1, 2\}$, $\{1, 3\}$, $\{2, 3\}$, $\{\ \}$, $\{1, 2, 3\}$.
  (b) $\{1, 2, 3\}$ is not a proper subset of S because $\{1, 2, 3\} = S$.
5. (a) $\{0, 10, 20, 30, 40, 50\}$.   **(b)** $\{5, 15, 25, 35, 45\}$.
  (c) Among the many possible answers are: $\{0, 5, 10, 15\}$, $\{5, 15, 25, 35\}$,
        $\{0, 10, 20, 30\}$, $\{0, 15, 30, 45\}$, and $\{35, 40, 45, 50\}$.
  (d) $\{0, 5, 10, 15, 20, 25, 30, 35, 40, 45, 50\}$. This subset is not a proper subset of
        M because it is equal to M.
7. (a) 8 subsets.   **(b)** 7 subsets.
9. $S = 2^x$; for $x = 5$, $S = 2^5 = 32$; for $x = 6$, $S = 2^6 = 64$.

2-2 Universal Set

1. Among the many correct answers are:
  (a) The set of all automobiles.
  (b) The set of all students at George Wythe High School.
  (c) The set of all whole numbers.
  (d) The set of integers.
3. Among the many correct answers are:
  (a) The set consisting of the first American astronaut to land on the moon.
  (b) The set of isosceles triangles.
  (c) The set of natural numbers less than 100.
  (d) The set of natural numbers less than 10.

2-3 Solution Sets

1. (a) $\{0, 1, 2, 3, 4, 5, 6\}$.   **(b)** $\{0, 1, 2, 3, 4, 5, 6, 7\}$.
  (c) $\{7, 8, 9, 10, \ldots\}$.   **(d)** $\{9\}$.
  (e) $\{0, 1, 2, 3, 4, 5\}$.   **(f)** $\{\ \}$.
3. (a) $\{2, 4, 6\}$.   **(b)** $\{\ \}$.
  (c) $\{4, 6, 8, 10, \ldots\}$.   **(d)** $\{\ \}$.
  (e) $\{\ \}$.   **(f)** $\{2, 4, 6, 8, 10, 12\}$.
5. (a) $\{1, 2, 3, 4, 5\}$.   **(b)** $\{0, 1, 2, 3, 4, 5\}$.
  (c) $\{0, 2, 4\}$.   **(d)** $\{1, 3, 5\}$.
7. (a) $(\triangle, \square) = \{(1, 2), (2, 1)\}$.   **(b)** $(\triangle, \square) = \{(0, 3), (1, 2), (2, 1), (0, 3)\}$.
  (c) $(\triangle, \square) = \{\ \}$.   **(d)** $(\triangle, \square) = \{\ \}$.

2-4 Union of Sets

1. $G \cup H = \{1, 2, 3, 4, 5, 6\}$; $n(G \cup H) = 6$.
3. $R \cup S = \{\%, @, \#, ¢, \&\}$; $n(R \cup S) = 5$.   **5.** $E \subseteq D$.
7. (a) The union is always the original set.
  (b) The union is always the original set.
  (c) The union is always the universal set.
  (d) The union is always the same for the two sets, regardless of order.

9. (a) The statement is possible. For example,
   if $X = \{a, b, c\}$ and $Y = \{b, c\}$, then $X \cup Y = \{a, b, c\} = X$.

   (b) The statement is possible. For example,
   if $X = \{a, b, c\}$ and $Y = \{b, c\}$, then $n(X) = 3, n(Y) = 2$, and $n(X \cup Y) = 3$.
   Thus, since $3 + 2 > 3$ we have an example of $n(X) + n(Y) > n(X \cup Y)$.

   (c) The statement is not possible because for all sets X and Y, $n(X \cup Y)$ will always be equal to or less than $n(X) + n(Y)$.

   (d) The statement is not possible because for all sets X and Y, $n(X \cup Y)$ will always be equal to or greater than $n(X)$.

2-5 Intersection of Sets

1. $A \cap B = \{b, c\}$; $n(A \cap B) = 2$.   3. $G \cap H = \{ \}$; $n(G \cap H) = 0$.

5. $N = M$.   7. $D = E$.

9. (a) $\{0, 1, 2, 3, \ldots\}$; that is, A.   (b) $\{ \}$.
   (c) $\{10, 20, 30, \ldots\}$.   (d) $\{0, 2, 4, 5, 6, 8, 10, 12, 14, 15, 16, \ldots\}$.
   (e) $\{1, 3, 5, 7, \ldots\}$; that is, C.   (f) $\{0, 1, 2, 3, \ldots\}$; that is, A.
   (g) $\{0, 1, 2, 3, \ldots\}$; that is, A.   (h) $\{5, 10, 15, \ldots\}$; that is, D.
   (i) $\{0, 1, 2, 3, \ldots\}$; that is, A.   (j) $\{0, 2, 4, 6, \ldots\}$; that is, B.
   (k) $\{5, 15, 25, \ldots\}$.   (l) $\{1, 3, 5, 7, 9, 10, 11, 13, 15, \ldots\}$.

11. (a) The intersection is always the original set.
    (b) The intersection is always the empty set.
    (c) The intersection is always the original set.
    (d) The intersection is always the same for the two sets, regardless of order.

13. (a) The statement is possible. For example, if $W = \{p, q, r\}$ and $K = \{p, q, r, s\}$, then $n(W) = 3$, and $n(W \cap K) = 3$.

    (b) The statement is possible. For example, if $W = \{p, q, r\}$ and $K = \{p, q, r, s\}$, then $n(W) = 3, n(K) = 4$, and $n(W \cap K) = 3$. Thus, since $3 + 4 > 3$, we have an example of $n(W) + n(K) > n(W \cap K)$.

    (c) The statement is not possible because for all sets W and K, $n(W) + n(K)$ will be equal to or greater than $n(W \cap K)$.

    (d) The statement is not possible because for all sets W and K, $W \cap K = K \cap W$, and $n(W \cap K) = n(K \cap W)$.

2-6 Disjoint Sets

1. $G \cap H = \{ \}$; $n(G \cap H) = 0$.   3. $R \cap S = \{12\}$; $n(R \cap S) = 1$.

5. G and H, X and Y.

7. (a) The intersection of sets W and Z is the empty set.
   (b) $n(W \cap Z) = 0$.

2-7 Complement of a Set

1. $A' = \{1, 3\}$.

3. (a) $\{1, 2, 3\}$; that is, A.   (b) $\{1, 2, 3, 4, 5, 6, 7, 8, 9\}$; that is, U.

(c) $\{1, 2, 4, 6, 8\}$.   **(d)** $\{1, 2, 4, 6, 8\}$; that is, B'.

(e) $\{1, 2, 3, 4, 5, 6, 7, 8, 9\}$; that is, U.   **(f)** $\{1, 2, 4\}$; that is, $(B \cup C)'$.

(g) $\{1, 2, 3, 4, 5, 6, 7, 8, 9\}$; that is, U.   **(h)** $\{1, 2, 4\}$; that is, $B' \cap C'$.

5. (a) $\{\ \}$.   **(b)** $\{1, 3, 9\}$.   **(c)** $\{1, 3, 9\}$.   **(d)** $\{\ \}$.

2-8 Cartesian Product of Sets

1. (a) $A \times B = \{(1, e), (1, f), (2, e), (2, f)\}$.   **(b)** $n(A \times B) = 4$.

(c) $B \times A = \{(e, 1), (e, 2), (f, 1), (f, 2)\}$.   **(d)** $n(B \times A) = 4$.

3. (a) $C \times D = \{(\text{red, white}), (\text{red, blue})\}$.   **(b)** $n(C \times D) = 2$.

(c) $D \times C = \{(\text{white, red}), (\text{white, blue})\}$.   **(d)** $n(D \times C) = 2$.

2-9 Venn Diagrams

1. (a) $A \cup B = \{1, 2, 3, 4, 5, 6, 7, 8\}$.   **(b)** $A \cap B = \{\ \}$.

3. (a) $M \cup N = \{\text{Louise, Ann, Lucille}\}$.   **(b)** $M \cap N = \{\text{Louise}\}$.

5. (a) $R \cup S$ **(b)** $R \cap S$

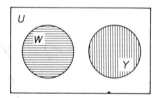   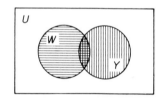

Set R is shaded horizontally and set S is shaded vertically; $R \cup S$ is the subset of U that is shaded in either or both directions.

Set R is shaded horizontally and set S is shaded vertically; $R \cap S$ is the subset of U that is shaded both horizontally and vertically.

7. (a) $W \cap Y = \varnothing$ **(b)** $W \cap Y \ne \varnothing$

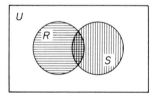   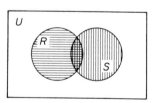

Set W is shaded horizontally and set Y is shaded vertically; $W \cap Y$ is the subset of U that is shaded both horizontally and vertically. Thus $W \cap Y = \varnothing$.

Set W is shaded horizontally and set Y is shaded vertically; $W \cap Y$ is the subset of U that is shaded both horizontally and vertically. Thus $W \cap Y \ne \varnothing$.

(c) $W \subset Y$

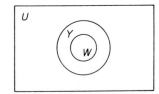

(d) $Y \subset W$

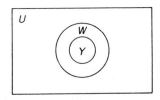

Set W is contained entirely within set Y, and there is part of Y that does not contain W. Thus $W \cap Y = W$.

Set Y is contained entirely within set W, and there is part of W that does not contain Y. Thus $W \cap Y = Y$.

(e) $W = Y$

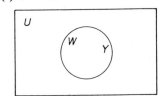

(f) $W \cup Y = Y$

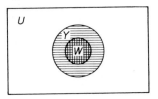

The same set has two names. Thus $W \cap Y = W = Y$.

Set Y is shaded horizontally and set W is shaded vertically; $W \cup Y$ is shaded in either or both directions. Thus $W \cap Y = Y$.

9. $(A \cap B) \cup C$

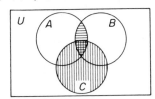

Set $A \cap B$ is shaded horizontally and set C is shaded vertically; $(A \cap B) \cup C$ is the subset of U that is shaded in either or both directions.

11. (a) C'

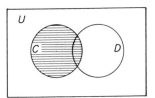

(b) D'

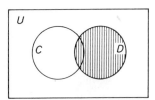

Set C is shaded horizontally; C' is the subset of U that is not shaded.

(c) $(C \cup D)'$

Set D is shaded vertically; D' is the subset of U that is not shaded.

(d) $C' \cup D'$

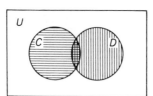

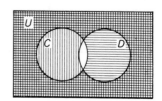

Set $C \cup D$ is the subset of U that is shaded in either or both directions; $(C \cup D)'$ is the subset of U that is not shaded.

(e) $(C \cap D)'$

Set C' is shaded horizontally and set D' is shaded vertically; $C' \cup D'$ is the subset of U that is shaded in either or both directions.

(f) $C' \cap D'$

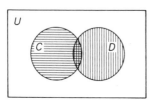

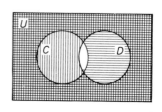

Set $C \cap D$ is shaded in both directions; $(C \cap D)'$ is the subset of U that is not shaded in both directions.

13. (a) $R \cup (S \cup T)$

Set C' is shaded horizontally and set D' is shaded vertically; $C' \cap D'$ is the subset of U that is shaded in both directions.

(b) $(R \cup S) \cup T$

$R \cup (S \cup T)$

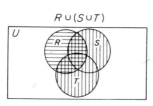

$(R \cup S) \cup T$

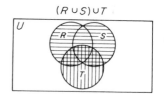

Set R is shaded horizontally and $(S \cup T)$ is shaded vertically; $R \cup (S \cup T)$ is the subset of U that is shaded in either or both directions.

Set $(R \cup S)$ is shaded horizontally and T is shaded vertically; $(R \cup S) \cup T$ is the subset of U that is shaded in either or both directions.

(c) $R \cap (T \cup S)$

$$R \cap (T \cup S)$$

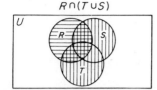

(d) $R \cap (T \cap S)$

$$R \cap (T \cap S)$$

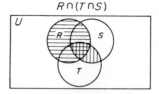

Set R is shaded horizontally and $(T \cup S)$ is shaded vertically; $R \cap (T \cup S)$ is the subset of U that is shaded in both directions.

(e) $(R \cap T) \cup (R \cap S)$

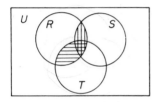

Set $R \cap T$ is shaded horizontally and $R \cap S$ is shaded vertically; $(R \cap T) \cup (R \cap S)$ is the subset of U that is shaded in either or both directions.

Set R is shaded horizontally and $(T \cap S)$ is shaded vertically; $R \cap (T \cap S)$ is the subset of U that is shaded in both directions.

(f) $(R \cup T) \cap (R \cup S)$

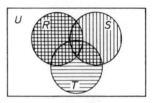

Set $(R \cup T)$ is shaded horizontally and $(R \cup S)$ is shaded vertically; $(R \cup T) \cap (R \cup S)$ is the subset of U that is shaded in both directions.

CHAPTER 3—SYSTEMS OF NUMERATION

3-1 Roman Numeration System

1. (a) XIII.   **(b)** LII.   **(c)** LXXIV.   **(d)** XCVI.
   (e) CXLIX.   **(f)** CCCLXXXVII.   **(g)** MCMLXXI.   **(h)** C̄CDXCVIII.

3. (a) XVI.   **(b)** XIX.   **(c)** CCXX.   **(d)** DCCXXXIV.

5. (a)   XII
            LIII
      ⎯⎯⎯⎯⎯⎯
      LXIIIII; this becomes LXV.

(b)   LXXIV
             XCV
      ⎯⎯⎯⎯⎯⎯
      XCLXXVIV; this becomes CLXIX.

(c)   CXLIX
             CCCLXXXVII
      ⎯⎯⎯⎯⎯⎯⎯⎯⎯⎯⎯
      CCCCLXLXXXIXVII;
              this becomes DXXXVI.

(d)   MCMLXXI
              C̄CDXCVIII
      ⎯⎯⎯⎯⎯⎯⎯⎯⎯⎯⎯
      C̄MCMCDXCLXXVIIII;
              this becomes C̄MMCDLXIX.

7. (a) LXV = 65.   **(b)** CLXIX = 169.
   (c) DXXXVI = 536.   **(d)** C̄MMCDLXIX = 102,469.

9. $12 \times 53 = 636$.

11. (a) XCIX.   **(b)** CMXCIX.   **(c)** LXXXVIII.   **(d)** DCCCLXXXVIII.

3-2 Early Egyptian Numeration System

1. (a) ∩|||||| **(b)** ∩∩∩∩ ||| ∩∩∩∩   **(c)** 99|

   (d) 9999∩∩∩||||| **(e)** ⚱⚱⚱9|||
       9999∩∩∩∩|||| ⚱⚱999||||

   (f) ⌐⌐⌐ ⚱⚱9999∩∩∩|| **(g)** ≈≈ ⌐⌐⌐⌐ ⚱⚱⚱99∩∩∩||||
        9999∩∩∩|| 99∩∩∩∩||||

   (h) ⚱ ⌐⌐⌐⌐ 9||||

3. (a) ∩   **(b)** 9   **(c)** ⚱   **(d)** ⌐

5. (a) ∩∩∩∩ ∩∩∩∩∩   **(b)** 99999 9999

   (c) ⚱⚱⚱⚱ ⚱⚱⚱⚱⚱   **(d)** ⌐⌐⌐⌐ ⌐⌐⌐⌐⌐

7. (a) $16 + 83 = 99.$ **(b)** $201 + 879 = 1,080.$
   (c) $5,407 + 32,864 = 38,271.$ **(d)** $253,478 + 1,040,104 = 1,293,582.$

9.
              ∩∩∩∩∩∩∩|||
                 ∩||||||
              ∩∩∩∩∩∩∩|||
              ∩∩∩∩∩∩∩|||
              ∩∩∩∩∩∩∩|||
              ∩∩∩∩∩∩∩|||
              ∩∩∩∩∩∩∩|||
              ∩∩∩∩∩∩∩|||
            99999999∩∩∩
  999∩∩∩∩∩∩∩∩∩∩∩∩∩∩∩∩∩|||||
  999∩∩∩∩∩∩∩∩∩∩∩∩∩∩∩∩∩|||||
  99∩∩∩∩∩∩∩∩∩∩∩∩∩∩∩∩∩|||||; this becomes ⚱999∩∩||||||||

11.   ⚱999∩∩|||||||| = 1328

3-3 Early Babylonian Numeration System

1. (a) ▼▼▼ ▼▼▼ ▼▼▼   **(b)** << << ▼▼▼ ▼▼▼   **(c)** <<< ▼▼▼ ▼▼▼ ▼▼▼   **(d)** ▼ ▼

  (e) ▼<▼▼   **(f)** ▼▼ ▼ ▼▼ ▼▼ ▼▼ ▼▼   **(g)** ▼ <▼▼ <▼▼   **(h)** ▼▼▼<▼▼▼ ▼▼▼▼<▼▼ ▼▼<▼▼

3. (a) ▼▼   **(b)** <   **(c)** ▼   **(d)** ▼▼

5. (a)
       ▼▼▼
       ▼▼▼
       ▼▼▼
   <<
   << ▼▼▼
   ―――――
   <<▼▼▼▼
   <<▼▼▼▼ ; this becomes <<< ▼▼
     ▼▼▼▼   <<

(b)
          ▼▼▼ ▼▼
     <<<▼▼▼ ▼▼
     << ▼▼▼
   ▼   ▼
   ―――――
   ▼<<<▼▼▼▼
   << ▼▼▼▼ ; this becomes ▼▼
     ▼▼

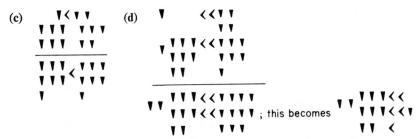

7. (a) ⟨⟨⟨ ▼▼ = 52
⟨⟨

(b) ▼▼ = 120

(c) ▼▼▼ ⟨▼▼▼
▼▼▼ ⟨ ▼▼▼ = 437
▼ ▼

(d) ▼▼ ▼▼▼ ⟨⟨
▼▼▼ ⟨⟨ ▼ = 7731
▼▼

3-4 Our Decimal System of Numeration

1.	Roman	Egyptian	Babylonian	Our Decimal
Place value	No	No	Yes	Yes
Base	No	Yes (ten)	Yes (sixty)	Yes (ten)
Additive	Yes	Yes	Yes	Yes
Subtractive	Yes	No	No	No
Multiplicative	Yes	No	No	No
Repetitive	Yes	Yes	No	No
Use of Zero	No	No	No	Yes

3. 145.   **5.** 99.

7. 0.

9. (a) $256 = (2 \times 100) + (5 \times 10) + (6 \times 1)$.

   (b) $5,764 = (5 \times 1,000) + (7 \times 100) + (6 \times 10) + (4 \times 1)$.

   (c) $97,438 = (9 \times 10,000) + (7 \times 1,000) + (4 \times 100) + (3 \times 10) + (8 \times 1)$.

   (d) $385,061 = (3 \times 100,000) + (8 \times 10,000) + (5 \times 1,000) + (0 \times 100) + (6 \times 10) + (1 \times 1)$.

3-5 Base of a System of Numeration

1. (a) $328 = (3 \times 100) + (2 \times 10) + (8 \times 1)$.

   (b) $9,657 = (9 \times 1,000) + (6 \times 100) + (5 \times 10) + (7 \times 1)$.

   (c) $19,175 = (1 \times 10,000) + (9 \times 1,000) + (1 \times 100) + (7 \times 10) + (5 \times 1)$.

(d) 263,849 = (2 × 100,000) + (6 × 10,000) + (3 × 1,000) + (8 × 100) + (4 × 10) + (9 × 1).

3. (a)    **(b)**   **(c)**

(d)   **(e)**

5. (a) 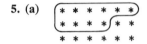   **(b)** 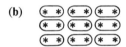   **(c)**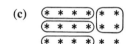

(d) 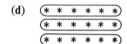   **(e)**

7. (a) 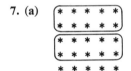   **(b)** 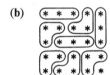   **(c)**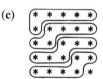

(d) 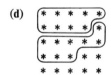   **(e)**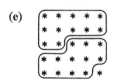

9. (a) Ten.   **(b)** Twelve.   **(c)** Nine.   **(d)** Five.
(e) Two.   **(f)** Seven.   **(g)** Three.   **(h)** Eight.
(i) Four.

11. I. Additive, subtractive, repetitive.
    II. Additive, symbol for zero (⊗), place value, base (3) with digits read left to right.
    III. Multiplicative, repetitive.
    IV. Additive, place value, base (4).

3-6 Powers of Numbers

1. (a) 8^3.   (b) 9^4.   (c) 3^2.   (d) 5^3.

3. The exponent and the number of zeros is the same; for example, $10^2 = 100$ and $10^3 = 1,000$.

5. (a) The base is 3, the exponent is 6, and the power is 3^6.
(b) The base is 7, the exponent is 2, and the power is 7^2.

3-7 Multiplication and Division of Powers

1. (a) 2^7.   (b) 3^3.   (c) 6^5.
(d) 4^8.   (e) 1.   (f) 5^{12}.

3. (a) 3^{14}.   (b) 3^9.   (c) 25^2, or 5^4.

(d) 4^{-6}.   (e) $2^6 \div 2 = 2^5$.   (f) $\dfrac{1}{21^6} = 21^{-6}$.

5. (a) One-tenth.   (b) One-hundredth.
(c) One-thousandth.   (d) One ten-thousandth.

CHAPTER 4—WHOLE NUMBERS: ADDITION AND SUBTRACTION

4-1 Addition of Whole Numbers

1. (a) 4.   (b) 2.   (c) 3.   (d) 0.   (e) 5.
3. (a) 6.   (b) 5.   (c) 8.   (d) 3.   (e) 7.
(f) 5.   (g) 2.   (h) 6.

5. Among the many possible correct answers are:
(a) $A = \{a, b\}$, $B = \{c, d, e, f, g\}$, $A \cup B = \{a, b, c, d, e, f, g\}$, $n(A) + n(B) = n(A \cup B)$; that is $2 + 5 = 7$.
(b) $G = \{\triangle, \square, \Diamond\}$, $H = \{m, n, p, q, r, s\}$, $G \cup H = \{\triangle, \square, \Diamond, m, n, p, q, r, s\}$, $n(G) + n(H) = n(G \cup H)$; that is $3 + 6 = 9$.
(c) $R = \{g, h, j, k\}$, $S = \{s, t, w\}$, $R \cup S = \{g, h, j, k, s, t, w\}$, $n(R) + n(S) = n(R \cup S)$; that is $4 + 3 = 7$.
(d) $M = \{\triangle, \Diamond, \square, \triangledown, \bigcirc, \varhexagon\}$, $N = \{\ \}$, $M \cup N = \{\triangle, \Diamond, \square, \triangledown, \bigcirc, \varhexagon\}$, $n(M) + n(N) = n(M \cup N)$; that is $6 + 0 = 6$.
(e) $E = \{e, f, g, h\}$, $K = \{k, l, m, n\}$, $E \cup K = \{e, f, g, h, k, l, m, n\}$, $n(E) + n(K) = n(E \cup K)$; that is $4 + 4 = 8$.
(f) $C = \{\ \}$, $D = \{\ \}$, $C \cup D = \{\ \}$, $n(C) + n(D) = n(C \cup D)$; that is, $0 + 0 = 0$.

7. (a)

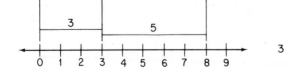

$3 + 5 = 8$

(b)

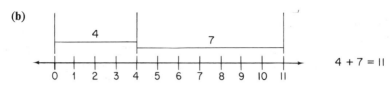

$$4 + 7 = 11$$

(c)

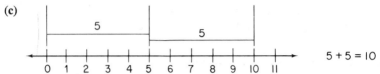

$$5 + 5 = 10$$

4-2 Closure and Uniqueness Properties

1. (a) Closed.   **(b)** Not closed.   **(c)** Not closed.   **(d)** Closed.
   (e) Closed.   **(f)** Closed.   **(g)** Closed.

4-3 The Commutative Property of Addition

1. (a) Not commutative.   **(b)** Commutative.   **(c)** Commutative.
   (d) Not commutative.   **(e)** Not commutative.

3. (a) $A = \{a, b\}$, $D = \{d, e, f, g\}$, $A \cup D = \{a, b, d, e, f, g\}$, $D \cup A = \{d, e, f, g, a, b\}$. Since $A \cup D = D \cup A$, we have $n(A) + n(D) = n(A \cup D)$ and $n(D) + n(A) = n(A \cup D)$. Thus we have $2 + 4 = 4 + 2$.

(b) $G = \{a, b, c, d, e, f, g\}$, $H = \{h, i, j, k, l, m, n, p\}$, $G \cup H = \{a, b, c, d, e, f, g, h, i. j, k, l, m, n, p\}$, $H \cup G = \{h, i, j, k, l, m, n, p, a, b, c, d, e, f, g\}$. Since $G \cup H = H \cup G$, we have $n(G) + n(H) = n(G \cup H)$ and $n(H) + n(G) = n(G \cup H)$. Thus we have $7 + 8 = 8 + 7$.

5. (a)

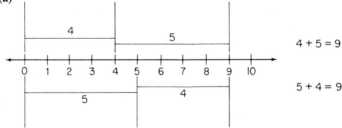

$$4 + 5 = 9$$

$$5 + 4 = 9$$

(b)

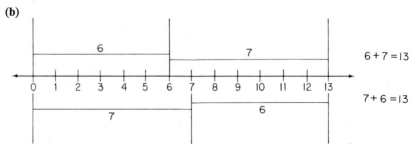

$$6 + 7 = 13$$

$$7 + 6 = 13$$

4-4 The Associative Property of Addition

1. (a) $C \cup D = \{g, h, j, k, l, m, n\}$, $n(C \cup D) = 7$, $n(E) = 5$,
$n(C \cup D) + n(E) = 7 + 5$,
$(C \cup D) \cup E = \{g, h, j, k, l, m, n, e, f, a, b, w\}$, $n((C \cup D) \cup E) = 12$.
Thus we have $7 + 5 = 12$.

(b) $n(C) = 3$, $D \cup E = \{k, l, m, n, e, f, a, b, w\}$, $n(D \cup E) = 9$,
$n(C) + n(D \cup E) = 3 + 9$,
$C \cup (D \cup E) = \{g, h, j, k, l, m, n, e, f, a, b, w\}$, $n(C \cup (D \cup E)) = 12$.
Thus we have $3 + 9 = 12$.

(c) From **1(a)**, $n((C \cup D) \cup E) = 12$ and from **1(b)** $n(C \cup (D \cup E)) = 12$.
Thus we have $12 = 12$.

(d) From **1(a)**, $n(C \cup D) = 7$, $n(E) = 5$, and $n(C \cup D) + n(E) = 7 + 5$.
From **1(b)**, $n(C) = 3$, $n(D \cup E) = 9$, and $n(C) + n(D \cup E) = 3 + 9$.
Thus we have $7 + 5 = 3 + 9$.

3. $(4 + 6) + 3 = 4 + (6 + 3)$.

5. (a) $(15 + 3) + 42 = 18 + 42 = 60$.
$15 + (3 + 42) = 15 + 45 = 60$.

(b) $(22 + 7) + 18 = 29 + 18 = 47$.
$22 + (7 + 18) = 22 + 25 = 47$.

(c) $(109 + 23) + 11 = 132 + 11 = 143$.
$109 + (23 + 11) = 109 + 34 = 143$.

(d) $(237 + 165) + 35 = 402 + 35 = 437$.
$237 + (165 + 35) = 237 + 200 + 437$.

7. (a) $27 + (36 + 13) + 64 = 27 + (13 + 36) + 64$
$= (27 + 13) + (36 + 64)$
$= 40 + 100$
$= 140$.

(b) $135 + (81 + 15) + 9 = 135 + (15 + 81) + 9$
$= (135 + 15) + (81 + 9)$
$= 150 + 90$
$= 240$.

(c) $58 + (24 + 32) + 16 = 58 + (32 + 24) + 16$
$= (58 + 32) + (24 + 16)$
$= 90 + 40$
$= 130$.

(d) $19 + (17 + 11) + 83 = 19 + (11 + 17) + 83$
$= (19 + 11) + (17 + 83)$
$= 30 + 100$
$= 130$.

9. (a) False.   **(b)** True.   **(c)** True.   **(d)** False.

11. (a) $(a + c) + b = a + (c + b)$ Associative property of addition
       $a + (c + b) = a + (b + c)$ Commutative property of addition

(b) $(c + b) + a = a + (c + b)$ Commutative property of addition
       $a + (c + b) = a + (b + c)$ Commutative property of addition

(c) $(c + a) + b = (a + c) + b$ Commutative property of addition
$(a + c) + b = a + (c + b)$ Associative property of addition
$a + (c + b) = a + (b + c)$ Commutative property of addition

4-5 Identity Element for Addition

1. (a) $n(R) = 3$, $n(S) = 0$, $R \cup S = \{r, s, t\}$, $n(R \cup S) = 3$. Thus, we have $3 + 0 = 3$.
   (b) $n(R \cup S) = 3$, $n(R) = 3$. Thus, we have $3 = 3$.
   (c) $n(R) = 3$, $n(S) = 0$. Thus, we have $3 + 0 = 3$.
   (d) $n(R) = 3$, $n(S) = 0$. Thus, we have $3 + 0 = 0 + 3 = 3$.
3. (a) $0 + 8 = 8$.   (b) $27 + 0 = 27$.
   (c) $0 + 0 = 0$.   (d) $x + 0 = 0 + x = x$, $x \in W$.

4-6 The Table of Basic Addition Facts

1. (a) One-more pattern.   (b) One-more pattern.   (c) Two-more pattern.
   (d) Each block in a particular diagonal row contains the same numeral.
   (e) One-less pattern.   (f) One-less pattern.
3. (a) The sum of two odd numbers is an even number.
   (b) The sum of two even numbers is an even number.
   (c) The sum of an odd number and an even number is an odd number.
   (d) The sum of an even number and an odd number is an odd number.
5. (a) $9 + 6 = 9 + (1 + 5) = (9 + 1) + 5 = 10 + 5 = 15$.
   (b) $8 + 5 = 8 + (2 + 3) = (8 + 2) + 3 = 10 + 3 = 13$.
   (c) $7 + 6 = 7 + (3 + 3) = (7 + 3) + 3 = 10 + 3 = 13$.
7. (a) $400 + 110 + (10 + 2)$ (b) $249 = 200 + 40 + 9$
   $400 + (110 + 10) + 2$ $\underline{+657 = 600 + 50 + 7}$
   $400 + 120 + 2$ $800 + 90 + 16$
   $400 + (100 + 20) + 2$ Place value $800 + 90 + (10 + 6)$
   $(400 + 100) + 20 + 2$ Associative property $800 + (90 + 10) + 6$
   $500 + 20 + 2$ Addition $800 + 100 + 6$
   522 Associative property $(800 + 100) + 6$
                            Addition $900 + 6$
                            Place-value notation 906
9. No. The addition table displays only two addends. Three or more addends are required for illustration of the associative property of addition.

4-7 Understanding the Addition Algorithm

1. (a) 2 (b) 20 (c) 24 (d) $\overset{1}{27}$
         $\underline{+3}$ $\underline{+30}$ $\underline{+35}$ $\underline{+38}$
          5 50 59 65

3. (a)
$$\begin{array}{r} 300 \\ +400 \\ \hline 700 \end{array}$$
(b)
$$\begin{array}{r} 320 \\ +460 \\ \hline 780 \end{array}$$
(c)
$$\begin{array}{r} \overset{1}{3}90 \\ +450 \\ \hline 840 \end{array}$$
(d)
$$\begin{array}{r} \overset{1\,1}{3}97 \\ +456 \\ \hline 853 \end{array}$$

5. (a)
$$\begin{array}{r} 27 \\ +34 \\ \hline 11 \\ 50 \\ \hline 61 \end{array}$$
(b)
$$\begin{array}{r} 278 \\ +165 \\ \hline 13 \\ 130 \\ 300 \\ \hline 443 \end{array}$$
(c)
$$\begin{array}{r} 2597 \\ +4036 \\ \hline 13 \\ 120 \\ 500 \\ 6000 \\ \hline 6633 \end{array}$$
(d)
$$\begin{array}{r} 396 \\ 222 \\ +137 \\ \hline 15 \\ 140 \\ 600 \\ \hline 755 \end{array}$$

4-8 Subtraction of Whole Numbers

1. (a) $13 - 6 = 7$ or $13 - 7 = 6$.
(c) $5 - 1 = 4$ or $5 - 4 = 1$.
(b) $59 - 38 = 21$ or $59 - 21 = 38$.
(d) $m - h = k$ or $m - k = h$.

3. (a) $3 + 4 = 7$; $7 - 4 = 3$.
(b) $2 + 6 = 8$; $8 - 6 = 2$.

5. (a) $7 - 3 = 4$.
(b) $10 - 4 = 6$.

7. (a) $4 - 3 = 1$.
(b) $6 - 2 = 4$.

4-9 Inverse Operations: Addition and Subtraction

1. (a) Opening a door.
(c) Taking off your shoes.
(b) Withdrawing $6.00 from the bank.
(d) Removing a pie from the oven.

3. Subtract 6 from the sum; that is, $(n + 6) - 6 = n$.

4-10 Properties of Subtraction of Whole Numbers

1. Among the many possible counterexamples are:
  (a) $9 - 4 \neq 4 - 9$, because $5 \neq {}^{-}5$.
  (b) $(9 - 5) - 2 \neq 9 - (5 - 2)$, because $2 \neq 6$.
  (c) $8 - 9 = {}^{-}1$, and the difference ${}^{-}1$ does not exist in the set of whole numbers.

3. (a) False. **(b)** True. **(c)** True. **(d)** False. **(e)** True. **(f)** False.

4-11 Understanding the Subtraction Algorithm

1. (a)
$$\begin{array}{r} 7 \\ -4 \\ \hline 3 \end{array}$$
(b)
$$\begin{array}{r} 70 \\ -40 \\ \hline 30 \end{array}$$
(c)
$$\begin{array}{r} 76 \\ -42 \\ \hline 34 \end{array}$$
(d)
$$\begin{array}{r} \overset{6\,1}{7\!\!\!/5} \\ -48 \\ \hline 27 \end{array}$$

3. (a)
$$\begin{array}{r} 500 \\ -200 \\ \hline 300 \end{array}$$
(b)
$$\begin{array}{r} 560 \\ -210 \\ \hline 350 \end{array}$$
(c)
$$\begin{array}{r} \overset{4\,1}{5\!\!\!/2}0 \\ -230 \\ \hline 290 \end{array}$$
(d)
$$\begin{array}{r} \overset{4\,1\,1}{5\!\!\!/2\!\!\!/7} \\ -239 \\ \hline 288 \end{array}$$

5. (a)
$$\begin{array}{rcccc} 67 = & 60 + 7 = & 50 + 17 \\ -19 = & -(10 + 9) = & -(10 + 9) \\ \hline & & 40 + 8 & = 48. \end{array}$$

(b) $50 = \quad 50 + 0 = \quad 40 + 10$
$\underline{-13 = -(10 + 3) = -(10 + \quad 3)}$
$\qquad\qquad\qquad\qquad 30 + \quad 7 \ = 37.$

(c) $236 = \quad 200 + 30 + 6 = \quad 200 + 20 + 16$
$\underline{-158 = -(100 + 50 + 8) = -(100 + 50 + \quad 8)}$
$\qquad\qquad\qquad\qquad\qquad = \quad 100 + 120 + 16$
$\qquad\qquad\qquad\qquad\qquad \underline{= -(100 + \quad 50 + \quad 8)}$
$\qquad\qquad\qquad\qquad\qquad\qquad\qquad\quad 70 + \quad 8 \ = 78.$

(d) $444 = \quad 400 + 40 + 4 = \quad 400 + 30 + 14$
$\underline{-355 = -(300 + 50 + 5) = -(300 + 50 + \quad 5)}$
$\qquad\qquad\qquad\qquad\qquad = \quad 300 + 130 + 14$
$\qquad\qquad\qquad\qquad\qquad \underline{= -(300 + \quad 50 + \quad 5)}$
$\qquad\qquad\qquad\qquad\qquad\qquad\qquad\quad 80 + \quad 9 \ \doteq 89.$

7. Yes, by using a missing addend interpretation. For example, $7 - 2 = n$ may be thought of as $n + 2 = 7$, where the addend 2 and the sum 7 are located respectively in a margin and interior of the table, and the missing addend is located in the other margin.

Yes, there are limitations. The solution for $a - b$ may be found in the table only if $b \leq a$.

CHAPTER 5—WHOLE NUMBERS: MULTIPLICATION AND DIVISION

5-1 Multiplication of Whole Numbers

1. **(a)** $3 + 3 + 3 + 3 = 12.$ **(b)** $1 + 1 + 1 + 1 + 1 = 5.$
$\qquad\quad 4 \times 3 = 12.$ $\qquad\qquad 5 \times 1 = 5.$

3. **(a)** $7 + 7 = 14.$ **(b)** $1 + 1 + 1 + 1 + 1 + 1 = 6.$
(c) $0 + 0 + 0 = 0.$ **(d)** $3 + 3 + 3 + 3 + 3 + 3 + 3 + 3 = 24.$
(e) Not possible.   **(f)** Not possible.

5. **(a)**

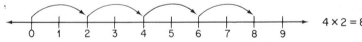
$4 \times 2 = 8$

(b)

$5 \times 1 = 5$

(c) Not possible to show $0 \times 3 = 0$ on a number line.
(d) Not possible to show $7 \times 0 = 0$ on a number line.

7. **(a)**    **(b)**   **(c)** Not possible to show $3 \times 0 = 0.$

(d)   * * * * * * * * *
           * * * * * * * * *

9.   Shirts   Pants

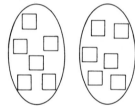

Jack can make 10 different outfits.

5-2 Closure and Uniqueness Properties

1. (a) Closed.   **(b)** Closed.   **(c)** Closed.   **(d)** Not closed.
   (e) Closed.   **(f)** Not closed.   **(g)** Closed.

5-3 The Commutative Property of Multiplication

1.   2 sets of 6 = 12   6 sets of 2 = 12

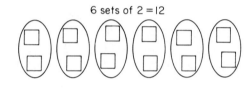

Thus 2 x 6 = 6 x 2 = 12

3. (a) No.   **(b)** Yes.

5.   8 x 4 = 32

Thus 8 x 4 = 4 x 8 = 32

7. (a) $5 \times 3 = \underline{3} \times \underline{5}$.   **(b)** $7 \times 9 = \underline{9} \times \underline{7}$.
Among the many correct answers are:
(c) $143 \times \underline{2} = \underline{2} \times 143$.   **(d)** $\underline{10} \times 4 = 4 \times \underline{10}$.

5-4 The Associative Property of Multiplication

1. (a) $(2 \times 5) \times 46 = 10 \times 46 = 460$.   **(b)** $(3 \times 6) \times 9 = 18 \times 9 = 162$.
   $2 \times (5 \times 46) = 2 \times 230 = 460$. $3 \times (6 \times 9) = 3 \times 54 = 162$.

(c) $(50 \times 20) \times 7 = 1,000 \times 7 = 7,000$.
$50 \times (20 \times 7) = 50 \times 140 = 7,000$.

(d) $(4 \times 8) \times 25 = 32 \times 25 = 800$.
$4 \times (8 \times 25) = 4 \times 200 = 800$.

3. (a) $10 \times 13 = 130$; yes.   (b) $100 \times 17 = 1,700$; yes.
(c) $100 \times 16 = 1,600$; yes.   (d) $1,000 \times 7 = 7,000$; yes.

5. $(3 \times 4) \times (5 \times 6) = 12 \times 30 = 360$.
$(3 \times 6) \times (4 \times 5) = 18 \times 20 = 360$.
$(3 \times 5) \times (4 \times 6) = 15 \times 24 = 360$.
$3 \times (6 \times 5) \times 4 = 3 \times 30 \times 4 = 360$.
All the products are equal.

5-5 Identity Element for Multiplication

1. (a) $n(A) = 1; n(B) = 3; n(A \times B) = 3$;
$n(A) \times n(B) = n(A \times B)$ since $1 \times 3 = 3$.

(b) $n(A \times B) = 3; n(B) = 3$;
$n(A \times B) = n(B)$ since $3 = 3$.

(c) $n(A) = 1; n(B) = 3$;
$n(A) \times n(B) = n(B)$ since $1 \times 3 = 3$.

(d) $n(A) = 1; n(B) = 3$;
$n(A) \times n(B) = n(B) \times n(A) = n(B)$ since $1 \times 3 = 3 \times 1 = 3$.

3. (a) $8 \times 1 = 8$.   (b) $10 \times 1 = 10$.
$1 \times 8 = 8$. $1 \times 10 = 10$.

5. (a) $1 \times 16 = \underline{16.}$ (b) $\underline{23} \times 1 = 23$.
(c) $1 \times \underline{1} = 1$.   (d) $\underline{0} \times 1 = 0$.
(e) $d \times \underline{1} = \underline{1} \times d = d; d \in W$.

5-6 Property of Multiplication by Zero

1. (a) $n(D) = 4; n(H) = 0; n(D \times H) = 0$;
$n(D) \times n(H) = n(D \times H)$ since $4 \times 0 = 0$.

(b) $n(D \times H) = 0; n(H) = 0$;
$n(D \times H) = n(H)$ since $0 = 0$.

(c) $n(D) = 4; n(H) = 0$;
$n(D) \times n(H) = n(H)$ since $4 \times 0 = 0$.

(d) $n(D) = 4; n(H) = 0$;
$n(D) \times n(H) = n(H) \times n(D) = n(H)$ since $4 \times 0 = 0 \times 4 = 0$.

3. (a) 0 sets of 7 cannot be pictured.

(b) 0×7 cannot be shown effectively on a number line.

(c) 0×7 cannot be shown effectively with an array.

(d) To show 0×7 using Cartesian product of sets, let $A = \{\ \}$ and $B = \{a, b, c, d, e, f, g\}$. Thus $n(A) = 0$ and $n(B) = 7$. The set of ordered pairs defined as $A \times B = \{\ \}$. Thus $n(A \times B) = 0$ and $n(A) \times n(B) = n(A \times B)$; that is $0 \times 7 = 0$.

5-7 Distributive Property of Multiplication over Addition

1. (a) $7 \times (8 + 29) = (7 \times \underline{8}) + (\underline{7 \times 29})$.

(b) $(5 \times 3) + (5 \times \underline{26}) = \underline{5} \times (\underline{3} + 26)$.

(c) $128 \times (64 + 216) = \underline{(128 \times 64)} + (128 \times 216)$.

(d) $(785 \times 38) + (785 \times 59) = \underline{785} \times (\underline{38 + 59})$.

3. (a) $(78 \times 43) + (78 \times 57) = 78 \times (43 + 57) = 78 \times 100 = 7{,}800$.

(b) $(39 \times 5) + (39 \times 5) = 39 \times (5 + 5) = 39 \times 10 = 390$, or
$(39 \times 5) + (39 \times 5) = (39 + 39) \times 5 = 78 \times 5 = 390$.

5. $(30 \times 13) + (37 \times 30) = (30 \times 13) + (30 \times 37)$ — Commutative property of multiplication.

$(30 \times 13) + (30 \times 37) = 30 \times (13 + 37)$ — Distributive property of multiplication over addition.

$30 \times (13 + 37) = 30 \times 50$ — Addition.

$30 \times 50 = 1{,}500$. — Multiplication.

7. Yes in both cases.

9. (a) $(3 + 5) \times (4 + 2) = 3 \times (4 + 2) + 5 \times (4 + 2)$ — Distributive property of multiplication over addition.

$3 \times (4 + 2) + 5 \times (4 + 2) = (3 \times 4) + (3 \times 2) + (5 \times 4) + (5 \times 2)$ — Distributive property of multiplication over addition.

(b) $(d + e) \times (g + h) = d \times (g + h) + e \times (g + h)$ — Distributive property of multiplication over addition.

$d \times (g + h) + e \times (g + h) = (d \times g) + (d \times h) + (e \times g) + (e \times h)$ — Distributive property of multiplication over addition.

5-8 The Table of Basic Multiplication Facts

1. The property of multiplication by zero.

3. The differences between these consecutive squares form a pattern.

$$1 - 0 = 1$$
$$4 - 1 = 3$$
$$9 - 4 = 5$$
$$16 - 9 = 7$$
$$25 - 16 = 9$$
$$36 - 25 = 11$$
$$49 - 36 = 13$$
$$64 - 49 = 15$$
$$81 - 64 = 17$$

These differences are consecutive odd whole numbers. The next five numbers represented on the main diagonal would be as follows.

First: $81 + 19 = 100$
Second: $100 + 21 = 121$
Third: $121 + 23 = 144$
Fourth: $144 + 25 = 169$
Fifth: $169 + 27 = 196$

5. (a) The product of two odd numbers is an odd number.
   (b) The product of two even numbers is an even number.
   (c) The product of an odd number and an even number is an even number.
   (d) The product of an even number and an odd number is an even number.

7. Row: 0, 700, 1400, 2100; Column: 0, 700, 1400, 2100.

9. No. The associative property of multiplication involves at least 3 factors and only 2 factors may be observed directly from the multiplication table.

5-9 Understanding the Multiplication Algorithm

1. (a)
$$\begin{array}{r} 4 \\ \times 2 \\ \hline 8 \end{array}$$

(b)
$$\begin{array}{r} 40 \\ \times\ 2 \\ \hline 80 \end{array}$$

(c)
$$\begin{array}{r} 43 \\ \times\ 2 \\ \hline 86 \end{array}$$

(d)
$$\begin{array}{r} 40 \\ \times 20 \\ \hline 800 \end{array}$$

(e)
$$\begin{array}{r} 43 \\ \times 21 \\ \hline 43 \\ 86 \\ \hline 903 \end{array}$$

3. (a)
$$\begin{array}{r} 200 \\ \times\ \ 4 \\ \hline 800 \end{array}$$

(b)
$$\begin{array}{r} \overset{1}{230} \\ \times\ \ 4 \\ \hline 920 \end{array}$$

(c)
$$\begin{array}{r} \overset{12}{236} \\ \times\ \ 4 \\ \hline 944 \end{array}$$

(d)
$$\begin{array}{r} 200 \\ \times\ 40 \\ \hline 8000 \end{array}$$

(e)
$$\begin{array}{r} \overset{1}{230} \\ \times\ 40 \\ \hline 9200 \end{array}$$

(f)
$$\begin{array}{r} \overset{12}{236} \\ \times\ 40 \\ \hline 9440 \end{array}$$

(g)
$$\begin{array}{r} \overset{12}{\overset{1}{236}} \\ \times\ 42 \\ \hline 472 \\ 944 \\ \hline 9912 \end{array}$$

(h)
$$\begin{array}{r} \overset{24}{236} \\ \times\ 78 \\ \hline 1888 \\ 1652 \\ \hline 18408 \end{array}$$

5. (a)
$$\begin{array}{r} 38 \\ \times 29 \\ \hline 72 \\ 270 \\ 160 \\ 600 \\ \hline 1102 \end{array}$$

(b)
$$\begin{array}{r} 75 \\ \times 14 \\ \hline 20 \\ 280 \\ 50 \\ 700 \\ \hline 1050 \end{array}$$

(c)
$$\begin{array}{r} 298 \\ \times\ 36 \\ \hline 48 \\ 540 \\ 1200 \\ 240 \\ 2700 \\ 6000 \\ \hline 10728 \end{array}$$

(d)
$$\begin{array}{r} 134 \\ \times 218 \\ \hline 32 \\ 240 \\ 800 \\ 40 \\ 300 \\ 1000 \\ 800 \\ 6000 \\ 20000 \\ \hline 29212 \end{array}$$

7. (b)

$$23 \times 45 = 1035$$

(c)

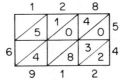

$$54 \times 128 = 6912$$

5-10 Division of Whole Numbers

1.

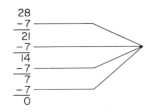

7 is subtracted 4 times; $28 \div 7 = 4$

3. (a) $51 \div 17 = d$ or $51 \div d = 17$.   **(b)** $39 \div b = g$ or $39 \div g = b$.
(c) $n \div 13 = 26$ or $n \div 26 = 13$.   **(d)** $t \div s = r$ or $t \div r = s$.
5. (a) $48 \div 16 = 3$.   **(b)** $98 \div 14 = 7$.
7. (a)

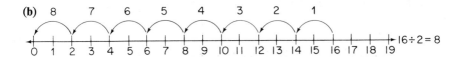

(b)

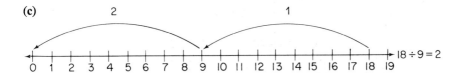

(c)

9. (a) * * * * * *
　　　 * * * * * *
　　　　$12 \div 6 = 2$

(b) • • • • •
　　　• • • • •
　　　• • • • •
　　　• • • • •
　　　• • • • •
　　　$25 \div 5 = 5$

(c) * * * * * * * *
　　　　$8 \div 8 = 1$

5-11 Inverse Operations: Multiplication and Division

1. (a) True.　　　　**(b)** True.　　　　**(c)** False.　　　　**(d)** True.

3. Multiply the quotient by 23; that is, if $n \div 23 = q$, then $q \times 23 = n$.

5-12 Properties of Division of Whole Numbers

1. Among the many possible counterexamples are:
　(a) $25 \div 5 \neq 5 \div 25$, because $5 \neq \frac{1}{5}$.
　(b) $48 \div (16 \div 8) \neq (48 \div 16) \div 8$, because $24 \neq \frac{3}{8}$.
　(c) $1 \div 3 = \frac{1}{3}$, and the quotient $\frac{1}{3}$ does not exist in the set of whole numbers.

3. (a) True.　　　　**(b)** False.　　　　**(c)** True.　　　　**(d)** True.
　(e) False.　　　　**(f)** True.　　　　**(g)** True.　　　　**(h)** True.

5-13 Division by Zero

1. True.

3. False; since $\dfrac{0}{0}$ is undefined.

5. False; since $\dfrac{n}{0}$ is undefined.

5-14 Understanding the Division Algorithm

1. (a) $23 \div 8 = 2 \, r \, 7$. Thus $23 = 8 \times 2 + 7$.
　(b) $85 \div 9 = 9 \, r \, 4$. Thus $85 = 9 \times 9 + 4$.
　(c) $392 \div 8 = 49 \, r \, 0$. Thus $392 = 8 \times 49 + 0$.
　(d) $408 \div 5 = 81 \, r \, 3$. Thus $408 = 5 \times 81 + 3$.
　(e) $426 \div 42 = 10 \, r \, 6$. Thus $426 = 42 \times 10 + 6$.
　(f) $923 \div 27 = 34 \, r \, 5$. Thus $923 = 27 \times 34 + 5$.
　(g) $51 \div 79 = 0 \, r \, 51$. Thus $51 = 79 \times 0 + 51$.
　(h) $95 \div 19 = 5 \, r \, 0$. Thus $95 = 19 \times 5 + 0$.

3. (a) $5 \times 3 = 15$.　　　　**(b)** $4 \times 19 = 76$.　　　　**(c)** $8 \times 216 = 1,728$.

CHAPTER 6—ELEMENTARY NUMBER THEORY

6-1 Multiples and Divisibility

1. 60; 84; 516; 278,190; and 0 are divisible by 2 since the one's digit in each case is a 0, 2, 4, 6, or 8.

3. (a) {0, 4, 8, 12, . . .}.   (b) {0, 6, 12, 18, . . .}.
    (c) {0, 8, 16, 24, . . .}.   (d) {0, 9, 18, 27, . . .}.

5. Among the many possible correct answers are:
    (a) 132, 544, 620, 712, 856.   (b) 131, 543, 619, 711, 855.

7. A number is divisible by 8 if and only if the last three digits of its numeral represent a number that is divisible by 8.

9. Yes it is true. Any number that is divisible by both 2 and 3 has 2 and 3 as factors. Thus it has 6 (2 × 3) as a factor and is divisible by 6.

11. Marty's rule does appear to work.

13. (a) If a number is not divisible by 3, there is no need to test for divisibility by 9, since 3 is a factor of 9.
    (b) If a number is not divisible by 6, there is still a need to test for divisibility by 9, since 6 is not a factor of 9.

6-2 Prime and Composite Numbers

1. 101, 103, 107, 109, 113, 127, 131, 137, 139, 149, 151, 157, 163, 167, 173, 179, 181, 191, 193, 197, 199.

3. Zero is neither a prime number nor a composite number.

5. (a) {2, 3, 4, 5, 6, 7, 8, 9, . . .}.   (b) {0, 1, 2, 3, 4, 5, 6, 7, . . .}.

7. (a) 5.   (b) 7.   (c) 13.   (d) 17.   (e) 23.

6-3 Fundamental Theorem of Arithmetic

1. (a) $2 \times 3 \times 7$.   (b) $3 \times 3 \times 11$.   (c) $2 \times 2 \times 37$.   (d) 5×71.

3. (a) 1, 2, 4, 8, 16, 32, 64.   (b) 4, 8, 16, 32, 64.   (c) 2.

5. 2×9; 3×6; $2 \times 3 \times 3$.

7.

$4 = 2 + 2$	$12 = 5 + 7$	$20 = 7 + 13$
$6 = 3 + 3$	$14 = 7 + 7$	$22 = 11 + 11$
$8 = 3 + 5$	$16 = 5 + 11$	$24 = 11 + 13$
$10 = 5 + 5$	$18 = 7 + 11$	

It appears that this could be done for all other even numbers (except 2).

6-4 Least Common Multiple

1. (a) Multiples of 6 = {6, 12, 18, 24, . . .};
       Multiples of 8 = {8, 16, 24, 32, . . .};

Common multiples of 6 and 8 = {24, 48, 72, . . .} ;
L.C.M. of 6 and 8 is 24.
  (b) Multiples of 9 = {9, 18, 27, 36, 45, . . .} ;
Multiples of 15 = {15, 30, 45, 60, . . .} ;
Common multiples of 9 and 15 = {45, 90, 135, . . .} ;
L.C.M. of 9 and 15 is 45.
  (c) Multiples of 12 = {12, 24, 36, 48, 60, 72, 84, 96, . . .} ;
Multiples of 32 = {32, 64, 96, 128, . . .} ;
Common multiples of 12 and 32 = {96, 192, 288, . . .} ;
L.C.M. of 12 and 32 is 96.
  (d) Multiples of 48 = {48, 96, 144, 192, . . .} ;
Multiples of 64 = {64, 128, 192, . . .} ;
Common multiples of 48 and 64 = {192, 384, 576, . . .} ;
L.C.M. of 48 and 64 is 192.

3. The L.C.M. of two prime numbers is the product of the two numbers.

5. (a) 150.   **(b)** 84.   **(c)** 90.

7. Among the many possible correct pairs of numbers are:
21 and 10, 16 and 35, 9 and 20, 36 and 25.

9. (a) $4 = 2 \times 2$
$18 = 2 \times 3 \times 3$
$20 = 2 \times 2 \times 5$
L.C.M. of 4, 18, and 20 is $2 \times 2 \times 3 \times 3 \times 5$; that is, 180.
  (b) $2 = 2 \times 1$
$5 = 5 \times 1$
$11 = 11 \times 1$
L.C.M. of 2, 5, and 11 is $2 \times 5 \times 11$; that is, 110.
  (c) $32 = 2 \times 2 \times 2 \times 2 \times 2$
$396 = 2 \times 2 \times 3 \times 3 \times 11$
L.C.M. of 32 and 396 is $2 \times 2 \times 2 \times 2 \times 2 \times 3 \times 3 \times 11$; that is, 3,168.
  (d) $239 = 1 \times 239$
$785 = 5 \times 157$
L.C.M. of 239 and 785 is $5 \times 157 \times 239$; that is, 186,615.

6-5 Greatest Common Factor

1. (a) Factors of 12 = {1, 2, 3, 4, 6, 12} ;
Factors of 16 = {1, 2, 4, 8, 16} ;
Common Factors of 12 and 16 = {1, 2, 4} ;
G.C.F. of 12 and 16 = 4.
  (b) Factors of 32 = {1, 2, 4, 8, 16, 32} ;
Factors of 48 = {1, 2, 3, 4, 6, 8, 12, 16, 24, 48} ;
Common Factors of 32 and 48 = {1, 2, 4, 8, 16} ;
G.C.F. of 32 and 48 = 16.
  (c) Factors of 64 = {1, 2, 4, 8, 16, 32, 64} ;
Factors of 72 = {1, 2, 3, 4, 6, 8, 9, 12, 18, 24, 36, 72} ;

Common Factors of 64 and 72 = {1, 2, 4, 8};
G.C.F. of 64 and 72 = 8.
  (d) Factors of 256 = {1, 2, 4, 8, 16, 32, 64, 128, 256};
Factors of 87 = {1, 3, 29, 87};
Common Factors of 256 and 87 = {1};
G.C.F. of 256 and 87 = 1.

3. The G.C.F. of any two prime numbers is always 1.

5. (a) 1.   (b) 1.   (c) 1.

7. The G.C.F. of any two relatively prime numbers is always 1.

9. (a) 24.   (b) 30.   (c) 180.   (d) 24,510.

11. Yes; $24 \times 4 = 96$ and $8 \times 12 = 96$.

13. 18.

15. Yes. Consider the pair of numbers 6 and 6. G.C.F. is 6. L.C.M. is 6.

17. Yes.

6-6 Perfect, Abundant, and Deficient Numbers

1. (a) 1, 2, 5.   (b) 1, 2, 4, 8.   (c) 1, 2, 3, 4, 6, 8, 12.
  (d) 1, 2, 4, 7, 14.   (e) 1, 2, 3, 4, 6, 9, 12, 18. (f) 1, 3, 5, 9, 15.
  (g) 1, 3, 31.   (h) 1, 3, 9, 13, 39.   (i) 1, 2, 4, 8, 31, 62, 124.
  (j) 1, 2, 11, 17, 22, 34, 187.

3. (a) Among the many possible correct answers are:
        11, 13, 17, 19, 23.
  (b) The numbers 11, 13, 17, 19 and 23 are all deficient numbers.
  (c) Yes, it is true. All prime numbers are deficient numbers because the only
        proper divisor of a prime number is 1. Since 1 is always less than the prime
        number under consideration, the prime number is a deficient number.

5. Among the many possible correct answers are:
  (a) 496.   (b) 18, 48, 72, 96, 100.   (c) 14, 21, 26, 32, 64.

7. 3 and 5, 5 and 7, 11 and 13, 17 and 19, 29 and 31.

9. Among the many possible tests of the conjecture are:
        $8 = 3 + 5; 10 = 3 + 7; 12 = 5 + 7; 14 = 7 + 7; 16 = 11 + 5;$
        $18 = 7 + 11; 20 = 7 + 13; 22 = 11 + 11; 24 = 11 + 13; 26 = 13 + 13.$

CHAPTER 7—INTEGERS: OPERATIONS AND PROPERTIES

7-1 The Set of Integers

1. Positive four plus positive nine is equal to positive thirteen.

3. Negative six plus positive eight is equal to positive two.

5. Zero plus negative fourteen is equal to negative fourteen.

7. Positive seven minus positive five is equal to positive two.

9. Negative three minus positive five is equal to negative eight.

11.

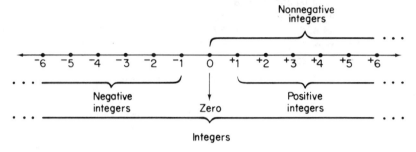

13. (a) False; zero is not included in the union whereas zero is an integer.
   (b) True. (At this level of mathematical sophistication.)
   (c) True.
   (d) False; since the set of integers is infinite in both directions, it should be represented as $\{\ldots, {}^-4, {}^-3, {}^-2, {}^-1, 0, 1, 2, 3, 4, \ldots\}$.

7-2 Additive Inverse

1. (a) $^+3$; that is, 3.   **(b)** $^-5$.   **(c)** $^-17$.   **(d)** $^+26$; that is, 26.
   (e) 0.   **(f)** ^-b.   **(g)** ^+b; that is, b.   **(h)** $^+(a + b)$; that is, $a + b$.
   (i) $^-(a - b)$; that is, $^-a + b$, or $b - a$.   **(j)** $^-(a \cdot b)$; that is, ^-ab.

3. (a)

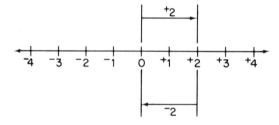

(b)

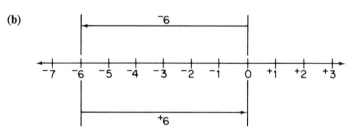

5. (a) $^-({}^-5)$ means the opposite of $^-5$ (also referred to as the negative of $^-5$).
   (b) $^-({}^-5) = {}^+5$ (We may also think $^-({}^-5) = 5$.)

7-3 Addition of Integers

1. (a)

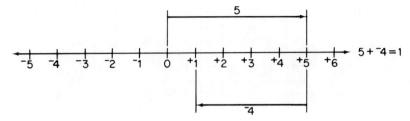

$$5 + {}^-4 = 1$$

(b)

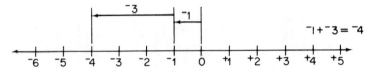

$${}^-1 + {}^-3 = {}^-4$$

(c)

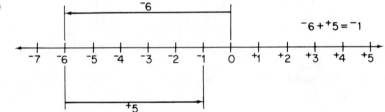

$${}^-6 + {}^+5 = {}^-1$$

(d)

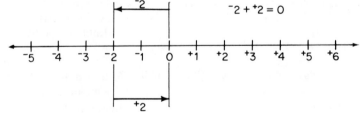

$${}^-2 + {}^+2 = 0$$

3.

+	$^-5$	$^-4$	$^-3$	$^-2$	$^-1$	0	1	2	3	4	5
$^-5$	$^-10$	$^-9$	$^-8$	$^-7$	$^-6$	$^-5$	$^-4$	$^-3$	$^-2$	$^-1$	0
$^-4$	$^-9$	$^-8$	$^-7$	$^-6$	$^-5$	$^-4$	$^-3$	$^-2$	$^-1$	0	1
$^-3$	$^-8$	$^-7$	$^-6$	$^-5$	$^-4$	$^-3$	$^-2$	$^-1$	0	1	2
$^-2$	$^-7$	$^-6$	$^-5$	$^-4$	$^-3$	$^-2$	$^-1$	0	1	2	3
$^-1$	$^-6$	$^-5$	$^-4$	$^-3$	$^-2$	$^-1$	0	1	2	3	4
0	$^-5$	$^-4$	$^-3$	$^-2$	$^-1$	0	1	2	3	4	5
1	$^-4$	$^-3$	$^-2$	$^-1$	0	1	2	3	4	5	6
2	$^-3$	$^-2$	$^-1$	0	1	2	3	4	5	6	7
3	$^-2$	$^-1$	0	1	2	3	4	5	6	7	8
4	$^-1$	0	1	2	3	4	5	6	7	8	9
5	0	1	2	3	4	5	6	7	8	9	10

5. For whole numbers a and b when $a \geq b$, $a + {}^-b = a - b$ if $a > b$, and $a + {}^-b = 0$ if $a = b$.

7. (a) $^-8 + 7 = \triangle$

$^-8 + 7 = ({}^-1 + {}^-7) + 7$ Renamed $^-8$ as $^-1 + {}^-7$.

$({}^-1 + {}^-7) + 7 = {}^-1 + ({}^-7 + 7)$ Associative property.

$^-1 + ({}^-7 + 7) = {}^-1 + 0$ Additive inverse property.

$^-1 + 0 = {}^-1$ Additive identity property.

Thus $^-8 + 7 = {}^-1$.

(b) $^-4 + {}^-9 = \triangle$

$({}^-4 + {}^-9) + 9 = \triangle + 9$ If $a = b$, then $a + n = b + n$, where a, b, and n are integers.

$^-4 + ({}^-9 + 9) = \triangle + 9$ Associative property.

$^-4 + 0 = \triangle + 9$ Additive inverse property.

$^-4 = \triangle + 9$ Additive identity property.

$^-4 + 4 = \triangle + 9 + 4$ If $a = b$, then $a + n = b + n$, where a, b, and n are integers.

$0 = \triangle + 9 + 4$ Additive inverse property.

$0 = \triangle + (9 + 4)$ Associative property.

Thus we conclude that $\triangle$ is the additive inverse of $9 + 4$; that is, $^-(9 + 4)$. Since $\triangle = {}^-4 + {}^-9$ in the beginning, we conclude that

$\triangle = {}^-4 + {}^-9 = {}^-(9 + 4) = {}^-13$.

(c) $5 + {}^-2 = \triangle$

$5 + {}^-2 = (3 + 2) + {}^-2$ Renamed 5 as $3 + 2$.

$(3 + 2) + {}^-2 = 3 + (2 + {}^-2)$ Associative property.
$3 + (2 + {}^-2) = 3 + 0$ Additive inverse property.
$3 + 0 = 3$ Additive identity property.
Thus $5 + {}^-2 = 3$.

7-4 Subtraction of Integers

1. (a) $16 - 7 = 9$. (b) ${}^-3 - {}^-9 = 6$. (c) ${}^-5 - {}^-1 = {}^-4$.
   $16 - 9 = 7$. ${}^-3 - 6 = {}^-9$. ${}^-5 - {}^-4 = {}^-1$.
3. (a) 6. (b) ${}^-x$. (c) ${}^-4$. (d) 1.
   (e) ${}^-(a + b)$. (f) ${}^-(4 + {}^-3)$; that is, ${}^-4 + 3$.
   (g) ${}^-({}^-13 + {}^-4)$; that is, $13 + 4$. (h) ${}^-({}^-a + 2)$; that is, $a - 2$.
5. (a) ${}^-5 - 4 = {}^-9$ and ${}^-5 + \underline{{}^-4} = {}^-9$. (b) $12 - {}^-1 = 13$ and $12 + \underline{1} = 13$.
   (c) ${}^-7 - 8 = {}^-15$ and ${}^-7 + \underline{{}^-8} = {}^-15$. (d) ${}^-6 - {}^-5 = {}^-1$ and ${}^-6 + \underline{5} = {}^-1$.
   (e) $x - y = z$ and $x + \underline{{}^-y} = z$, where x, y, and z are integers.
7. (a) Closed.
   (b) Not closed; ${}^+6 - {}^+6 = 0$ and 0 is not a positive integer.
   (c) Not closed; ${}^-5 - {}^-5 = 0$ and 0 is not a negative integer.
   (d) Not closed; ${}^+3 - {}^+5 = {}^-2$ and ${}^-2$ is not a nonnegative integer.
   (e) Closed.
   (f) Not closed; ${}^+7 - {}^-3 = {}^+10$ and ${}^+10$ is not an odd integer.
9. No; $({}^-6 - {}^+3) - {}^-8 \neq {}^-6 - ({}^+3 - {}^-8)$ because ${}^-1 \neq {}^-17$.
11. (a) Yes. (b) No. (c) No.

7-5 Multiplication of Integers

1. (a)

4
−16 ⟵ (⁻4 × 4)
−12
−8 ⟵ (⁻2 × 4)
−4
0
4 ⟵ (1 × 4)
8
12
16 ⟵ (4 × 4)

(b)

5
−25 ⟵ (⁻5 × 5)
−20
−15
−10
−5 ⟵ (⁻1 × 5)
0
5
10 ⟵ (2 × 5)
15
20 ⟵ (4 × 5)

(c)

9
−45
−36 ⟵ (⁻4 × 9)
−27
−18
−9 ⟵ (⁻1 × 9)
0 ⟵ (0 × 9)
9
18
27 ⟵ (3 × 9)
36

3. (a) ${}^-21$. (b) ${}^-24$. (c) ${}^-16$. (d) ${}^-9$.
   (e) ${}^+20$. (f) ${}^+27$. (g) ${}^+5$. (h) ${}^+42$.
   (i) 0. (j) 0. (k) 0. (l) 0.
   (m) ${}^-140$. (n) ${}^+540$. (o) ${}^-1,368$. (p) ${}^-1$.
5. Among the many possible patterns observed are:
   When one factor is 0, the product is 0.
   When one factor is 1, the product is the other factor.

When one factor is ⁻1, the product is the additive inverse (or negative of) the
other factor.
When both factors are negative, the product is positive.
When both factors are positive, the product is positive.
When one factor is negative and one factor is positive, the product is negative.

7-6 Division of Integers

1. (a) $^-54 \div 9 = \; ^-6.$ **(b)** $35 \div \; ^-5 = \; ^-7.$ **(c)** $^-32 \div \; ^-8 = 4.$
    $^-54 \div \; ^-6 = 9.$ $35 \div \; ^-7 = \; ^-5.$ $^-32 \div 4 = \; ^-8.$

3. (a) True.   **(b)** True.   **(c)** False.   **(d)** False.

5. You must multiply the quotient by ⁻4 to obtain j; that is, $j \div \; ^-4 = n$ if and only
    if $n \times \; ^-4 = j$.

7. (a) ⁻4.   **(b)** 1.   **(c)** ⁻1.   **(d)** 0.   **(e)** 1.
    (f) ⁻10.   **(g)** 9.   **(h)** ⁻8.   **(i)** ⁻9.

9. (a) Yes.   **(b)** No.   **(c)** No.   **(d)** No.

11. (a) No; there is no integer j such that $3 \div 0 = j$ and $j \times 0 = 3$.
    (b) Yes; division by zero is excluded or not defined for the set of integers.

13. (a) Distributive property of division over addition from the right (12a and 12b).
    (b) Distributive property of division over subtraction from the right (12c and
    12d).
    (c) Counterexample of distributive property of division over addition from
    the left (12e); counterexample of distributive property of division over
    subtraction from the left (12f).

Chapter 8—Rational Numbers: Addition and Subtraction

8-1 The Set of Rational Numbers

1. (a) Yes.   **(b)** Yes.   **(c)** Yes.   **(d)** No.
    (e) Yes.   **(f)** Yes.   **(g)** Yes.   **(h)** Yes.
    (i) Yes.   **(j)** Yes.   **(k)** Yes.   **(l)** Yes.
    (m) Yes.   **(n)** Yes.   **(o)** Yes.   **(p)** No.

8-2 Fractional Numbers

1. (a) Unlike.   **(b)** Like.   **(c)** Like.   **(d)** Unlike.

3. (a)   **(b)**

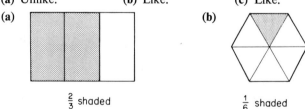

$\frac{2}{3}$ shaded $\frac{1}{6}$ shaded

(c)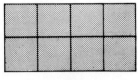

$\frac{8}{8}$ shaded

(d)

$\frac{0}{5}$ shaded

5.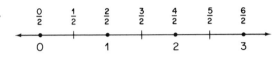

$\frac{0}{2} = 0$ $\frac{2}{2} = 1$ $\frac{4}{2} = 2$ $\frac{6}{2} = 3$

7. (a)

$\frac{1}{3}$ shaded $\frac{2}{6}$ shaded   Thus $\frac{1}{3} = \frac{2}{6}$

(b)

$\frac{5}{10}$ shaded $\frac{1}{2}$ shaded   Thus $\frac{5}{10} = \frac{1}{2}$

(c)

$\frac{1}{4}$ shaded $\frac{2}{8}$ shaded   Thus $\frac{1}{4} = \frac{2}{8}$

(d)

$\frac{1}{3}$ shaded $\frac{3}{9}$ shaded   Thus $\frac{1}{3} = \frac{3}{9}$

8-3 Equality and Inequality of Rational Numbers

1. (a) True.   **(b)** False.   **(c)** False.
   (d) True.   **(e)** True.   **(f)** False.
3. (a) $>$.   **(b)** $<$.   **(c)** $<$.   **(d)** $=$.

8-4 Numerals—the Names for Numbers

1. (a)

      $\frac{1}{2}$ shaded $\frac{3}{6}$ shaded   Thus $\frac{1}{2} = \frac{3}{6}$

  (b)

      $\frac{3}{4}$ shaded $\frac{6}{8}$ shaded   Thus $\frac{3}{4} = \frac{6}{8}$

  (c)

      $\frac{1}{3}$ shaded $\frac{3}{9}$ shaded   Thus $\frac{1}{3} = \frac{3}{9}$

3. (a)

$\frac{1}{2}$ shaded $\frac{2}{4}$ shaded   Thus $\frac{1}{2} = \frac{2}{4}$

(b)

$\frac{1}{3}$ shaded $\frac{2}{6}$ shaded   Thus $\frac{1}{3} = \frac{2}{6}$

(c)

$\frac{3}{4}$ shaded $\frac{6}{8}$ shaded   Thus $\frac{3}{4} = \frac{6}{8}$

5. (a) $\square = {}^-9.$ **(b)** $\square = 52.$ **(c)** $\square = 60.$
     (d) $\square = {}^-5.$ **(e)** $\square = 0.$ **(f)** $\square = 8.$

7. (a) $\frac{4}{5}.$ **(b)** $\frac{3}{5}.$ **(c)** $\frac{3}{7}.$
     (d) $\frac{97}{101}.$ **(e)** $0.$ **(f)** $\frac{11}{15}.$

8-5 Addition of Rational Numbers

1. (a)

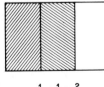

$$\frac{1}{3} + \frac{1}{3} = \frac{2}{3}$$

(b)

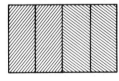

$$\frac{2}{4} + \frac{2}{4} = \frac{4}{4} = 1$$

(c)

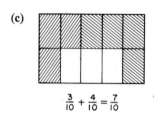

$$\frac{3}{10} + \frac{4}{10} = \frac{7}{10}$$

3. (a) $\frac{5}{7}$.   **(b)** $\frac{5}{8}$.   **(c)** $\frac{13}{7}$ (also $1\frac{6}{7}$).
   (d) $\frac{1}{13}$.   **(e)** $\frac{-3}{4}$.   **(f)** $\frac{14}{17}$.
5. (a) $\frac{41}{24}$ (also $1\frac{17}{24}$).   **(b)** $\frac{15}{12}$ (also $1\frac{1}{4}$).   **(c)** $\frac{44}{45}$.   **(d)** $\frac{26}{21}$ (also $1\frac{5}{21}$).
   (e) $\frac{23}{18}$ (also $1\frac{5}{18}$).   **(f)** $\frac{83}{15}$ (also $5\frac{8}{15}$).   **(g)** $\frac{-11}{48}$.   **(h)** $\frac{68}{9}$ (also $7\frac{5}{9}$).
   (i) $\frac{29}{30}$.

8-6 Properties of Addition of Rational Numbers

1. (a) $\frac{81}{91}$.   **(b)** $\frac{81}{91}$.
3. Commutative property of addition for rational numbers.
5. Zero.
7. The answers are the same.
9. (a) $\frac{5}{3}$ (also $1\frac{2}{3}$).   **(b)** $\frac{3}{4}$.   **(c)** 0.
   (d) $\frac{16}{17}$.   **(e)** $\frac{16}{17}$.   **(f)** $\frac{7}{9}$.
11. Commutative property of addition for rational numbers.

8-7 Subtraction of Rational Numbers

1. (a) $\frac{6}{7} - \frac{2}{7} = \frac{6}{7} + \frac{-2}{7} = \frac{4}{7}$.

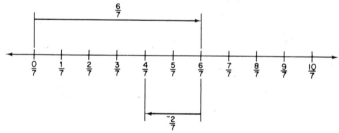

(b) $\frac{-2}{5} - \frac{-4}{5} = \frac{-2}{5} + \frac{4}{5} = \frac{2}{5}$.

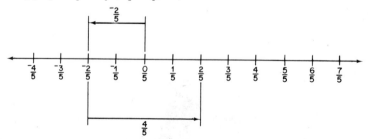

(c) $\frac{9}{4} - \frac{5}{4} = \frac{9}{4} + \frac{-5}{4} = \frac{4}{4} = 1.$

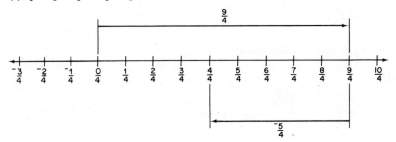

(d) $\frac{-1}{3} - \frac{-1}{3} = \frac{-1}{3} + \frac{1}{3} = 0.$

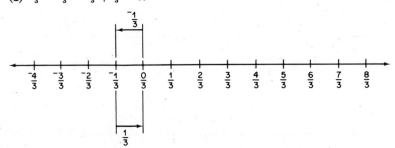

3. (a) 0.   **(b)** $\frac{-3}{8}$.   **(c)** $\frac{13}{4}$ (also $3\frac{1}{4}$).   **(d)** 0.

   (e) $\frac{-10}{15}$ (also $\frac{-2}{3}$).   **(f)** $\frac{16}{91}$.   **(g)** $\frac{-1}{30}$.   **(h)** $\frac{-1}{96}$.

8-8 Properties of Subtraction of Rational Numbers

1. (a) $n = \frac{5}{14}$.   **(b)** $n = \frac{-5}{14}$.   **(c)** No.   **(d)** No.

3. (a) $\frac{7}{8}$.   **(b)** $\frac{6}{5}$.   **(c)** $\frac{3}{7}$.   **(d)** $\frac{-4}{9}$.

5. (a) $\frac{0}{13} = 0$.   **(b)** $\frac{0}{8} = 0$.   **(c)** $\frac{0}{3} = 0$.   **(d)** $\frac{0}{8} = 0$.

7. (a) True.   **(b)** False.   **(c)** False.   **(d)** False.

8-9 Mixed Numerals and Computation

1. (a) $\frac{11}{3}$.   **(b)** $\frac{25}{4}$.   **(c)** $\frac{21}{8}$.

   (d) $\frac{8}{5}$.   **(e)** $\frac{31}{6}$.   **(f)** $\frac{62}{7}$.

   (g) $\frac{15}{2}$.   **(h)** $\frac{119}{6}$.   **(i)** $\frac{119}{8}$.

3. (a) $-2\frac{7}{12}$.   **(b)** $1\frac{1}{40}$.   **(c)** $14\frac{1}{42}$.

   (d) 0.   **(e)** $12\frac{1}{15}$.   **(f)** $-1\frac{9}{56}$.

CHAPTER 9—RATIONAL NUMBERS: MULTIPLICATION AND DIVISION

9-1 Multiplication of Rational Numbers

1. (a)

$$3 \times \frac{1}{7} = \frac{3}{7}$$

(b)

$$4 \times \frac{1}{4} = \frac{4}{4}$$

(c)

$$6 \times \frac{1}{5} = \frac{6}{5}$$

3. (a) $\frac{2}{4} + \frac{2}{4} + \frac{2}{4} = \frac{6}{4}$. **(b)** $\frac{7}{16} + \frac{7}{16} = \frac{14}{16}$.

    $3 \times \frac{2}{4} = \frac{6}{4}$. $2 \times \frac{7}{16} = \frac{14}{16}$.

5. (a)

$$\frac{3}{4} \times \frac{2}{7} = \frac{6}{28}$$

(b)

$$\frac{1}{2} \times \frac{2}{3} = \frac{2}{6}$$

(c)

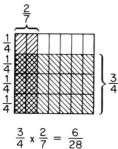

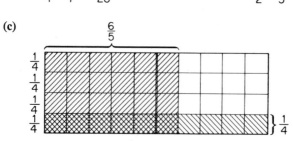

$$\frac{1}{4} \times \frac{6}{5} = \frac{6}{20}$$

9-2 Properties of Multiplication of Rational Numbers

1. (a) $(\frac{1}{3} \times \frac{1}{5}) \times \frac{3}{8} = \frac{1}{15} \times \frac{3}{8} = \frac{3}{120}.$ **(b)** $\frac{1}{3} \times (\frac{1}{5} \times \frac{3}{8}) = \frac{1}{3} \times \frac{3}{40} = \frac{3}{120}.$

3. Associative property.

5. (a) $\frac{1}{2} \times \frac{3}{4} = \frac{3}{8}.$ **(b)** $\frac{1}{2} = \frac{2}{4}; \frac{3}{4} = \frac{6}{8}.$ **(c)** $\frac{2}{4} \times \frac{6}{8} = \frac{12}{32}.$ **(d)** Yes.   **(e)** Yes.

7. (a) $\frac{2}{3} \times 1 = \frac{2}{3}.$

     (b) $\frac{4}{7} \times \frac{2}{2} = \frac{8}{14} = \frac{4}{7}.$

     (c) $\frac{3}{3} \times \frac{1}{8} = \frac{3}{24} = \frac{1}{8}.$

     (d) $6 \times \frac{5}{5} = \frac{30}{5} = 6.$

9. No; if $a = 0$, then $\dfrac{a}{a} = \dfrac{0}{0}$ has no meaning.

11. Zero property for multiplication.

13. Among the possible correct answers:

You can say $\dfrac{c}{d} = 1$ and $\dfrac{a}{b}$ equals any rational number;

You can say $\dfrac{a}{b} = 0$ and $\dfrac{c}{d}$ equals any rational number.

9-3 Multiplicative Inverse

1. (a) $\dfrac{1}{8}.$ **(b)** $\dfrac{4}{5}.$ **(c)** $\dfrac{-8}{-7}.$ **(d)** Not possible since $\frac{3}{0}$ is not defined.

   (e) $\dfrac{-9}{-9}.$ **(f)** $\dfrac{-5}{12}.$ **(g)** $\dfrac{6}{6}.$

   (h) Not possible, since $\dfrac{7}{-2+2} = \dfrac{7}{0}$ and $\dfrac{7}{0}$ is not defined.

   (i) Not possible, since $\dfrac{1}{(\frac{1}{4} + \frac{-1}{4})} = \dfrac{1}{0}$ and $\dfrac{1}{0}$ is not defined.

   (j) $\dfrac{11}{-3 + -2} = \dfrac{11}{-5}.$

3. (a) $\dfrac{6}{7} \times \dfrac{7}{6} = 1.$ **(b)** $\dfrac{-13}{14} \times \dfrac{14}{-13} = 1.$

   (c) $\dfrac{-2}{-9} \times \dfrac{-9}{-2} = 1.$ **(d)** There is no number $\square$ such that $\square \times 0 = 1.$

   (e) $\dfrac{6}{6} \times \dfrac{6}{6} = 1.$ **(f)** $\dfrac{97}{141} \times \dfrac{141}{97} = 1.$

5. 0.

9-4 Division of Rational Numbers

1. (a) $\dfrac{5}{2}.$ **(b)** 1.   **(c)** $\dfrac{121}{169}.$ **(d)** $\dfrac{eh}{fg}.$

   (e) $\dfrac{1}{7}.$ **(f)** $\dfrac{26}{5}.$ **(g)** $\dfrac{-7}{6}.$ **(h)** 27.

3. (a) Yes.   **(b)** Yes.

5. (a) True.   **(b)** True.

    (c) False; $\frac{5}{9} \div 9 = \square$ if and only if $\square \times 9 = \frac{5}{9}$.

    (d) False; since it is possible for g to equal 0 and $\frac{0}{0}$ is not defined.

9-5 Properties of Division of Rational Numbers

1. (a) True.   **(b)** True.   **(c)** True.

    (d) True.   **(e)** True.   **(f)** False; division by zero

                                                           is not defined.

    (g) False; division is not in general commutative, and $\frac{11}{16} \neq \frac{16}{11}$.

    (h) False; $\frac{1}{4} \div (\frac{1}{5} \div \frac{1}{6}) = \frac{5}{24}$ and $(\frac{1}{4} \div \frac{1}{5}) \div \frac{1}{6} = \frac{30}{4}$; thus

         $\frac{1}{4} \div (\frac{1}{5} \div \frac{1}{6}) \not> (\frac{1}{4} \div \frac{1}{5}) \div \frac{1}{6}$.

3. (a) In general, for any rational number $\frac{a}{b}, \frac{a}{b} \div 1 = \frac{a}{b}$. For example, $\frac{6}{7} \div 1 = \frac{6}{7}$.

    (b) In general, for any rational number $\frac{a}{b}, 1 \div \frac{a}{b} \neq \frac{a}{b}$. For example, $1 \div \frac{6}{7} \neq \frac{6}{7}$.

5. Multiply by $\frac{3}{-2}$. If $\frac{p}{q} \div \frac{-2}{3} = n$, then $n \times \frac{3}{-2} = \frac{p}{q}$.

7. No; when $\frac{s}{t} = 0, 0 \div 0 \neq 1$.

9-6 Mixed Numerals and Computation

1. (a) $\frac{14}{3}$.   **(b)** $\frac{51}{8}$.   **(c)** $\frac{47}{6}$.

    (d) $\frac{21}{4}$.   **(e)** $\frac{5}{4}$.   **(f)** $\frac{28}{5}$.

    (g) $\frac{182}{19}$.   **(h)** $\frac{35}{8}$.   **(i)** $\frac{-26}{3}$.

3. (a) $\frac{112}{153}$.   **(b)** $\frac{94}{63}$.   **(c)** $\frac{25}{112}$.

    (d) $\frac{208}{95}$.   **(e)** $\frac{-39}{133}$.   **(f)** $\frac{85}{56}$.

9-7 Density of the Rational Numbers

1. No.

3. It is impossible to identify a next rational number after any given rational number, since if a rational number $n > \frac{1}{2}$ is selected, there will be infinitely many other rational numbers between $\frac{1}{2}$ and n.

5. (a) $\frac{2}{5} + \frac{3}{5} = \frac{5}{5}$ and $\frac{5}{5} \div 2 = \frac{5}{5} \times \frac{1}{2} = \frac{5}{10}$,

         $\frac{2}{5} + \frac{5}{10} = \frac{9}{10}$ and $\frac{9}{10} \div 2 = \frac{9}{10} \times \frac{1}{2} = \frac{9}{20}$,

         $\frac{2}{5} + \frac{9}{20} = \frac{17}{20}$ and $\frac{17}{20} \div 2 = \frac{17}{20} \times \frac{1}{2} = \frac{17}{40}$,

         $\frac{2}{5} + \frac{17}{40} = \frac{33}{40}$ and $\frac{33}{40} \div 2 = \frac{33}{40} \times \frac{1}{2} = \frac{33}{80}$,

         $\frac{2}{5} + \frac{33}{80} = \frac{65}{80}$ and $\frac{65}{80} \div 2 = \frac{65}{80} \times \frac{1}{2} = \frac{65}{160}$.

    Among the rational numbers between $\frac{2}{5}$ and $\frac{3}{5}$ are $\frac{5}{10}, \frac{9}{20}, \frac{17}{40}, \frac{33}{80}$, and $\frac{65}{160}$.

    (b) $\frac{7}{10} + \frac{8}{10} = \frac{15}{10}$ and $\frac{15}{10} \div 2 = \frac{15}{10} \times \frac{1}{2} = \frac{15}{20}$,

         $\frac{7}{10} + \frac{15}{20} = \frac{29}{20}$ and $\frac{29}{20} \div 2 = \frac{29}{20} \times \frac{1}{2} = \frac{29}{40}$,

         $\frac{7}{10} + \frac{29}{40} = \frac{57}{40}$ and $\frac{57}{40} \div 2 = \frac{57}{40} \times \frac{1}{2} = \frac{57}{80}$,

         $\frac{7}{10} + \frac{57}{80} = \frac{113}{80}$ and $\frac{113}{80} \div 2 = \frac{113}{80} \times \frac{1}{2} = \frac{113}{160}$,

         $\frac{7}{10} + \frac{113}{160} = \frac{225}{160}$ and $\frac{225}{160} \div 2 = \frac{225}{160} \times \frac{1}{2} = \frac{225}{320}$.

    Among the rational numbers between $\frac{7}{10}$ and $\frac{8}{10}$ are $\frac{15}{20}, \frac{29}{40}, \frac{57}{80}, \frac{113}{160}$, and $\frac{225}{320}$.

(c) $\frac{-3}{4} + ^-1 = \frac{-7}{4}$ and $\frac{-7}{4} \div 2 = \frac{-7}{4} \times \frac{1}{2} = \frac{-7}{8}$,

   $\frac{-3}{4} + \frac{-7}{8} = \frac{-13}{8}$ and $\frac{-13}{8} \div 2 = \frac{-13}{8} \times \frac{1}{2} = \frac{-13}{16}$,

   $\frac{-3}{4} + \frac{-13}{16} = \frac{-25}{16}$ and $\frac{-25}{16} \div 2 = \frac{-25}{16} \times \frac{1}{2} = \frac{-25}{32}$,

   $\frac{-3}{4} + \frac{-25}{32} = \frac{-49}{32}$ and $\frac{-49}{32} \div 2 = \frac{-49}{32} \times \frac{1}{2} = \frac{-49}{64}$,

   $\frac{-3}{4} + \frac{-49}{64} = \frac{-97}{64}$ and $\frac{-97}{64} \div 2 = \frac{-97}{64} \times \frac{1}{2} = \frac{-97}{128}$.

Among the rational numbers between $\frac{-3}{4}$ and $^-1$ are $\frac{-7}{8}$, $\frac{-13}{16}$, $\frac{-25}{32}$, $\frac{-49}{64}$, and $\frac{-97}{128}$.

7. 0.

9. Yes; it may be determined as $\left(\dfrac{a}{b} + \dfrac{c}{d} \right) \div 2$.

Chapter 10—Ratio, Percent, and Decimals

10-1 Ratio and Proportion

1. (a) $\frac{4}{9}$.   (b) $\frac{30}{55}$.   (c) $\frac{32}{56}$.   (d) $\frac{53}{17}$.

3. (a) $\frac{1}{3}$.   (b) $\frac{1}{6}$.   (c) $\frac{3}{2}$.   (d) $\frac{1}{30}$.

5. (a) 45.   (b) 39.   (c) $\frac{5}{8}$.   (d) $\frac{7}{3}$.

7. (a) False.   (b) True.   (c) False.   (d) True.

9.

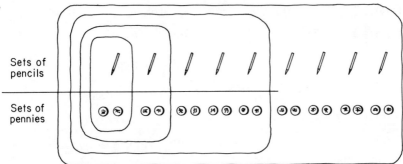

Sets of pencils

Sets of pennies

(a) 1 pencil for 2 cents.   (b) 2 pencils for 4 cents.
(c) 9 pencils for 18 cents.

11.

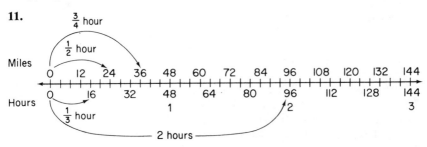

(a) $\frac{1}{3}$ hour → 16 miles.   (b) $\frac{1}{2}$ hour → 24 miles.

(c) $\frac{3}{4}$ hour → 36 miles.   (d) 2 hours → 96 miles.

10-2 Percent

1. (a) 75%.   (b) 100%.   (c) 160%.   (d) 350%.

 (e) 40%.   (f) 200%.   (g) 50%.   (h) $87\frac{1}{2}$%.

3. 65.   5. 5.

7. (a) $\frac{350}{2000} = 17\frac{1}{2}$%.   (b) $\frac{41}{328} = 12\frac{1}{2}$%.

10-3 Decimal Notation

1. (a) 57,293.641. (b) 300.0704. (c) 0.00006. (d) 0.2956.

3. (a) Two hundred forty-six. (b) Two and forty-six hundredths.

 (c) One hundred seventy-eight and two hundred ninety-six thousandths.

 (d) Nine tenths.

5. (a) $\dfrac{1}{100}$.   (b) $\dfrac{1}{10,000}$.   (c) $\dfrac{1}{1,000,000}$.

10-4 Terminating and Repeating Decimals

1. (a) 0.7. (b) 0.78. (c) 0.65. (d) 0.375.

3. (a) $\dfrac{4}{7} = 0.\overline{571428}$.   (b) $\dfrac{10}{15} = 0.\overline{6}$.

 (c) $\dfrac{8}{11} = 0.\overline{72}$.   (d) $\dfrac{29}{37} = 0.\overline{783}$.

5. (a) $\dfrac{5}{13} = 0.\overline{384615}$.   (b) Repeating decimal.

 (c) The cycle is 384615.

7. (a) $\dfrac{23}{100}$.   (b) $\dfrac{165}{1,000}$; that is, $\dfrac{23}{200}$.

 (c) $3\dfrac{738}{10,000}$; that is $3\dfrac{369}{5,000}$.   (d) $\dfrac{9}{1,000}$.

9. $\frac{13}{40}$, $\frac{7}{8}$, and $\frac{9}{25}$ will have decimal representations that terminate.

10-5 Operations with Numbers Expressed as Terminating Decimals

1. (a)
$$\begin{array}{r} 7.8 \\ 2.75 \\ +\ 6. \\ \hline 16.55 \end{array}$$
 $\dfrac{780}{100} + \dfrac{275}{100} + \dfrac{600}{100} = \dfrac{780 + 275 + 600}{100} = \dfrac{1,655}{100} = 16.55.$

 (b)
$$\begin{array}{r} 0.123 \\ +1.1 \\ \hline 1.223 \end{array}$$
 $\dfrac{123}{1,000} + \dfrac{1,100}{1,000} = \dfrac{123 + 1,100}{1,000} = \dfrac{1,223}{1,000} = 1.223.$

(c)

$$\begin{array}{r} 16.88 \\ -\ 9.56 \\ \hline 7.32 \end{array}$$

$\dfrac{1,688}{100} - \dfrac{956}{100} = \dfrac{1,688 - 956}{100} = \dfrac{732}{100} = 7.32.$

(d)

$$\begin{array}{r} 0.700 \\ -0.289 \\ \hline 0.411 \end{array}$$

$\dfrac{700}{1,000} - \dfrac{289}{1,000} = \dfrac{700 - 289}{1,000} = \dfrac{411}{1,000} = 0.411.$

3. (a) $\dfrac{6.5}{15}.$ **(b)** $\dfrac{7,890,000}{36}.$ **(c)** $\dfrac{478.96}{2,231}.$ **(d)** $\dfrac{59,200}{1}.$

5. (a) $9\%.$ **(b)** $0.5\%.$ **(c)** $278\%.$ **(d)** $75\%.$

10-6 Scientific Notation

1. (a) $9.78 \times 10^{-5}.$ **(b)** $5.64 \times 10^6.$ **(c)** $1.265 \times 10^0.$
(d) $4.31 \times 10^{-13}.$ **(e)** $9.6783 \times 10^7.$ **(f)** $1.0005 \times 10^{-2}.$

3. $2.7724 \times 10^{16}.$

5. Among the many possible answers are:

1 ton $\overset{m}{=} 2.0 \times 10^3$ pounds.
1 mile $\overset{m}{=} 5.28 \times 10^3$ feet.
1 year $\overset{m}{=} 3.65 \times 10^2$ days.

CHAPTER 11—REAL NUMBER SYSTEM

11-1 Irrational Numbers

1. Among the many positive irrational numbers less than 10 are: $\sqrt[3]{2}, \sqrt{5}, \sqrt{6}, \sqrt{7}, \sqrt[3]{11}, 3\sqrt{8}, \frac{2}{3}\sqrt{2}, 2\pi, \sqrt[6]{10}, \sqrt[5]{3}.$

3. Assume that $\sqrt{3} = \dfrac{a}{b}$ where a and b are integers that are not both divisible by 3 and $b \neq 0$. If $\dfrac{a}{b} = \sqrt{3}$, then $\dfrac{a^2}{b^2} = 3$. Since a and b are integers, a^2 and b^2 are also integers and $a^2 = 3b^2$. The integer $3b^2$ is divisible by 3 since any integer is divisible by 3 if it is equal to the product of 3 and another integer. Since a^2 is equal to $3b^2$, 3 is a factor of a^2 and, consequently, 3 is a factor of a. Thus, $a = 3k$, where k is an integer. Then the statement $a^2 = 3b^2$ may be written as $(3k)^2 = 3b^2$, and $(3k) \times (3k) = 3b^2$, and $k \times 3k = b^2$, and $3k^2 = b^2$. Thus b is an integer whose square is divisible by 3, and b must be divisible by 3. This is contrary to the assumption that a and b are not both divisible by 3. Thus, our assumption that $\sqrt{3}$ is rational is false, and $\sqrt{3}$ is irrational.

11-2 Approximating $\sqrt{2}$ as a Decimal

1. (a) $\sqrt{5}.$ **(b)** $\sqrt{7}.$ **(c)** $\sqrt{13}.$ **(d)** $\sqrt{16} = 4.$

3. (a) 2.2.   **(b)** 2.8.   **(c)** 3.2.   **(d)** 5.4.

5. (a) Using a trial and error procedure:

$$1^2 = 1 \qquad\qquad (2.1)^2 = 4.41$$
$$2^2 = 4 \qquad\qquad (2.2)^2 = 4.84$$
$$3^2 = 9 \qquad\qquad (2.3)^2 = 5.29 \text{ and so on.}$$
$$\text{Thus } 2 < n < 3. \qquad\qquad \text{Thus } 2.2 < n < 2.3.$$

We determine that a rational number n for $n^2 = 5$ could never be found since we continue to get a decimal representation.

Using indirect reasoning:

Assume that $n = \dfrac{a}{b}$ where a and b are integers that are not both divisible by 5 and $b \neq 0$. Then $n^2 = \left(\dfrac{a}{b}\right)^2 = 5$ and $\dfrac{a^2}{b^2} = 5$. Since a and b are integers, a^2 and b^2 are also integers and $a^2 = 5b^2$. The integer $5b^2$ is divisible by 5 since any integer is divisible by 5 if it is equal to the product of 5 and another integer. Since a^2 is equal to $5b^2$, 5 is a factor of a^2 and, consequently, 5 is a factor of a. Thus $a = 5k$, where k is an integer. Then the statement $a^2 = 5b^2$ may be written as $(5k)^2 = 5b^2$, and $(5k) \times (5k) = 5b^2$, and $k \times (5k) = b^2$, and $5k^2 = b^2$. Thus b is an integer whose square is divisible by 5, and b must be divisible by 5. This is contrary to the assumption that a and b are not both divisible by 5. Thus our assumption that n is rational is false, and there is no rational number n such that $n^2 = 5$.

(b), (c), (d) could be shown similarly.

7. $\sqrt{2} \times \sqrt{2} = \sqrt{4} = 2$, and 2 is not an irrational number.

11-3 Real Numbers

1. (a) Irrational.   **(b)** Rational; $\sqrt{9} = 3$.
  (c) Rational.   **(d)** Rational.
  (e) Irrational.   **(f)** Rational.
  (g) Irrational.   **(h)** Rational; $\sqrt{16} = 4$.

3. $\sqrt{9} = 3.\bar{0}$; $\frac{17}{26} = 0.6\overline{538461}$; $0.172 = 0.172\bar{0}$; $\frac{22}{7} = 3.\overline{142857}$; $\sqrt{16} = 4.\bar{0}$.

5. $\sqrt{3} = 1.7320\ldots$

Thus, $2 + \sqrt{3} = 2 + 1.7320\ldots = 3.7320\ldots$

Since the sum $3.7320\ldots$ is neither a repeating nor a terminating decimal, the sum is an irrational number.

We may also use indirect reasoning to show that $2 + \sqrt{3}$ is irrational:

We know that $\sqrt{3}$ is not a rational number.

Assume that $2 + \sqrt{3}$ is a rational number.

Then there exist integers a and b ($b \neq 0$) such that $2 + \sqrt{3} = \dfrac{a}{b}$. Then

$\sqrt{3} = \dfrac{a}{b} - 2$, and this may be written as $\sqrt{3} = \dfrac{a - 2b}{b}$. Since a and b are

integers ($b \neq 0$), we know that $\dfrac{a - 2b}{b}$ is a rational number. This contradicts the fact that $\sqrt{3}$ is not a rational number. Thus our original assumption is false and $2 + \sqrt{3}$ must be an irrational number.

7. No.

9. $\sqrt{3} = 1.7320\ldots$
Thus $^-2\sqrt{3} = ^-2 \times 1.7320\ldots = ^-3.4640\ldots$
Since the product $^-3.4640\ldots$ is neither a repeating nor a terminating decimal, the product is an irrational number.
We may also use indirect reasoning to show that $^-2\sqrt{3}$ is irrational:
We know that $\sqrt{3}$ is not a rational number.
Assume that $^-2\sqrt{3}$ is a rational number.
Then there exist integers a and b ($b \neq 0$) such that $^-2\sqrt{3} = \dfrac{a}{b}$. Then $\sqrt{3} = \dfrac{a}{b}\left(-\dfrac{1}{2}\right)$, and this may be written as $\sqrt{3} = \dfrac{^-a}{2b}$. Since a and b are integers ($b \neq 0$), we know that $\dfrac{^-a}{2b}$ is a rational number. This contradicts the fact that $\sqrt{3}$ is not a rational number. Thus our original assumption is false and $^-2\sqrt{3}$ must be an irrational number.

11-4 Properties of Real Numbers

1. (a) Not dense.   (b) Not dense.   (c) Dense.
(d) Dense.   (e) Dense.   (f) Dense.

3. The real numbers are complete.

5. (a) All the sets of numbers have an additive identity, namely zero.
(b) All the sets of numbers have a multiplicative identity, namely one.

7. (a) $\dfrac{1}{\sqrt{5}}$.   (b) $\dfrac{1}{^-\sqrt{2}}$ or $\dfrac{^-1}{\sqrt{2}}$.   (c) $\sqrt{11}$.   (d) $\dfrac{1}{\sqrt{6} + \sqrt{7}}$.

CHAPTER 12—NONDECIMAL SYSTEMS OF NUMERATION

12-2 Base Five Numeration

1. (a) 32_{five}.   (b) 241_{five}.   (c) 304_{five}.   (d) 123_{five}.

3.

0_{five} 1_{five} 2_{five} 3_{five} 4_{five} 10_{five} 11_{five} 12_{five} 13_{five} 14_{five} 20_{five} 21_{five}

5. (a) $413_{\text{five}} = (4 \times 5^2) + (1 \times 5^1) + (3 \times 5^0)$.
(b) $2{,}243_{\text{five}} = (2 \times 5^3) + (2 \times 5^2) + (4 \times 5^1) + (3 \times 5^0)$.
(c) $42.3_{\text{five}} = (4 \times 5^1) + (2 \times 5^0) + (3 \times 5^{-1})$.
(d) $1.32_{\text{five}} = (1 \times 5^0) + (3 \times 5^{-1}) + (2 \times 5^{-2})$.

12-3 Computation in Base Five Notation

1. (a) 42_{five}.   (b) 341_{five}.   (c) $10,000_{five}$.   (d) $1,212_{five}$.

3. (a) 11_{five}.   (b) 4_{five}.   (c) 123_{five}.   (d) $1,344_{five}$.

5. (a) 310_{five}.   (b) 413_{five}.   (c) $34,001_{five}$.   (d) $100,322_{five}$.

7. (a) 14_{five}.   (b) 432_{five}.   (c) 21_{five}.   (d) $3,422_{five}$.

9.

+	E	F	G	H	J
E	E	F	G	H	J
F	F	G	H	J	FE
G	G	H	J	FE	FF
H	H	J	FE	FF	FG
J	J	FE	FF	FG	FH

×	E	F	G	H	J
E	E	E	E	E	E
F	E	F	G	H	J
G	E	G	J	FF	FH
H	E	H	FF	FJ	GG
J	E	J	FH	GG	HF

11. (a) Yes.   (b) No.   (c) Yes.   (d) No.

13. (a) Yes.   (b) No.   (c) Yes.   (d) No.

15. Among the many correct answers are:
   (a) $G \times (F + H) = (G \times F) + (G \times H)$.
   (b) $F \times (H - G) = (F \times H) - (F \times G)$.

12-4 Change of Base

1. (a) 38.   (b) 125.   (c) 14,876.   (d) 15,625.

12-5 Base Nine Numeration

1. (a)

1 set of nine and 3 ones, written 13_{nine}.

(b) ★ ★ ★ ★

★ ★ ★ ★

0 sets of nine and 8 ones, written 8_{nine}.

(c)

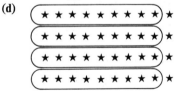

3 sets of nine and 1 one, written 31_{nine}.

(d)

4 sets of nine and 4 ones, written 44_{nine}.

3.

Six Thousand Five Hundred Sixty-ones	Seven Hundred Twenty-nines	Eighty-ones	Nines	Ones
$9 \times (9 \times 9 \times 9)$	$9 \times (9 \times 9)$	9×9	9×1	1
9^4	9^3	9^2	9^1	9^0

5.

+	0	1	2	3	4	5	6	7	8
0	0	1	2	3	4	5	6	7	8
1	1	2	3	4	5	6	7	8	10
2	2	3	4	5	6	7	8	10	11
3	3	4	5	6	7	8	10	11	12
4	4	5	6	7	8	10	11	12	13
5	5	6	7	8	10	11	12	13	14
6	6	7	8	10	11	12	13	14	15
7	7	8	10	11	12	13	14	15	16
8	8	10	11	12	13	14	15	16	17

×	0	1	2	3	4	5	6	7	8
0	0	0	0	0	0	0	0	0	0
1	0	1	2	3	4	5	6	7	8
2	0	2	4	6	8	11	13	15	17
3	0	3	6	10	13	16	20	23	26
4	0	4	8	13	17	22	26	31	35
5	0	5	11	16	22	27	33	38	44
6	0	6	13	20	26	33	40	46	53
7	0	7	15	23	31	38	46	54	62
8	0	8	17	26	35	44	53	62	71

7. (a) 22_{nine}.   **(b)** 136_{nine}.   **(c)** 180_{nine}.   **(d)** 1_{nine}.

9. (a) 48_{nine}.   **(b)** 503_{nine}.   **(c)** 121_{nine}.   **(d)** $1{,}583_{nine}$.

12-6 Base Two Numeration

1. (a) 21.   **(b)** 31.   **(c)** 38.   **(d)** 2,064.

3. (a) $1{,}001_{two}$.   **(b)** $1{,}101{,}001_{two}$.   **(c)** $1{,}001_{two}$.   **(d)** 101_{two}.

     (e) $1{,}110{,}100_{two}$.   **(f)** $1{,}101_{two}$.   **(g)** $10_{two}r100_{two}$.

12-7 Base Twelve Numeration

1. (a) 233.   **(b)** 23.   **(c)** 15,372.   **(d)** 18,995.

3. (a) $1E_{twelve}$.   **(b)** 65_{twelve}.   **(c)** $1{,}849_{twelve}$.   **(d)** $1{,}TT8_{twelve}$.

5. (a) $E3_{twelve}$.   **(b)** $20{,}T56_{twelve}$.   **(c)** $3{,}213{,}ET3_{twelve}$.   **(d)** $TT0{,}E22_{twelve}$.

7. (a) $1E_{twelve}$.   **(b)** 91_{twelve}.   **(c)** 207_{twelve}.   **(d)** $2{,}00E_{twelve}$.

9. (a) True.   **(b)** False.   **(c)** False.   **(d)** False.

CHAPTER 13—INFORMAL NONMETRIC GEOMETRY

13-2 Points

1. (a) A.

     (b) Since a point has no size and cannot be seen, neither point is larger than the other.

3. (a) False (a dot is a picture of a point).   **(b)** False (space is a set of points).

     (c) True.   **(d)** True.

     (e) True.

13-3 Curves

1. Because a curve is a path which connects two locations and the path consists of an infinite set of locations called points.

3. Pictures **(a)** and **(b)** represent closed curves.

5. (a) {(simple, closed), (simple, not closed), (not simple, closed), (not simple, not closed)}.

     (b) In Exercise 2, picture **(a)** is simple, closed; picture **(b)** is not simple, closed; picture **(c)** is simple, not closed; and picture **(d)** is not simple, not closed.

7. (a) You do not have to go through either C or D. For example:

(b) You do not have to go through *C*. You do not have to go through *D*. You must, however, go through at least one of *C* and *D*.

13-4 Surfaces and Regions

1. The curve in picture (a) separates the surface that contains it. The curve in picture (b) does not separate its surface because no pair of points of the surface exist such that any curve joining these two points also contains a point of the given curve.

3. (a) Yes.
   (b) Among the possible answers are:
      None of the points of the given curve are contained in the drawn curve.

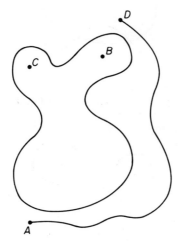

Two points of the given curve are contained in the drawn curve.

Four points of the given curve are contained in the drawn curve.

(c) Among the possible answers are:
Three points of the given curve are contained in the drawn curve.

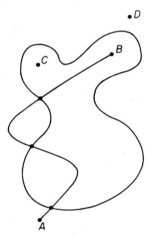

Five points of the given curve are contained in the drawn curve.

(d) Among the possible answers are:
Two points of the given curve are contained in the drawn curve.

Four points of the given curve are contained in the drawn curve.

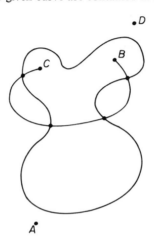

(e) *A* and *D* are exterior to the given curve.
   A is exterior and *B* is interior with respect to the given curve.
   B and *C* are interior to the given curve.
   When both points are exterior, the curve joining the two points contains an
   even number of points of the given curve; that is, 0, 2, 4, 6, 8, . . . points.
   When one point is interior and one point is exterior, the curve joining the two
   points contains an odd number of points of the given curve; that is, 1, 3, 5,
   7, 9, . . . points.
   When both points are interior, the curve joining the two points contains an
   even number of points of the given curve; that is, 0, 2, 4, 6, 8, . . . points.

13-6 Lines, Line Segments, and Rays

1. (a) 0.   **(b)** 2.   **(c)** 1.   **(d)** 0.
3. Among the possible answers are:
   (a) $\overleftrightarrow{AB}$, $\overleftrightarrow{AC}$, and $\overleftrightarrow{BC}$.
   (b) $\overline{AB}$, $\overline{AC}$, and $\overline{BC}$.
   (c) $\overrightarrow{AB}$, $\overrightarrow{AC}$, and $\overrightarrow{BC}$.
5. (a)   **(b)**

(c)   **(d)**

7. (a)

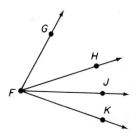

(b) An infinite number.

9. (a) True.   **(b)** True.   **(c)** False.   **(d)** True.

13-7 Points, Lines, and Planes

1. (a)   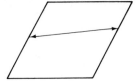   **(b)**

(c)   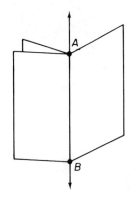   **(d)**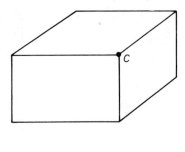

3. Among the many correct answers are:

(a) $\overleftrightarrow{RT}$ and $\overleftrightarrow{SY}$.   (b) $\overleftrightarrow{RX}$ and $\overleftrightarrow{TZ}$.   (c) $\overleftrightarrow{RT}$ and $\overleftrightarrow{ST}$.

(d) Plane RST and plane XYZ (only one correct answer in this case).

(e) Plane RST and plane TSY.   (f) $\overleftrightarrow{TS}$ and plane RSY.

(g) $\overleftrightarrow{RS}$ and plane RSY.   (h) $\overleftrightarrow{TZ}$ and plane RSY.

(i) $\overleftrightarrow{RS}$, $\overleftrightarrow{TS}$, and $\overleftrightarrow{YS}$.   (j) Plane RST, plane RXT, and plane SYT.

5. (a) True.   **(b)** False (the three points must not be on the same line).

(c) False.   **(d)** False.

7. (a) Yes.   **(b)** No ($\overleftrightarrow{AC}$ intersects the plane ABC in every point of the line AC).

13-8 Angles, Polygons

1. The drawings **(a)**, **(c)**, and **(d)** represent polygons.
3. (a) The vertices are A, B, C, D, and E.
    (b) The sides are $\overline{AB}$, $\overline{BC}$, $\overline{CD}$, $\overline{DE}$, and $\overline{EA}$.
    (c) The possible diagonals are $\overline{AC}$, $\overline{AD}$, $\overline{BE}$, $\overline{BD}$, and $\overline{CE}$.
    (d) The angles are $\angle ABC$, $\angle BCD$, $\angle CDE$, $\angle DEA$, and $\angle EAB$.

5.

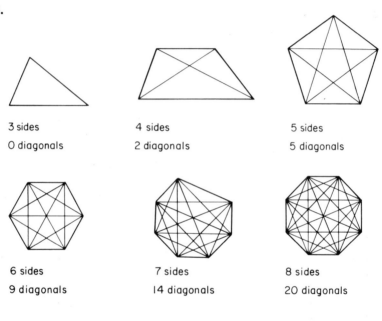

| 3 sides | 4 sides | 5 sides |
| O diagonals | 2 diagonals | 5 diagonals |

| 6 sides | 7 sides | 8 sides |
| 9 diagonals | 14 diagonals | 20 diagonals |

| 9 sides | 10 sides |
| 27 diagonals | 35 diagonals |

Yes; for the number of sides, n, the number of diagonals, d, is $\dfrac{n \times (n-3)}{2}$.

7. (a) B.   **(b)** $\varnothing$.   **(c)** D.   **(d)** A.
   (e) $\varnothing$.   **(f)** $\angle ABC$.   **(g)** $\overline{AC}$.   **(h)** $\overleftrightarrow{AC}$ (also named $\overleftrightarrow{EC}$ or $\overleftrightarrow{EA}$).

CHAPTER 14—INFORMAL METRIC GEOMETRY

14-1 Congruence and Similarity

1. The curves represented in pictures **(a)** and **(e)** are congruent; the curves shown in pictures **(b)** and **(c)** are congruent.

3. The curves represented in pictures **(a)**, **(b)**, **(c)**, and **(e)** of Exercise 1 are the same length.

5. (a)

(b) Among the possible correct drawings are:

(c)

7. Yes, the two models represent congruent triangles.

14-2 Measurement of Segments

1. $\overline{XY} \cong \overline{AL} \cong \overline{MN}$.

3. $\overline{CD}$ is 2 units, $\overline{EF}$ is 3 units, $\overline{GH}$ is $1\frac{1}{2}$ units.

14-3 Circles

1. Among the many correct answers are:
  (a) $\overline{RS}$, $\overline{MS}$, and $\overline{SN}$.
  (c) $\overline{RT}$ and $\overline{MN}$ (only one correct answer in this case).
  (b) $\overparen{RM}$, $\overparen{RV}$, $\overparen{RMW}$, and $\overparen{RNT}$.
  (d) R, V, N, T, and W.

3. (a) True. **(b)** False. **(c)** False (may not be diameters of the same circle or of circles with congruent radii). **(d)** False. **(e)** True. **(f)** True. **(g)** True. **(h)** False.

5. (a) Yes.   **(b)** Yes.   **(c)** Yes (since the radius has not been given).

14-4 Congruence and Measurement of Plane Angles

1. $\angle ABC \cong \angle MON$.

3. $\angle RQT$, $\angle NAP$, and $\angle DEG$ are right angles.

5. ∠*GHI* is 10 units, ∠*JKL* is 28 units, and ∠*MNO* is ½ unit.

7. Yes.

14-5 Classification of Triangles and Quadrilaterals

1. (a) Scalene.   **(b)** Equilateral and   **(c)** Isosceles.
                                      isosceles.

   (d) Isosceles.   **(e)** Isosceles.   **(f)** Scalene.

   (g) Equilateral and   **(h)** Scalene.
        isosceles.

3. (a)   **(b)**    **(c)**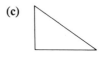

   (d)    **(e)**   **(f)**

5. (a) Always true.   **(b)** Sometimes true.   **(c)** Sometimes true.

   (d) Always true.   **(e)** Always true.   **(f)** Sometimes true.

14-6 Area

1. Area of region *ABCD* is 8 units.

3. Area of region *FGH* is estimated as 27 units.

5. (a) Yes.   **(b)** No.   **(c)** Yes; regions **(a)** and **(f)** are congruent;
                                                regions **(b)** and **(e)** are congruent.

   (d) No.   **(e)** Yes.   **(f)** Yes.

14-7 Volume

1. (c) Because two different units of volume were used.

3. A cubic inch is a standard unit of volume measure which is comprised of a cubical solid with edges 1 inch in length. A cubic foot is a standard unit of volume measure which is comprised of a cubical solid with edges 1 foot in length.

5. (a) $V = 6 \times 5 \times 10 = 300$ cubic units.

   (b) $V = 10 \times 10 \times 10 = 1,000$ cubic units.

   (c) $V = 3 \times 6 \times 0.5 = 9$ cubic units.

Index

A

C

O

P

R

S